Smart Energy Technologies Towards a Sustainable Future

This book offers a comprehensive exploration of renewable energy systems, elucidating the fundamental principles underpinning technologies such as solar photovoltaics, wind turbines, and biomass conversion.

Features:

- Discusses the role of computational fluid dynamics simulations in bioenergy production and for thermal solar collectors.
- Covers geothermal power plant design and performance optimization and solar-assisted heat pump technologies for building applications.
- Highlights the role of smart technologies such as artificial intelligence in sustainable energy technologies.
- Presents case studies on renewable energy sources, power demand, policies, and solar thermal desalination technologies.
- Explains valorization of biomass to fuel briquettes and environmental life-cycle analysis of renewable energy.

The text is primarily written for senior undergraduates, graduate students, and academic researchers in the fields of environmental engineering, renewable energy, clean technologies, mechanical engineering, electrical engineering, chemical engineering, and civil engineering.

Engineering Tribology, Manufacturing and Applied Energy

Series Editors

Yashvir Singh and Nishant Kumar Singh

About The Series:

Tribology is the study of parts when they are in contact with each other and have a considerable impact during their sliding or rotating motion. The series will publish the books highlighting the latest developments and research in tribology with its application in the manufacturing and energy sector. The series will include books covering established areas of tribology and emerging fields such as tribo-chemistry, surface engineering, nano-tribology, and bio-tribology. The series will focus on the recent developments and advancements in green tribology for example environmentally friendly composites. Aimed at senior undergraduate, graduate students, academic researchers and professionals, the proposed series will cover the latest techniques in the field of tribology with applications in diverse fields such as manufacturing, energy, and medical.

Modeling and Simulation of Fluid Flow and Heat Transfer
Edited by Reshu Gupta, Mukesh Kumar Awasthi

Smart Energy Technologies Towards a Sustainable Future
Edited by Dan Bahadur Pal and Ashish Kapoor

For more information on series page, please visit our website:

Smart Energy Technologies Towards a Sustainable Future

Edited by

Dan Bahadur Pal and Ashish Kapoor

CRC Press
Taylor & Francis Group
Boca Raton London New York

CRC Press is an imprint of the
Taylor & Francis Group, an **informa** business

First edition published 2026
by CRC Press
2385 NW Executive Center Drive, Suite 320, Boca Raton FL 33431

and by CRC Press
4 Park Square, Milton Park, Abingdon, Oxon, OX14 4RN

CRC Press is an imprint of Taylor & Francis Group, LLC

ISBN: 9781032947402 (hbk)
ISBN: 9781032947396 (pbk)
ISBN: 9781003581536 (ebk)

DOI: 10.1201/9781003581536

Typeset in Sabon
by Newgen Publishing UK

Contents

Preface

This book is dedicated to the comprehensive exploration of Smart Energy Technologies Towards a Sustainable Future. It aims to provide readers with an in-depth analysis of the principles, innovations, and applications that underpin various renewable energy sources. The focus encompasses solar, wind, hydro, geothermal, and biomass energy systems, delving into the intricacies of each technology. Furthermore, the project undertakes a meticulous examination of the environmental, economic, and social sustainability aspects associated with renewable energy. This includes assessing the environmental impact, economic viability, and societal implications of adopting renewable energy solutions on a global scale. In addition to highlighting the advantages of sustainable renewable energy, the project candidly addresses the challenges and obstacles that can hinder widespread adoption. It explores innovative solutions, emerging trends, and the role of policies and regulations in advancing the renewable energy sector. The primary aim of this book, "Sustainable Renewable Energy Technologies," is to serve as a comprehensive and authoritative resource for individuals seeking a profound understanding of the multifaceted world of renewable energy. It caters to a diverse audience, including students, researchers, policymakers, and industry professionals.

Chapter 1 introduces the future prospects of hydrogen generation, storage, transportation, and utilization, while Chapter 2 and Chapter 3 examine renewable energy sources, power demand, and policies, with a particular focus on India, and explore hydrogen as a clean energy solution from source to application. Chapters 4 and 5 expand on case studies and policy frameworks related to India's renewable energy landscape. Chapter 6 highlights innovations in lead-free piezoelectric materials for dielectric energy storage and harvesting, and Chapter 7 discusses advanced substrate treatment methods for improving biogas plant efficiency. Chapter 8 underscores the importance of life-cycle analysis in evaluating sustainable energy systems, while Chapter 9 explores smart integrated biorefineries for sustainable development. Further, Chapter 10 presents insights from computational fluid dynamics studies on microfluidic fuel cells, and Chapter 11 models the conversion of cellulose into hexane through process simulation. Chapter 12 focuses on sustainable materials for photovoltaic systems, followed by Chapter 13, which discusses carbon-based nanomaterials in biofuel cells. Finally, Chapter 14 concludes with biohydrogen production from food waste, emphasizing its potential for sustainable energy generation and waste valorization.

This book aims to simplify complex ideas and provide practical insights into how smart biomass can revolutionize biofuels, helping to meet the world's growing energy needs in an environmentally friendly way.

Acknowledgements

We sincerely thank all contributors and researchers whose expertise enriched this book, *Smart Energy Technologies Towards a Sustainable Future*. The editors are grateful to the Harcourt Butler Technical University, Kanpur, India, for technical facilities. We are grateful to the publishers for their support and to the academic community for providing the foundation of knowledge in this field.

Special thanks to our families, friends, and colleagues for their encouragement, and to our readers, whose curiosity inspires such efforts. We hope this book serves as a valuable resource for further exploration in Smart Energy Technologies Towards a Sustainable Future.

With sincere gratitude,
The Editors

About the editors

Dan Bahadur Pal is currently working as Associate Professor, Department of Chemical Engineering, Harcourt Butler Technical University, Kanpur, Uttar Pradesh, India. He received his M.Tech. in 2011 and Ph.D. in 2017 in the field of Chemical Engineering from the Indian Institute of Technology (BHU) Varanasi, Uttar Pradesh, India. Before that, he completed his B.Tech. in Chemical Engineering from UPTU, Lucknow, in 2006. Dr. Pal completed a doctoral degree in the field of nanotechnology and catalysis. These nanofibers have very promising potential to provide benefits to nanotechnologies, energy, environment, catalysts, sensors, etc. While pursuing research, he has published 105 publications in internationally refereed journals and 12 books and has contributed to 53 book chapters. From all the publications, he got 4,218 citations with a 29 h-index and 63i-10 indexes. Dr. Pal's research interests include nanotechnology, catalysis, energy and environment, and waste management, with a special focus on developing processes and materials by using waste as raw materials. He also prefers to work on bio-waste processing and value addition. He is the top 2% scientist in the field of Energy by Standford University list for the year 2024 & 2025.

Ashish Kapoor is currently working as Professor in the Department of Chemical Engineering at Harcourt Butler Technical University, Kanpur, Uttar Pradesh, India. He earned his B.Tech. and M.Tech. (dual degree program) in Chemical Engineering from the Indian Institute of Technology (IIT) Madras, Tamil Nadu, India, in 2004. He received his Ph.D. from the University of Illinois Urbana-Champaign (UIUC), Illinois, USA, in 2011. Before assuming his current position, Dr. Kapoor worked as a Product Engineer at Lam Research Corporation, California,

USA, and as a faculty member in the Department of Chemical Engineering at SRM Institute of Science and Technology, Kattankulathur, Tamil Nadu, India. With over 12 years of experience in industry and academia, he has authored more than 75 peer-reviewed international publications and has an h-index of 18. His research interests include microfluidics, lab-on-a-chip technologies, environmental remediation, biomass valorization, and modelling and simulation.

Contributors

Neeru Anand
University School of Chemical Technology, Guru Gobind Singh Indraprastha University, India

Trisanti Anindyawati
Research Center for Applied Microbiology, National Research and Innovation Agency (BRIN), Cibinong, Indonesia

Adarsh Kumar Arya
Department of chemical Engineering, Harcourt Butler Technical University Kanpur Uttar Pradesh, India

Shraddha Awasthi
Department of Environmental Studies, MLNCE, University of Delhi, New Delhi, India

Sahar Naveed Baig
Department of Biotechnology, University of Sargodha, Sargodha, Pakistan

Laxmi Deepak Bhatlu
Department of Chemical Engineering, KPR Institute of Engineering and Technology, Arasur, Coimbatore, India

Beata Bień
Department of Environmental Engineering and Biotechnology, Faculty of Infrastructure and Environment, Czestochowa University of Technology, Czestochowa, Poland

Krishna Srihari Bonasi
Department of Chemical Engineering, National Institute of Technology Srinagar, Srinagar, Jammu & Kashmir, India

Dilip Chorei
Department of Chemistry, Haflong Government College, Dima Hasao, Assam, India

Abhay Kumar Choudhary
Department of Chemical Engineering and Technology, National Institute of Technology Patna, Patna, Bihar, India

Apridah Cameliawati Djohan
Research Center for Applied Microbiology, National Research and Innovation Agency (BRIN), Cibinong, Indonesia

Anna Grobelak
Department of Environmental Engineering and Biotechnology, Faculty of Infrastructure and Environment, Czestochowa University of Technology, Czestochowa, Poland

Titin Haryati
Research Center for Applied Microbiology, National Research and Innovation Agency (BRIN), Cibinong, Indonesia

Muhammad Irfan
Department of Biotechnology, University of Sargodha, Sargodha, Pakistan
Research Centre for Applied Microbiology, National Research and Innovation Agency (BRIN), Cibinong, Indonesia

Sumit Kumar Jana
Department of Chemical Engineering, Birla Institute of Technology, Mesra, Ranchi, Jharkhand, India

Kainat
Department of Biotechnology, University of Sargodha, Sargodha Pakistan.

Hiyaman Kalita
Department of Chemistry, Gauhati University, Guwahati, Assam, India

Avanish Kumar
Department of Chemical Engineering, Harcourt Butler Technical University Kanpur Uttar Pradesh India

Brajesh Kumar
Department of Chemical Engineering, National Institute of Technology Srinagar, Jammu & Kashmir, Hazratbal, India

Gulshan Kumar
University School of Basic and Applied Sciences, Guru Gobind Singh Indraprastha University, Delhi, India

Krishna Kumar
University School of Basic and Applied Sciences, Guru Gobind Singh Indraprastha University, India

Vimal Kumar
Departments of Chemistry, Brahmanand College, Kanpur, Uttar Pradesh, India

Ika Rahmatul Layly
Research Center for Applied Microbiology, National Research and Innovation Agency (BRIN), Cibinong, Indonesia

Saurav Kumar Maity
University School of Basic and Applied Sciences, Guru Gobind Singh Indraprastha University, India

Magesh Kumar Marichamy
Department of Chemical Engineering, SRM Institute of Science and Technology, Chennai, Tamil Nadu, India

Ambneesh Mishra
Faculty of Biology and Environmental Protection, University of Lodz, Poland

Sankar Neogi
Department of Chemistry, Haflong Government College, Dima Hasao, Assam, India

Ramdendra Pal
Department of Chemical Engineering, Harcourt Butler Technical University, Kanpur, Uttar Pradesh, India

Vijayeta Pal
School of Materials Science and Technology, Indian Institute of Technology, Banaras Hindu University, Varanasi, Uttar Pradesh, India

Vimal Singh Pal
Department of Chemical Engineering, Harcourt Butler Technical University, Uttar Pradesh, India

Deoashish Panjiara
Department of Chemical and Petrochemical Engineering, Ganpat University, Mehsana, 384012, Gujarat, India

Sanchayita Rajkhowa
Department of Chemistry, Haflong Government College, Dima Hasao, Assam, India

Amit Kumar Rathoure
Department of Chemical Engineering, Harcourt Butler Technical University, Kanpur, Uttar Pradesh, India

Jyotirmoy Sarma
Department of Chemistry, Assam Don Bosco University, Tapesia Gardens, Sonapur, Assam, India

Balasubramanian Sivasamy
Department of Chemical Engineering, KPR Institute of Engineering and Technology, Coimbatore, Tamil Nadu, India

Daria Sławczyk
Department of Environmental Engineering and Biotechnology, Faculty of Infrastructure and Environment, Czestochowa University of Technology, Czestochowa, Poland

Bonasi Krishna Srihari
Department of Chemical Engineering, National Institute of Technology, Hazratbal, Jammu and Kashmir, Srinagar, India

Ahmad Thontowi
Research Center for Applied Microbiology, National Research and Innovation Agency (BRIN), Cibinong, Indonesia

Amit Kumar Tiwari
Department of Chemical Engineering, Birla Institute of Technology, Mesra, Ranchi, Jharkhand, India

Uplabdhi Tyagi
University School of Chemical Technology, Guru Gobind Singh Indraprastha University, Delhi, India

Neeraj Kumar Yadav
Department of Chemical Engineering, Government Polytechnic Mohammadpur Bahraich, Board of Technical Education, Uttar Pradesh, India

Sudeep Yadav
Department of Chemical Engineering, Bundelkhand Institute of Engineering and Technology, Jhansi, Uttar Pradesh, India

Chapter 1

Hydrogen generation, storage and transportation future prospects

Magesh Kumar Marichamy

1.1 INTRODUCTION

Hydrogen not only occupies a predominant role in meeting the future energy demand but also plays a vital role in fuelling cars, trucks, ships, airplane, etc. It is a green fuel because it produces only water vapour when burned or oxidized. Even though it is present in all living beings and non-living beings on earth in compound form, its presence for direct fuel use is very scarce. It could minimize carbon emissions and accelerate the policy of achieving net-zero discharge of carbon pollutants. Being the lightest element in nature, it occupies the first position, among all the elements, in the modern periodic table.<<end>> Hydrogen gas is combustible but does not support combustion and has one electron and one proton as sub-atomic particles. Hydrogen gas has an atomic number of 1, a relative atomic mass of 1.008 amu and an electronic configuration of $1s^1$ (Royal Society of Chemistry, 2024). Since the number of valence electrons is one, hydrogen can react readily with other elements and compounds either by losing the single electron to form H^+ ion or by gaining an electron to form H^- ion. Approximately 66.7% of all the atoms in our body are hydrogen, and the remaining 33.3% comprises other elements, namely carbon, oxygen, chlorine, sulfur, etc. (Grochala, 2015). Because of its lightness, it cannot be used directly for fuel applications; instead, it must be used in compressed, liquefied or solidified form. Philips' research laboratories fortuitously discovered that hydrogen gas can be retrieved in virgin form when stored in intermetallic compounds in the form of hydride (Van Vucht et al., 1970). This lays the foundation for the design and development of "electrochemical-based" hydrogen storage systems and the use of "nickel–metal hydride" batteries as an energy source in road vehicles. The first element, "hydrogen", started to occupy a supreme position in energy industries and thrive for sustained energy production methods. Apart from fulfilling energy demands, hydrogen is the primary raw material for the manufacture of industrial chemicals. Because of risk involvement in *hydrogen transport*, it is prepared in situ wherever required. It is produced predominantly by steam – reforming of natural gas and other hydrocarbons.

DOI: 10.1201/9781003581536-1

Hydrogen is also produced from carbon monoxide, which in turn is formed in the water-gas shift reaction (American Chemical Society, 2020). Apart from these processes, which depend on fossil fuels and other *hydrocarbons*, the potential for "green hydrogen" production is gaining momentum. The Government of India has set a target to attain energy independence by 2047, and green hydrogen has the ability to fulfil this policy. Production of green hydrogen from biomass is another viable option for achieving sustainability and circularity in energy generation and utilization. Biomass is gasified at high temperature to form hydrogen gas (Ministry of New and Renewable Energy, GOI, 2024). Technology for energy harnessing from hydrogen is neither simple nor cost-effective for large-scale production. There is a degree of uncertainty in using hydrogen as an energy sources in new applications due to ambiguity in government policies, sluggishness in infrastructure development and competition from other energy-producing companies. The investment for hydrogen production is categorized into four segments, namely, production, distribution, storage and end-use applications.

The potential of hydrogen as an energy source is exemplified by a drastic increase in investment. The International Energy Agency (IEA) has given permission to develop 1500 low-carbon projects. The market value for green hydrogen increased from $676 million in 2022 to $7300 million by 2027, accounting for an 11 times increase (International Energy Agency, 2024). Hydrogen, a credible future fuel, offers endless benefits to society, the economy and the environment. It reduces the export of crude oil from foreign countries and lessens carbon emissions from transport sectors. The energy-yielding capacity of hydrogen gas is 122 kJ/g, almost 2.75 times greater than conventional fossil fuels (Balat, 2008). All these positive attributes of hydrogen gas towards sustainable green energy provide a motivation to illustrate the production scenario, storage facilities and energy aspects in this chapter.

1.2 HYDROGEN: PROPERTIES AND CATEGORIES

Hydrogen is the most abundant and simplest of all the elements in the universe. A single atom of hydrogen has one electron and one proton, and most of the mass of the hydrogen atom is concentrated at the centre of the nucleus. Since it has an equal distribution of positive and negative charged particles, a hydrogen atom is considered electrically neutral. Even in liquid form, hydrogen does not have more density. One cubic metre volume of liquid water contains 111 kg of hydrogen, whereas the same volume of liquid hydrogen contains only 71 kg of hydrogen. The energy density is a measure of the packing efficiency of hydrogen atoms in a fuel. Hence, hydrocarbons having a complex structure have more energy density because more hydrogen atoms are packed in a single molecule. This is the reason for the poor energy density of hydrogen fuel, although its energy-to-mass ratio is the best of all fuels.

Hydrogen contributes to achieving zero emissions targets, making it the cleanest hydrogen among all types. Even though green hydrogen is reliable and eco-friendly, building the infrastructure for electricity generation using renewable sources is cost-intensive. Since a mammoth capital investment, overhead costs and production costs are involved, the overall costs for green hydrogen production are too inflated. In addition to the aforementioned hydrogen types, turquoise hydrogen and yellow hydrogen contribute in a small way to the global production scenario. Turquoise hydrogen is manufactured utilizing natural gas as feedstock under extreme temperature and pressure conditions, and the process follows multiple reaction mechanisms. The emitted carbon compounds are captured, stored and used for value-added processes like enhanced oil recovery, chemical manufacture, etc. Yellow hydrogen is produced using biomass, such as agricultural waste or forestry residue, and it is considered a low-carbon-emitting method.

1.3 HYDROGEN: AN ENERGY SOURCE AND *ENERGY CARRIER*

The major requirements of an ideal fuel are inexhaustibility, cleanliness, easy transport, higher energy density and non-dependence on foreign imports. Hydrogen satisfies all the criteria and is hence considered the best and cleanest option for fossil fuels. Undoubtedly, hydrogen offers impeccable benefits in terms of diverse supply and reduced emissions of carbon compounds and air contaminants. Hydrogen and fuel cells can meet the nation's energy demand with the ability for use in a wide range of applications, including industries, *transportation*, commercial activities, etc. The applications include distributed power systems, backup power, potable power, supplementary power for aircraft, ships, forklifts, freight vehicles, etc. The US Energy Department has proved that a hydrogen-based process has the potential to mitigate emissions as per the following details. In light-duty highway vehicles, more than a 70% reduction in emissions can be achieved if gasoline vehicles are disposed of and replaced with hydrogen-powered engines. In the auxiliary power unit (APU) system, more than 60% emissions are reduced compared to the engine idling condition. In combined heat and power (CHP) systems, when hydrogen is used as an energy source, more than 35%–50% reduction in emissions is achieved over conventional power plants (US Department of Energy, 2017).

Hydrogen is an energy carrier that transports energy in usable form from one place to another. Hydrogen, similar to electricity, carries energy, but it must be obtained from other sources. The diffusion coefficient of hydrogen in air at standard temperature and pressure is 0.61 cm^2/s, which shows fleeting tendency during leakage. This is one of the desired safety features in handling, storage and transport of hydrogen. The positively charged protons diffuse through the electrolyte solution to reach the cathode, where they combine chemically with oxygen and free electrons to produce heat energy and water (US Department of Energy, 2024). A typical fuel cell system consists of the components, namely, fuel cell stack, fuel cell processor, power conditioners, air compressors and humidifiers. The fuel cell stack is the heart of a fuel cell power generation plant. Electricity in direct current (DC) form is produced from electrochemical reactions that take place in a fuel cell. A single fuel cell generates less than 1 V of power, which is not sufficient for the majority of the applications. So, individual fuel cells are arranged in series to form a fuel cell stack consisting of hundreds of small fuel cells. Type of fuel cell, size of cell, operating temperature and gas pressure are the critical variables that affect the efficiency of fuel cells. The type of electrolyte employed in fuel cell operation forms the basis of fuel cell classification. The important types of fuel cells are proton exchange membrane (PEM) fuel cell, direct methanol fuel cell, alkaline fuel cell, phosphoric acid fuel cell, molten carbonate fuel cell and solid oxide fuel cell. In a PEM cell, a solid polymer is used as an electrolyte and a porous carbon containing platinum catalyst is used as an electrode. It is used mainly for transportation purposes like cars, buses and heavy duty trailers and for stationary power generation in small-wattage requirements. Molten carbonate fuel cells (MCFC) use molten carbonate salt mixture suspended in a ceramic lithium aluminium oxide matrix as electrolyte and operate at a high temperature of 650°C. It is now used for electrical utility, military and industrial operations. The major disadvantage of this cell is its short life span because of the high temperature condition and the corrosive nature of the electrolyte. The next energy producer is *solid oxide fuel cells* (SOFCs), which use a hard, non-porous ceramic compound as the electrolyte. A new technique based on a metal hydrate storage system satisfies safety features, economic considerations and industry expectations (Naveed Khan, 2025). A hydrate-based storage technique uses water as a host molecule and hydrogen as a guest molecule at low-temperature and high-pressure conditions. This process necessitates the use of promoters to moderate the process conditions and reduce the cost factors.

1.4 COMPARISON BETWEEN HYDROGEN AND HYDROCARBON FUELS IN ENERGY PERSPECTIVE

Before venturing into projects related to the development of sustainable energy technology, there is a need to understand the pros and cons of prospective fuel. Hydrogen is emerging as an ideal candidate to minimize carbon emissions and shift towards a sustainable energy future. It is necessary to understand the potential of hydrogen to attain the target of reducing environmental footprint. The global energy demand, coupled with environmental concerns, is increasing at an accelerated rate, which has intensified the process to search for sustainable energy sources. Hydrogen and fossil fuels represent two extreme poles in this regard, each with its own set of positive and negative attributes.

Hydrogen energy and fossil fuel energy potentials should be compared based on factors like production costs, usage efficiency and environmental damages. It is inferred that hydrogen energy outsmarts fossil fuel energy in factors like effective energy carrying, environmental concern mitigation and resource conservation (Awad and Veziroğlu, 1984). As the fossil fuel era nears extinction, humanity must find a way to manufacture future fuels. The selection criteria to be considered are convenience for transport, easy to convert to other forms at the user end, high energy output efficiency, environmental friendliness and safety handling. While using surface vehicles for transportation, the future fuel should be lightweight, occupy less space and cover extra mileage than prescribed. Table 1.1 presents the relevant data, i.e. energy output per unit mass of fuel and energy produced per unit volume of fuel for fossil fuels, which are currently used and/or under consideration as future fuel (liquid hydrogen and gas hydrogen) (Vezirg̃olu and Barbir, 1992).

The tabulated energy data infer that hydrogen is the lightest fuel for a fixed quantity of energy liberated, whereas methanol is the heaviest. Based on energy produced per unit volume, liquid hydrogen occupies more space compared with other fuels. When gaseous fuels are compared, natural gas occupies 32% more volume than hydrogen gas to liberate the same amount of energy. For comfortable transportation of fuels, the mass and volume-based energy characteristics of fuel are compared. During transportation through pipelines and surface vehicles, the acceleration force acting on fuel

Table 1.1 Energy output per unit mass of fuel and energy produced per unit volume of fuel for fossil fuels

Fuel	*Energy per unit mass (J/kg)*	*Energy per unit volume (J/m^3)*	*Motivity factor (Φ_m)*
Fuel oil	45.5	38.65	0.78
Gasoline	47.4	34.85	0.76
Jet fuel	46.5	35.3	0.75
LPG	48.8	24.4	0.62
LNG	50	23	0.61
Methyl alcohol	22.3	18.1	0.23
Ethanol	29.9	23.6	0.37
Natural gas	50	0.04	0.75
Liquid hydrogen	141.9	10.10	1
Gas hydrogen	141.9	0.013	1

is proportional to mass to achieve the least magnitude of mass per unit energy for a given fuel.

$$F_a \propto \frac{M}{E} \tag{1.1}$$

where F_a is the acceleration force needed to transport the fuel, M is the mass of fuel and E is the energy generated by the same fuel. Drag force acting on fluid influences the volume of liquid transported and is given by the expression

$$F_d \propto \left(\frac{V}{E}\right)^{0.6667} \tag{1.2}$$

where F_d is the drag force and V is fuel volume. For a fuel to be designated as the best transportation fuel, both F_a and F_d must be as small as possible. The sum of two forces should be as low as possible, and their product is expected to be low. Equations (1.1) and (1.2) are combined to avoid dimensional compatibility issues, and augmented form is reported as

$$F_a F_d \propto \left(\frac{M}{E}\right)\left(\frac{V}{E}\right)^{0.6667} \tag{1.3}$$

For good transportation fuel, the reciprocal of the product $(F_a F_d)$ should be as high as possible. This inverse product is normalized with respect to hydrogen properties, and the normalized form shall be called the "motivity factor (Φ_m)". It is defined as

$$\Phi_m = \frac{\left(\frac{E}{M}\right)\left(\frac{E}{V}\right)^{0.6667}}{\left(\frac{E_H}{M_H}\right)\left(\frac{E_H}{V_H}\right)^{0.6667}} \tag{1.4}$$

where the subscript "H" refers to hydrogen. Table 1.1 shows that hydrogen has the best motivity factor while methyl alcohol has the least motivity factor. The refinement of the motivity factor is based on the mass of the energy storage matrix (Utgikar et al., 2011). Another feature of hydrogen fuel is its versatility of use in different energy forms at the end-of-pipe. Figure 1.1 summarizes the different fuels and processes by which they are transformed to other energy forms.

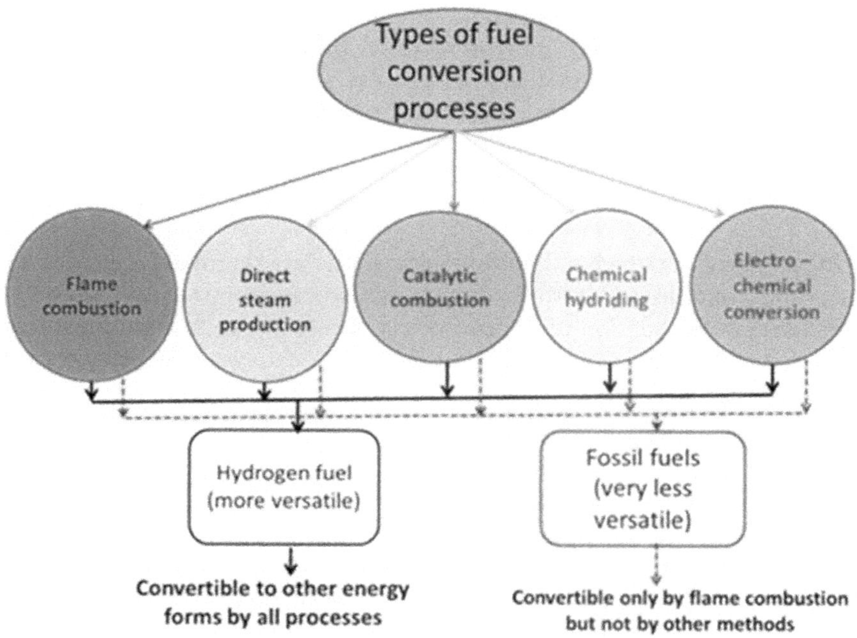

Figure 1.1 Different fuels and processes.

It is inferred from Figure 1.1 that, except for hydrogen, all fuels can be converted to other forms of energy using only the combustion process. For metamorphosing hydrogen to other energy types, five different methods are followed. In flame combustion, hydrogen is converted directly to steam, whereas it is converted to heat in the catalytic combustion process. In an *electrochemical process*, hydrogen energy is converted into electrical energy. Hence, hydrogen is much more multifaceted than fossil fuels in meeting the energy challenges. The *flame combustion* process is used to produce heat from hydrogen for direct applications in internal combustion (IC) engines, diesel engines, jet and rocket engines.

1.5 METHODS OF *HYDROGEN GENERATION*

The trend for fossil fuel usage declines drastically as new technologies are followed in the power sector and as the environmental concerns are on the rise. Hence, technically competent persons and skilled industrialists perceive the need to modernize existing energy generation units. In due course of time, fossil fuels will be washed away from the energy sector map if the energy benefits of hydrogen are realized in an efficient and economical way.

Currently, hydrogen is mainly produced from fossil fuels by reforming and pyrolysis processes. In the steam reforming process, steam and hydrocarbons react at high temperature and pressure to produce carbon dioxide and hydrogen. Feed stocks used in this process are liquefied petroleum gas (LPG), natural gas and naphtha. The major chemical reaction is

$$CH_4 + H_2O \xrightarrow{\text{heat supply}} CO + 3H_2 \tag{1.5}$$

In the pyrolysis process, liquid hydrocarbons are thermally degraded in anaerobic conditions to form hydrogen and other carbonaceous materials. The thermal degradation of hydrocarbons occurs at extreme temperatures and needs the support of a transition metal as a catalyst. The general pyrolysis stoichiometry is

$$C_nH_m \rightarrow nC + \frac{1}{2}mH_2 \tag{1.6}$$

Even though this method produces less quantity of hydrogen compared with other methods, it has the benefit of producing functional carbon nanomaterials in the shapes of tubes, spheres and fibres. The fossil gasification method involves thermochemical reactions in series between the compounds such as air, steam and fossil fuels. High-temperature gasification of coal forms products like hydrogen and carbon monoxide. This approach of coal degradation minimizes emissions and improves the energy density of fuel. Gasification of coal follows the stoichiometry

$$3C(\text{from coal}) + O_2 + 2H_2O \rightarrow H_2 + 3CO \tag{1.7}$$

Agricultural residues and woody biomass can replace hydrocarbons in hydrogen production because of their abundant availability and eco-friendly nature. Burning of biomass fuels does not affect the concentration of carbon dioxide in the atmosphere because the development and growth of biomass involve CO_2 consumption during photosynthesis. Carbon neutrality is maintained when energy is generated from biomass since biomass combustion does not contribute to excess carbon emission. Biomass is gasified at 500°C to produce a complex mixture of compounds containing carbon monoxide, carbon dioxide, hydrogen gas and other species. The reaction mechanism is

$$C_xH_yO_zN_aS_b + O_2 + N_2 + \text{steam} \rightarrow H_2 + CO_2 + \text{other compounds}(\text{biomass}) \tag{1.8}$$

Another method for producing hydrogen from biomass is pyrolysis, which occurs in the absence of oxygen. Biomass is more difficult to gasify because of the presence of unbound moisture content and the complex structure of lignin and cellulose compounds. The next method uses the electrolysis principle to split water, applying electricity produced from *renewable energy*. Among all the methods, it produces 95% pure hydrogen (Le et al., 2023). Water splitting for hydrogen production is classified as low-temperature electrolysis, high-temperature electrolysis and photolytic degradation. Low-temperature electrolysis involves the splitting of water into hydrogen and oxygen at room temperature. General stoichiometry for the process is

$$H_2O + \text{electric current} \rightarrow H_2 + \frac{1}{2}O_2 \tag{1.9}$$

In high-temperature electrolysis process, dissociation of water occurs at 750°C to form hydrogen and oxygen.

1.5.1 Technical specifications in each method

Hydrogen production on a commercial scale is achieved by the methods, namely, thermal processes, biological processes, electrolytic processes and photolytic processes. Technical specifications in fossil fuel-based methods are reported as: Steam reforming process uses light hydrocarbons as feedstock and operates in the temperature range of 900°C–1000°C. In the steam reforming process, natural gas (methane) reacts with steam at high pressure, 300–2500 kPa, utilizing nickel as a catalyst to produce a large amount of hydrogen, CO and CO_2. It is an endothermic type reaction, and heat must be supplied to sustain the reaction rate. The magnitude of the heat of reaction is positive. This first step is sequentially followed by the water-gas shift reaction, which converts carbon monoxide with steam to produce CO_2 and more hydrogen. A downstream process, pressure-swing adsorption, separates pure hydrogen from CO_2 and other impurities.

Partial oxidation reforming method uses hydrocarbons, heavy fuel oil and coal as feedstock, and the operating temperature should be more than 1000°C. The hydrocarbons are oxidized partially with a limited amount of oxygen, not sufficient to satisfy the stoichiometric requirements, to produce CO and hydrogen. Here, oxygen is the "limiting reactant" that decides the conversion and extent of reaction, whereas hydrocarbon gases are "excess reactant". Oxygen is supplied in less quantity than required by the stoichiometry, and combustion occurs in a restricted environment to produce carbon monoxide and hydrogen. It is an exothermic process in which heat is produced and has a negative magnitude of heat of reaction. The rate of reaction in partial oxidation process is much larger than the *steam reforming* method. The reaction mechanism is

$$CH_4 + 0.5O_2 \rightarrow CO + 2H_2 + \text{heat} \; \Delta H_r = (-\text{ve}) \qquad (1.10)$$

where ΔH_r is the heat of reaction, and a negative sign indicates the exothermic nature of the process.

In pyrolysis technology, hydrocarbon feed stocks are gasified in the absence of oxygen at 500°C–800°C to produce hydrogen gas. Pyrolysis of methane does not produce CO_2 gas; instead, it forms solid carbon (C), which can be collected and stored for other industrial and commercial applications. The word pyrolysis has roots in the Greek language, literally meaning "separation using fire". Pyro means "fire" and lysis means "separation". It is a thermochemical decomposition of hydrocarbons without oxygen and combustion reactions. Methane pyrolysis involves splitting the feedstock to form hydrogen gas and solid carbon, resulting in net-zero carbon emissions. Moreover, when methane pyrolysis is carried out using biogas and CO_2 is utilized from atmosphere, then this process does not affect carbon balance in the ecosystem.

Gasification process uses coal as raw material to produce hydrogen and other emissions at an operating temperature of 700°C–1200°C. Technical aspects of non-fossil fuel-based hydrogen production processes are explained based on operating conditions, energy yield efficiency and hydrogen yield. The water electrolysis method utilizes low temperature and high pressure to achieve 55%–80% energy efficiency and product gas yield of 111 g H_2/kg feedstock. The water thermolysis process produces oxygen as a byproduct but involves high capital costs, and energy efficiency obtained is 20%–50%. In the biomass pyrolysis process, the yield of hydrogen depends on the nature of the feedstock, the maximum energy efficiency attained is 35%–50% and the quantity of hydrogen produced is 25–65 g H_2/kg feedstock. The energy efficiency of the biomass gasification process is 30%–60% with a product yield of 40–190 g H_2/kg feedstock. Table 1.2 shows the comparison of the energy efficiency of different hydrogen production routes.

The steam methane reforming process has more potential for energy conversion from raw materials to products without losing much energy during

Table 1.2 Energy efficiency of different hydrogen production routes

Conversion technology	*Energy efficiency in conversion (%)*	*Energy efficiency with carbon capture and storage system (CCS) in %*
Coal gasification	60	43
Steam methane reforming	75	60
Biomass gasification	35–50	Not feasible
Thermochemical water splitting	20–45	Not feasible
Water electrolysis	50–70	Not feasible
Methane pyrolysis	58	58

transformation, but the inclusion of a carbon capture and storage system (CCS) and greenhouse gas control systems reduces the energy conservation efficiency to 60% from a whopping 75% efficiency. Additional infrastructure also adds to investment cost by more than 30% (Sánchez-Bastardo et al., 2021).

1.5.2 Challenges and innovations

Hydrogen has significant potential to neutralize the carbon destabilization status of fossil fuels. To switch from a fossil fuel energy system to a non-fossil fuel energy system, the major roadblock is infrastructure investment and capital cost. A challenge is to create the facilities for bulk-scale manufacture, storage and transportation of hydrogen. With existing facilities, it is very difficult to fill the supply–demand energy gap only with the support of hydrogen energy. Due to urbanization, population growth and industrialization, the energy demand is soaring sky high, and energy distributors find it difficult to accomplish the mission to attain net-zero energy system. Unfortunately, hydrogen cannot single-handedly meet the energy demand, but it can act as a buffer and bridge the gap between unreliable renewable sources and depleting fossil fuels. A system called "power-to-'hydrogen'-to-power" is practiced in real time, which utilizes renewable energy to produce and store hydrogen that can drive a gas turbine to produce green electricity without any carbon emissions. In the energy usage domain, the aviation sector is totally outdated and not ready to cope with the emerging green energy initiatives and still uses petroleum-based fuels. Aviation authorities have envisioned the future of hydrogen airplanes, fuelling stations and safety infrastructure. Challenges like higher combustion temperature, formation of contrails, storage considerations and fire-explosion risks need to be analysed before retrofitting the existing airplane fuel system. While enough production of hydrogen to meet the current demand is the first challenge, it is followed by challenges posed for storage space. Salt caverns, exhausted oil reservoirs and deep sea are the promising ways to store hydrogen. All the technologies used for hydrogen production have their own merits and limitations. High capital investment, storage problems, poor energy transformation efficiency and undesired byproduct formation are a few challenges to be elucidated before transformation is completed.

1.6 NECESSITIES AND METHODS OF HYDROGEN STORAGE

Hydrogen storage is an important aspect for the sustained transition from fossil fuels to non-polluting fuels. An economically beneficial and energy-constructive hydrogen storage system is mandatory for portable, backup and transportable energy applications. Advanced storage methods are

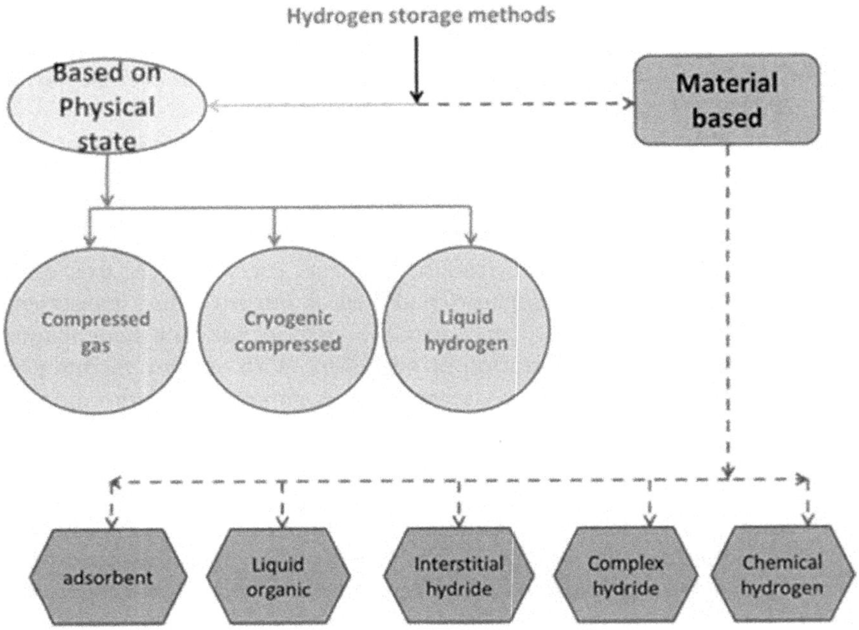

Figure 1.2 Hydrogen storage methods.

needed because hydrogen has the highest energy value per unit mass and the lowest energy per unit volume. Hydrogen can be stored in either gaseous or liquid form. Storage in gas form requires high pressure, whereas storage in liquid form requires cryogenic temperature. Hydrogen storage methods are summarized in Figure 1.2. Strong uncertainty and volatility of the hydrogen source consequently affect the demand and supply of constant power.

1.6.1 Physical and material-based storage

Hydrogen can be stored in different ways, depending on the nature of the end use and the characteristics of the storage system. A geological storage system refers to the storage of hydrogen gas in geological formations, including exhausted oil and gas reservoirs, salt caverns and non-mineable coal seams. This method has the advantage of storing a large volume of hydrogen in natural formations, requiring no additional infrastructure. The challenges include pukka characterization of natural sites using sophisticated analytical facilities and preventing unnoticed leakage of hydrogen through interstitial spaces between the formations. A standard method to store hydrogen is the use of high-pressure cylindrical tanks, where the stored gas is pressurized up to 690 bars. This method guarantees high energy density during storage and an easy transportation method. The inherent drawback of this method is the need for specialized materials of construction to fabricate pressure vessels to withstand very high pressure and explosion damage. Multistage compressors used to increase the pressure of hydrogen for storage add to the cost factor and cumbersome energy consumption. Hydrogen can also be stored in liquid form, but it requires refrigeration cooling facilities to bring down the temperature to –253°C. Liquid hydrogen is appropriate for long-term storage applications, including rocket fuel, space vehicles, and deep space travel capsules. Moreover, storage facilities for liquid hydrogen require apt material of construction that could bear small pressure values. One of the major drawbacks of this method is the difficulty in maintaining low temperature with insulating materials. Identifying suitable insulating material having poor thermal conductivity is the major criterion to maintain the phase and state of liquid hydrogen; otherwise, due to the fugacity of hydrogen molecules, leakage occurs frequently. Insulation material thickness is another important parameter that decides the rate of heat exchange between the surrounding atmosphere and the system (liquid hydrogen inside the container). To minimize heat loss rate, critical insulation thickness should be used.

In a material-based storage system, hydrogen is stored in solid materials such as metal hydrides, metal organic frameworks and carbon-based nanomaterials, where hydrogen is absorbed and stored inside the materials. This method is suitable for both portable and stationary energy requirements, where the standards to meet are high energy density and lightweight.

1.6.2 Engineering problems in storage

Hydrogen is the undisputable future fuel to drive the growth phase in industries, transportation, aviation, and sea transport. At present, hydrogen is stored in vessels at high pressure and delivered at normal pressure at the point of use. In such cases, hydrogen embrittlement (HE) is a major concern that needs to be addressed because it weakens the metal structure and the integrity of storage vessels. It occurs when hydrogen is absorbed by the material, leading to ductility reduction and decrease in fracture toughness. The risk of HE is reduced by the use of low hydrogen content welding accessories, providing multilayer surface coatings, and heating the storage vessel to pull out the adsorbed hydrogen molecules. Compressed hydrogen storage system is simple, easy to operate and does not require any complex infrastructure, but it is not suitable for long distance travel energy need and large volume storage requirements. Liquid hydrogen storage involves cooling the hydrogen gas to a very low temperature and storing it in insulated vessels. Liquid hydrogen has a low boiling point, which makes it difficult to store and handle. Metal hydrides, a solid material, absorb hydrogen in pores and release it wherever required through chemical reactions. But this method

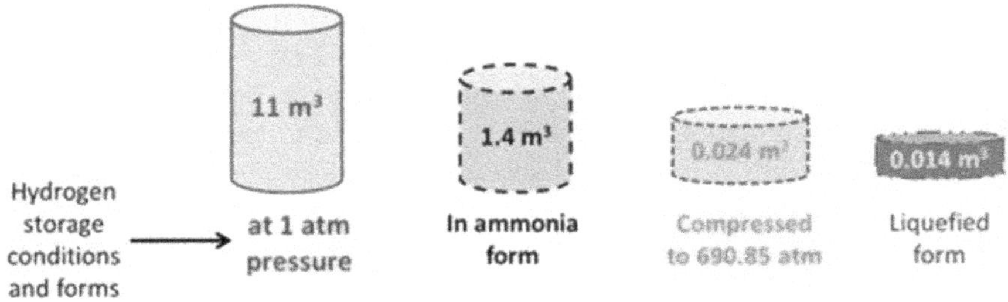

Figure 1.3 Hydrogen in different forms and conditions.

has low storage capacity for hydrogen. Storage using carbon nanotubes is an upcoming technology that can store a large volume of hydrogen at room temperature. The challenge lies with the development of perfect nanotubes and controlling the adsorption and desorption rate of hydrogen molecules inside the porous structure. Hydrogen storage poses a great challenge and puts a barrier to widespread applications as an alternate fuel and attains sustainability in energy demand. Storage container volume required to contain 1 kg of hydrogen in different forms and conditions is illustrated in Figure 1.3. Hydrogen storage challenges must be addressed to achieve the goal of zero-emission standards and energy industry self-reliance.

1.6.3 Safety measures in hydrogen storage

Hydrogen can be stored either in gas form or in liquid form, depending on use requirements. Gas storage requires high-pressure containers (operating pressure range 98.7–986.9 atm) while cryogenic temperature condition is required for liquid form storage. During compression of hydrogen, the pressure increases and hence suitable storage vessels constructed with materials resilient to high-pressure conditions are desired. Based on the material of construction, four categories of vessels are used to store hydrogen gas. Type I vessels are constructed from stainless steel material to withstand pressures of up to 200 bar. It is heavier because metals are fully used for fabrication and construction. Type II storage vessels are made of aluminium reinforced with glass fibres to increase toughness and ability to withstand pressure. Type III is made of metal fortified with organic material, primarily a combination of aluminium/carbon composite, to withstand pressure up to 700 bar. This vessel is more expensive than other types due to its composition. Type IV is a non-metal type container, carbon fibre with polymer lining. This type is very light in weight due to the absence of metals. Prolonged storage at mild conditions or short-duration storage at extreme conditions of hydrogen in any form are detrimental to the chemical and physical properties of storage vessels. Hydrogen can cause the material to become brittle, to lose tensile strength and fracture toughness and lead to fatigue of metals. Recommended materials for hydrogen storage vessels are austenitic stainless steel, aluminium alloys, ferritic steel alloys and copper alloys. The materials to be avoided are martensitic steels, ductile cast iron, nickel alloys and titanium alloys. With regards to hydrogen plant safety, the optimal location and safety features incorporated determine the overall safety. Instead of indoor storage, the outdoor storage is most preferred to meet any eventuality, hazards due to accidental leaks, fires and explosions. A few guidelines to consider while constructing and identifying optimal location of storage places include the provision of good ventilation and exhaust to dilute the accumulated hydrogen concentration, providing adequate separation distance between production unit and storage place, storage vessels not exposed to sunlight directly, only diffused radiations are permitted and refrigeration arrangements to maintain ambient temperature (should not exceed more than 50°C).

Ventilation and fresh air circulation are very important in places where hydrogen storage vessels are constructed. It guarantees quick dissipation of leaked hydrogen gas without forming a flammable fuel-air mixture. When hydrogen is leaked, it rises due to buoyancy and collects near the top of the room. Hence, a gas detector system, fire alarms and flame suppressors are to be fitted near the vicinity of the room ceiling. The general principle is that hydrogen rises up because of its low density and lightweight than air and occupies the highest point of the room for a short duration. Hence, the ideal location to monitor the gas concentration is the tallest portion of the room. Leaks are a major concern in hydrogen handling and storage because of the simple molecular weight and structure. Leak detectors are installed at joints, seals, storage vessels and valves to ascertain leaks in storage and handling. On-spot detection includes the use of a soap solution to apply in leak-suspected places to form soap bubbles due to the fleeting pressure of hydrogen gas. This method is a crude way to detect leaks of any gas. Once a leak is confirmed, sophisticated hydrogen detectors are used to measure the eluted hydrogen concentration. In addition to all these features, understanding the physical and chemical properties of hydrogen and training in safe hydrogen handling practices are a major component to ensure the safe use of hydrogen. Besides these, regular tests for storage leak, transport mishap and maintenance of hydrogen production systems show that hydrogen can be produced, stored and dispensed safely.

1.7 MODES OF HYDROGEN TRANSPORTATION

Hydrogen supply to the end-user can be achieved using high-pressure tankers or through pipelines. Hydrogen storage and transportation in high-pressure tankers face indistinguishable challenges as seen in high-pressure storage vessels. High-pressure hydrogen transport is achieved using road, rail or sea route transportation, which offers fewer challenges to reach various destinations with the existing transport networks and routes. Pipeline transport of hydrogen is suited for distributing large quantities of hydrogen. Already, pipelines are available to transport natural gas across the borders of countries. The same existing infrastructure can be used to transport hydrogen gas with restricted modifications.

Road transport is a common practice for distributing hydrogen over short distances. Compressed hydrogen can be stored in metal containers (like LPG gas cylinders) and moved to designated destinations via road route. Hydrogen stored in solid-state materials can be transported comfortably using roads. Rail transport is preferred for longer distances and massive volume of hydrogen. Ships are used to transport liquid hydrogen between countries for commercial and trade reasons. Facilities needed are the same as those of liquefied natural gas (LNG) tankers, but suitable thermal insulation is required in order to prevent evaporation loss. Pipeline transport is suited for long-range logistics, but the embrittlement process hampers the service life of pipelines.

1.7.1 Sustainable supply chain

After mitigating the challenges associated with production and storage of hydrogen, if a sustainable supply chain is not followed, then it is difficult to achieve the target of zero emissions within a short period. The production-supply chain of hydrogen is planned using blockchain technology, advanced analytics and modern software tools, and when followed meticulously, it can reduce environmental impacts and fuel wastage. A supply chain starts from production, storage and transportation and ends with delivery to users. The transportation cost can be reduced when hydrogen is produced, stored and used locally in a small region. But the hindrances are the lack in availability of meaningful quantities of pure hydrogen from Mother Earth, infrastructure requirements for large capacity production, specialized transportation requirements and safety issues in handling. The coordination and integrity between renewable energy producers, electrolyser dealers, transportation companies and end-users ensures a seamless supply chain. The long-term plan for a sustained supply chain includes in situ production of hydrogen, compact and concrete storage system, low-distance transport, marketing spots and frequent commercial use (Sgarbossa et al., 2022).

1.7.2 Safety measures during transportation

Hydrogen handling is associated with hazards like fire, explosion and asphyxiation. Hydrogen's individualities like high dispersion coefficient and low atomic mass cause hydrogen to fade away rapidly without causing an explosion in open areas/highly ventilated areas. Other hazards are reduced by means of proper training of operators and maintenance of equipment. Nitrogen, an inert gas, is used to purge hydrogen-related processes to prevent the formation of a flammable hydrogen/oxygen mixture. Sufficient ventilation can prevent asphyxiation of persons working in confined areas. Hydrogen burns with an invisible blue flame, which requires sophisticated flame detectors to prevent flame propagation.

Hydrogen gas cylinders must be stored in an upright position to prevent them from being knocked over during road transport. Protective valve caps should be in place to prevent leakage during storage and transportation. Storage in cylinders should be in accordance with CGA P-1 standards for "Safe Handling of Compressed Gases in Containers", and the cylinder should be kept away from sunny weather, rainfall and mechanical damage (Voltaix, Inc., 2021). Hydrogen gas cylinders storage and use place should be away from any oxidizing gases at least at a distance of 6.1 m. When the above protocols are followed during the production, storage and transportation of hydrogen fuel, then everyone's ambition to achieve self-sufficiency and zero carbon emission can be reached sustainably without devastating environmental components.

1.8 CONCLUSION

Hydrogen is an ideal fuel that encompasses the aspects of fuel qualities and environmental concerns. Decarbonization, zero emissions at the user point and energy production from eco-friendly sources are the prime objectives of human beings to prevent Mother Earth from irreversible damage. Hydrogen energy is pivotal in minimizing greenhouse gas emissions, plays a major role in developed nations' decarbonization policies and is promoted as the energy future for developing countries. The production of green hydrogen is energy-intensive and expensive because of the dependency on fossil fuels. Economic constraints pose a great challenge to immediately shifting from carbon-intensive fossil fuels to green hydrogen. In contrast, developed nations act as the pioneers to develop a policy framework and infrastructure establishment. Also, the technical challenges faced in hydrogen storage and transportation can be overcome by universal collaborative efforts to share emerging technologies, intellectual property rights commercialization, strategic investments and improved logistics – supply chain management. Policy makers, technocrats, industrialists and the common public should understand the concealed meaning of "one earth, one future" for life to flourish and sustain on planet earth.

REFERENCES

American Chemical Society. (2020). Molecule of the week archive, ACS, Washington DC, USA. www.acs.org/molecule-of-the-week/archive/h/hydrogen.html

Awad, A.H., Veziroğlu, T.N. (1984). Hydrogen versus synthetic fossil fuels. *International Journal of Hydrogen Energy*, 9(5), 355–366. https://doi.org/10.1016/0360-3199(84)90055-7

Balat, M. (2008). Potential importance of hydrogen as a future solution to environmental and transportation problems. *International Journal of Hydrogen Energy*, 33(15), 4013–4029. https://doi.org/10.1016/j.ijhydene.2008.05.047

Grochala, W. (2015). First there was hydrogen. *Nature Chemistry*, 7, 264. https://doi.org/10.1038/nchem.2186

International Energy Agency. (2024). Hydrogen production and infrastructure projects database, IEA, Paris Cedex, France. www.iea.org/data-and-statistics/data-product/hydrogen-production-and-infrastructure-projects-database

Le, P.A., Trung, V.D., Nguyen, P.L., Phung, T.V.B., Natsuki, J., Natsuki, T. (2023). The current status of hydrogen energy: an overview, *RSC Advances*, 13(40), 28262–28287. https://doi.org/10.1039/D3RA05158G

Ministry of New and Renewable Energy. (2024). Brief on green hydrogen, Government of India. https://mnre.gov.in/en/hydrogen-overview/

Naveed Khan, M. (2025). Hydrogen as future sustainable energy resource: an insight into technological advancements in hydrate-based hydrogen storage. *International Journal of Hydrogen Energy*, 97(6), 1386–1398, https://doi.org/10.1016/j.ijhydene.2024.12.020

Royal Society of Chemistry. (2024). www.rsc.org/periodic-table/element/1/hydrogen

Sánchez-Bastardo, N., Schlögl, R., Ruland, H. (2021). Methane pyrolysis for zero-emission hydrogen production: a potential bridge technology from fossil fuels to a renewable and sustainable hydrogen economy. *Industrial & Engineering Chemistry Research*, 60(32). https://doi.org/10.1021/acs.iecr.1c01679

Sgarbossa, F., Arena, S., Tang, O., Peron, M. (2022). Renewable hydrogen supply chains: a planning matrix and an agenda for future research. *International Journal of Production Economics*, 250, 108712. https://doi.org/10.1016/j.ijpe.2022.108712

US Department of Energy. (2017). www.energy.gov/eere/articles/hydrogen-clean-flexible-energy-carrier

US Department of Energy. (2024). www.energy.gov/eere/fuelcells/fuel-cells

Utgikar, V.P., Vijaykumar, J., Thyagarajan, K. (2011). Refinement of motivity factor in comparison of transportation fuels. *International Journal of Hydrogen Energy*, 36(5), 3302–3304. https://doi.org/10.1016/j.ijhydene.2010.12.097

Van Vucht, J.H.N., Kuijpers, F.A., Bruning, H.C.A.M. (1970). Reversible room-temperature absorption of large quantities of hydrogen by intermetallic compounds. *Philips Research Reports, U.S. Department of Energy, Office of Scientific and Technical Information*, 25(2), 133–40 CODEN: PRREA9; ISSN: 0031-7918. www.osti.gov/biblio/4129528

Veziroglu, T.N., Barbir, F. (1992). Hydrogen: the wonder fuel. *International Journal of Hydrogen Energy*, 17(6), 391–404. https://doi.org/10.1016/0360-3199(92)90183-W

Voltaix, Inc. (2021). MSDS Document Number – H000, Material Safety Data Sheet for Hydrogen (H_2), University of Illinois Urbana-Champaign. www-s.mechse.uiuc.edu/cleanroom/files/Current%20MSDS/Hydrogen.pdf

Chapter 2

Renewable energy sources, power demand, policies

An Indian perspective

Vimal Singh Pal, Vijayeta Pal, and Dan Bahadur Pal

2.1 INTRODUCTION

In India, the increasing population and expanding economy are driving up the energy demand. It is a challenge to fulfil this requirement while also addressing environmental considerations. Renewable energy has become a vital participant in India's pursuit of a sustainable and environmentally friendly energy solution (Charles et al., 2024). As the global population expands, there is a corresponding rise in the need for energy to meet the expanding demand for electricity in residential, commercial, and communal settings (Binyamin et al., 2024). Since the Industrial Revolution, several nations have increasingly relied on fossil fuels as their energy sources (coal, oil, and gas). The phenomena affect global climate and human health. One of the most significant issues of concern in the 21st century is global warming and climate change, primarily attributed to *greenhouse gas* (GHG) emissions due to the increased utilization of fossil fuels (Singh, 2024). Approximately 75% of worldwide GHG emissions are attributed to the combustion of fossil fuels for energy production. Fossil fuels are accountable for significant levels of localized air pollution, which poses a health concern resulting in a minimum of 5 million premature fatalities annually (Anjum et al., 2024). To cut emissions and mitigate the adverse effects of local air pollution, there is an urgent global imperative to transition expeditiously towards energy sources characterized by low carbon emissions, such as nuclear and renewable technologies. The development and proliferation of renewable energies play a crucial role in upholding sustainable energy levels and safeguarding the environment against the effects of climate change (Raihan et al., 2024c). The present world is confronted with the dual imperative of meeting the escalating energy requirements and mitigating GHG emissions while enhancing energy efficiency. Renewable energy offers an effective solution to the pressing challenge of meeting energy demands while reducing emissions (Batra, 2023). Furthermore, it is crucial to note that this factor also plays a significant role in the realm of energy security, in addition to the enhancement of environmental preservation and the augmentation of employment

DOI: 10.1201/9781003581536-2

opportunities throughout diverse nations. Thus, renewable energy sources are currently of great interest to nations worldwide due to their pollution-free nature, widespread availability, cost-effectiveness, and abundant reserves on Earth (Raihan et al., 2024c). Renewable energy technology necessitates the utilization of naturally occurring energy resources, for instance, solar radiation, wind, water, biomass, and *geothermal energy*, among others. Numerous countries widely recognize renewable energy as a pivotal and influential aspect of the latest energy technological advancements (Hassan et al., 2024). Consequently, these governments have established ambitious objectives for renewable energy as an integral component of their policy frameworks. The significance of low-carbon development has become increasingly crucial due to the progress in national policies and the advancement of renewable energy technologies (Siddik et al., 2023). Recently, there has been more focus on the utilization of clean energy sources for attaining the Sustainable Development Goals (SDGs) proposed by the United Nations that serve as a blueprint for the advancement of global human well-being, material conditions, and the preservation of the natural environment, with a projected timeline of achievement by the year 2030.

Renewable energy is one of the most critical factors contributing to the sustainable development of the country (Algarni et al., 2023). There are different sources of renewable energy available in the country, but the efficient usage of the same counts for the growth and development of the country. Though India is using renewable sources of energy, resources are still not sufficient to meet the demand of increasing population and to tackle the environmental conditions (Algarni et al., 2023).

India is one of the fastest-growing developing nations, characterized by strong economic growth. To sustain its economic growth, India needs to meet increasing energy demand from various energy sources. With India's financial and industrial development, energy consumption is also growing exponentially (Singh et al., 2024). It is also predicted that with the current trend of energy consumption, India will increase its global energy consumption demand from 5% in 2016 to 11% in 2040. India has achieved industrialization by increasing its share in the worldwide economy. India also needs to accommodate the issues of a growing population and environmental degradation. With the nation's increasing population, the gap between the demand and supply of energy is also anticipated to increase in the upcoming days (Yawale et al., 2023). India's annual population growth rate is 1.18%, responsible for almost 18% of the global population. India's population in 2022 was 1.417 billion, which would be 1.512 billion in 2030, 1.605 billion in 2040, and 1.658 billion in 2050 (International Energy Agency (IEA), 2022). Every year, India's population increases more than that of any nation worldwide. With the increasing population and economic growth, the energy demand of India is also growing exponentially, and it is expected to be the fastest energy-consuming nation among powerful nations by 2040 (Singh et al., 2024).

India, with a diverse and culturally rich community of 1.3 billion people, is the third-largest energy-consuming country in the world, driven by rising wages and improved standards of living. India experiences a tremendous rise in its energy demand due to its growing economy, population, urbanization, and industrialization (Ministry of Power, Government of India, 2023). As a direct consequence, India faces a perplexing energy crisis because of its limited conventional energy sources in comparison to the required energy needs (Husain et al., 2024). Since 2000, energy use has skyrocketed, with coal, oil, and solid biomass still providing 80% of the demand. Hence, it becomes imperative that India starts considering and investing in alternate sources of energy (Ministry of New and Renewable Energy (MNRE), 2023d).

India's growth rate for renewable energy has been among the highest in the world. Globally, it ranks fourth in terms of both RE and wind power capacity and stands at the fifth position when it comes to solar power capacity (Giri et al., 2024). India ranked third in the 2021 Renewable Energy Country Attractiveness Index, attracting numerous foreign investors, and currently has a total installed renewable energy capacity exceeding 160 GW (including large hydro) (Ministry of New and Renewable Energy (MNRE), 2023d). By the end of 2022, the government intends to reach a cumulative RE capacity of 175 GW and boost it further to 500 GW by 2030 (Ministry of New and Renewable Energy (MNRE), 2022c). The importance of renewable energy is as follows:

1. **Navigating Climate Change:** The adoption of renewable energy sources is a crucial step in mitigating the impacts of climate change (Girgibo et al., 2024). By reducing reliance on fossil fuels, India can contribute to global efforts aimed at curbing GHG emissions.
2. **Energy Security:** Diversifying the energy mix with renewables enhances energy security. By tapping into abundant and indigenous sources like sunlight, wind, and water, India can reduce its dependence on imported fossil fuels. This strengthens its energy resilience (Ministry of New and Renewable Energy (MNRE), 2022d).
3. **Job Creation:** The renewable energy sector has the potential to create a significant number of jobs (Hassan et al., 2024). From manufacturing solar panels to constructing and maintaining power plants, the entire value chain offers employment opportunities. This fosters economic growth.
4. **Rural Electrification:** Renewable energy can play a pivotal role in electrifying rural areas (Mishra et al., 2023). This is primarily in the form of decentralized solar power. Off-grid solar solutions provide

a sustainable and cost-effective means of bringing electricity to remote communities. This improves living standards and fosters development.

5. **Technological Advancements:** Investing in sustainable energy stimulates innovation and technological advancements (Asghar et al., 2024). As India continues to develop its clean energy infrastructure, a ripple effect of innovation extends beyond the energy sector. This is influencing various industries.
6. **Cost Competitiveness:** The cost of renewable energy technologies, especially solar power, has witnessed a significant decline in recent years (Hassan et al., 2024). This cost competitiveness makes renewable energy an attractive option. This is not just from an environmental standpoint but also from an economic perspective.

Even though India has achieved a fast and remarkable economic growth, energy is still scarce. Strong economic growth in India is escalating the energy demand, and more energy sources are required to cover this demand (Nassar and Khaleel, 2024). At the same time, due to the increasing population and environmental deterioration, the country faces the challenge of sustainable development.

The estimated potential of wind power in the country during 1995 was found to be 20,000 MW (20 GW), solar energy was 5×10^{15} kWh/pa, bioenergy was 17,000 MW, bagasse cogeneration was 8,000 MW, and small hydropower was 10,000 MW. For 2006, the renewable potential was estimated as 85,000 MW with wind 4,500 MW, solar 35 MW, biomass/bioenergy 25,000 MW, and small hydropower of 15,000 MW (Ministry of New and Renewable Energy (MNRE), 1995). According to the annual report of the Ministry of New and Renewable Energy (MNRE) for 2017–2018, the estimated potential of wind power was 302.251 GW (at 100-m mast height), of small hydropower 19.749 GW, biomass power 17.536 GW, bagasse cogeneration 5 GW, waste to energy (WTE) 2.554 GW, and solar 748.990 GW. The estimated total renewable potential amounted to 1,096.080 GW, assuming 3% wasteland (Ministry of New and Renewable Energy (MNRE), 2018). India is a tropical country and receives significant radiation, and hence the solar potential is very high (Kumar et al., 2023).

2.2 HISTORICAL CONTEXT OF INDIA'S ENERGY SECTOR

Energy Status before Independence: Before 1947, when India became independent, the energy sector was basically based on conventional fuels such as biomass, wood, and coal (Singh, 2024). A lot of the country's energy consumption was through biomass in the rural areas and through coal in the cities. Most of the energy consumed at the time was used for domestic purposes – to cook, heat, and light (Ministry of Power, Government of India, 2022e). Most villagers used wood, dung, and farm wastes as their sources of energy (Hoyos et al., 2024). Coal came into use in towns and industries during the British period.

Coal was an essential fuel in the Indian energy system during those times, considering how coal supported the growth of industrial and transport sectors (Ministry of Coal, Government of India, 2022). Actually, the British first introduced coal mining in India in the late 18th century, which grew exponentially in the mid-19th century (Dé, 2024). In fact, coal was the staple of the Indian energy system for power through the railways, steam engines, and recently emerging industries like textiles and steel production (Ministry of Coal, Government of India, 2023). However, despite abundant coal deposits, energy availability was very low, and the vast majority of the population, especially the rural population, suffered from the lack of modern energy facilities (Al Mubarak et al., 2024).

By the turn of the century, electricity production was generally small, with localized grids serving both urban and industrial areas. The first hydroelectric power plant was commissioned at Darjeeling in 1897, but this was the start of electricity generation in the country (Sarkar, 2024). General electrification remained at a low level, and only a few cities could adopt the newly emerging energy facilities. Rural India continued to rely mainly on traditional sources of energy.

2.2.1 Post-independence development

From 1947, which was the year of independence, India began moving towards industrialization and modernization (Ramesh, 2024). In the process, the energy sector was exploited grossly. The Indian government did all this to help increase the energy infrastructure, especially regarding thermal power plants that utilize coal for economic growth (Ministry of Power, Government of India, 2022g). With the overall objective of providing all the energy that the industry and growing towns needed, in the early years after independence, the principal fuel utilized was coal, for electricity generation purposes as well as for industrial use (Singh, 2024).

Hydropower was one of the most reliable renewable energies in the 1950s and 1960s. Moreover, it constituted a high percentage of power generation. However, by the 1970s, coal had acquired a position of prime dependency for India, and the tempo of development of hydropower projects declined (Central Electricity Authority (CEA), 2022a). Since its independence, India has followed a policy of self-reliance and dominance of the public sector. Government companies dominated both generation and distribution of power (Senawiang et al., 2024). Availability increased in urban areas, but access to electricity is quite limited in rural areas. People are still more dependent on fuel resources like wood and biomass in rural areas.

2.2.2 An era of liberalization and energy reform

After all, the Indian energy sector did undergo a significant transformation with the economic liberalization of 1991 (Ahmed and Chakraborty, 2024). The sector remained very weak then, with less investment and bad services. The government-dominated industry could not cope with the continuously increasing demand for electricity, resulting in both power shortages and erratic supplies. Thus, the 1991 reforms were aimed at solving the economic problems of the country. Roughly in the same time frame, this period ushered in privatization, deregulation, as well as private investment in the energy sector (Ministry of Power, Government of India, 2022f).

This also included liberalization for private companies to enter the energy sector, mainly in the generation of power. Significant reforms of the Electricity Act 2003 have included bifurcation of state electricity boards into different units, opening up the sector to private investment and competition (Vardhan et al., 2024). The very law included making electric supply available to all the villages, encouraging renewable sources, and providing open access to electricity markets. Private investment into independent power producers and both conventional and non-conventional energy projects mushroomed in these years (Ministry of Power, Government of India, 2022c).

Liberalization in the thermal power sector triggered many private coal-based power plants that somewhat eased the power shortage of the country (Singh, 2023). During this time, natural gas also emerged as a power generation alternative, with domestic gas fields being developed alongside investments in liquefied natural gas (LNG) infrastructure. However, the country faced several problems, including technological obsolescence, power losses during transmission, and high coal consumption, all of which were becoming increasingly damaging to the environment (Ministry of Power, Government of India, 2022a).

India gradually progressed in this century for the energy of tomorrow with renewable energy (Raihan et al., 2024c). The government developed various policies that urged the private sector investment in hydropower, wind power, and solar power. Although the whole process was not hectic in the 1990s for renewable energy, the path for the future was fully constructed then itself (Kaur and Rehman, 2024). For clean energy, the government made a ministry known as the Ministry of New and Renewable Energy (MNRE), and most state governments offered incentives to projects based on renewable energy.

2.2.3 India's shift towards renewable energy

Indian energy policy started changing in the early 2000s, when environmental issues relating to climate change started coming into relevance. It was realized that the vast dependence on fossil energy would accelerate pollution and aggravate global warming, even as it weakened energy security (Hassan et al., 2024). Thereby, India focused on renewable energy. International agreements, especially on the Kyoto Protocol and the Paris Agreement, have dramatically altered the Indian energy policy.

The environmental damage caused by fossil fuels and the falling cost of renewable energy technologies prompted India to move quickly towards renewable energy (Mondal et al., 2024). Emphasis was placed on solar energy. The Jawaharlal Nehru National Solar Mission came into operation in 2010 with a target of achieving 100,000 gigawatts of solar capacity by 2022 (Ministry of New and Renewable Energy (MNRE), 2010). That meant the first major milestone in India's shift into life under renewable energy – the country used to rely a lot on traditional means of energy and slowly integrated solar power into its policy on energy (Kumar et al., 2024). This is part of a global trend characterized by increasing investments in clean energy technologies and cooperation on climate change issues (Adelekan et al., 2024). Its commitment to the International Solar Alliance, for example, has reflected India's global leadership role in renewable energy.

The energy sector of India has undergone significant changes in the last hundred years. From dependence on traditional fuels and coal after independence to the development of thermal and hydropower, and then again toward renewable energy in the 21st century, the country's energy sector has, in a way, undergone continuous change (Singh, 2024).

2.3 INDIA'S CURRENT ENERGY STRUCTURE AND ELECTRICITY DEMAND

Energy Consumption Patterns: A characteristic of India's energy consumption is rapid economic growth, industrialization, and urbanization, and it shows widespread increase in the country's energy demand from various sectors, with primary consumers being residential, industrial, commercial, and agricultural sectors (Majeed et al., 2023). All these sectors affect India's total energy demand.

Residential Sector: The consumption of electricity by the residential sector accounts for about 25% of the total electricity consumption in India. It includes the kinds of electricity used in houses: lighting, heating, cooling, cooking, and household appliances (Central Electricity Authority (CEA), 2022b). Urban houses consume more electricity than rural houses because cities use more electrical appliances and there is better access to electricity (Osman et al., 2024). In short, it has improved dramatically over the past decade and is expected to continue raising electricity demand in the future.

Industrial Sector: The industrial sector accounts for the largest proportion of the total energy consumption in the country, which is about 40%–45%. Highly energy-consuming industries include those pertaining to steel, cement, chemical manufacturing, and textiles; electricity, coal, and natural gas are the main fuels on which these industries depend (Ministry of Power, Government of India, 2022b). With the industrial development of the country, this sector's energy demand has increased, and this sector has been the highest user of energy (Wang et al., 2024).

Commercial Sector: This includes offices, shops, hotels, and hospitals that account for around 8%–10% of the overall electricity consumption. Electricity is primarily consumed in the form of air conditioning, lighting, elevators, and office equipment (Central Electricity Authority (CEA), 2022c). Its energy demand is on the rise with growing service sectors in the Indian urban areas (Kumar et al., 2023). In this category, there is a growing need for cooling.

Agricultural Sector: The agricultural sector accounts for around 15%–20% of the aggregate energy consumed in India. Electricity is mainly used for irrigation because water pumps are widely used (Ministry of Coal, Government of India, 2022a). This sector is one of prime importance to the Indian economy and its energy demand depends on water requirement, weather, and availability of electricity in rural areas (Raihan et al., 2024c). Supply of cheap electricity to farmers in many states has increased the energy demand in this sector.

Supremacy of Fossil Fuels: Renewable energy, though gaining importance, still has fossil fuels (coal, oil, and natural gas) dominating India's energy system (Eweade et al., 2024). These are still the primary sources for power generation, transportation systems, and industries. Their use in the country can be justified as these fuels are readily available, a lot of infrastructure is already in place to produce them, and their prices have historically been low as well.

Coal: India has the largest energy source in the form of coal. Coal contributes to 55% of the electricity generation capacity in India (Central Electricity Authority (CEA), 2022a). The Indian power production is made stable and reliable by coal-fired power plants (Kumar et al., 2024). India is the second largest producer as well as consumer of coal in the world, and its major reserves can be found in Jharkhand, Odisha, and Chhattisgarh (Ministry of Coal, Government of India, 2022a). However, the utilization of coal has severe environmental and health consequences, contributing significantly to air pollution and greenhouse gas emissions, thereby prompting discussions on the need for clean energy (Raihan et al., 2024c).

Oil: Though, in India, oil is mainly used by the transport sector, it is also utilized for industrial processes as well as the generation of power (Prajapati et al., 2024). The country imports more than 80% of its oil requirements, and therefore, the import bill is relatively high. This has posed a challenge for the energy security of the country as well as its macro-economic balance. The other important reason is that the transport sector is a source of air pollution to a large extent, mainly in cities, so efforts are being made towards the production of electric vehicles (EVs) and minimum oil use (Ministry of Power, Government of India, 2022d).

Natural Gas: It constitutes a mere 6%–8% of India's energy basket. It is used for power generation, industrial work, and fertilizer production. It burns cleaner than both coal and oil, which is why the country is trying to consume more of it. The government has new targets to build more LNG terminals and gas pipelines to take the country towards a gas-based economy (Mehrotra et al., 2024).

2.3.1 Contribution of renewable energy

In the past two or three years, India has taken some significant steps to increase the share of renewable energy in its power generation capacity. By 2024, renewable energy – Sun, wind, hydropower, and biomass – will account for nearly 40% of the total installed power capacity of India (Gupta et al., 2024). Table 2.1 shows the year-wise achievements of installed renewable energy capacity.

Solar Energy: In India, solar energy is the most critical component of its renewable energy policy. India receives more than 300 sunny days in a year and hence has tremendous solar power potential (Vijayalaxmi et al., 2024). India launched the Jawaharlal Nehru National Solar Mission (JNNSM) in 2010, which aims to achieve a cumulative solar power capacity of 100 GW by 2022, and it has almost reached the target (Srivastava et al., 202). Huge solar farms, rooftop solar projects, and decentralized solar plants have encouraged the rapid generation of solar power. Table 2.2 shows the targets and achievements in Phase I, Phase II, and Phase III and Table 2.3 shows the application-wise status of the installations under the off-grid and decentralized solar PV applications.

Brief on Off-grid Solar PV Programme:

Wind Energy: Wind energy is an essential constituent of India's renewable energy too (Husain et al., 2024). Due to fairly strong wind sources, states like Tamil Nadu, Gujarat, and Maharashtra have made India one of the world's biggest wind power markets. By 2024, India will have installed over 40 GW of wind power capacity (Sikarwar et al., 2024). Newer technologies such as larger wind turbines and better grid systems make wind power efficient while operational. Table 2.4 shows the potential of wind energy in India.

Hydropower: Hydropower has been a part of India's energy system for generations, especially in northern Indian states where rivers have a lot of resources (Utturkar et al., 2024).

Biomass: Among renewable energies, this sector is only the smallest with biomass energy, which is derived from organic matter, such as agricultural waste and forest residues (Ali et al., 2024). Figure 2.1 shows the

Table 2.1 Year-wise achievements of installed renewable energy capacity (MW) (excluding large hydro power)

Sector	*Cumulative achievements (till-31.03. 2014)*	*2014-15*	*2015-16*	*2016-17*	*2017-18*	*2018-19*	*2019-20*	*2020-21*	*2021-22*	*2022-23*	*2023-24*	*Cumulative achievements (till-31.10.2024)*
Wind power	21,042.58	2,311.77	3,423.05	5,502.37	1,865.23	1,480.97	2,117.79	1,503.3	1,110.53	2,275.55	3,253.38	47,716.72
Solar power	2,821.91	1,171.62	3,130.36	5,658.63	9,563.69	6,750.97	6,510.06	5,628.8	12,760.5	12,783.8	15,033.24	92,119.18
Small hydro power	3,803.68	251.68	218.11	106.38	105.95	107.34	90.01	103.65	62.09	95.4	58.95	5,077.25
Biomass (bagasse) cogeneration	7,419.23	295.67	304.85	161.95	519.1	402.7	97	173.37	59.69	0.00	0.00	9,806.42
Biomass (non-bagasse) cogeneration	531.82	60.05	59.24	2.2	9.5	12	0.00	97.24	0.00	42.4	107.34	921.79
Waste to power	90.58	0.00	0.00	23.5	24.22	0.00	9.34	21	54.5	25	1.60	249.74
Waste to energy (off-grid)	139.79	9.71	5.69	11.77	5.55	6.58	19.11	20.75	34.66	52.28	30.17	356.00
Total	**35,849.59**	**4,100.5**	**7,141.3**	**11,466.81**	**12,093.24**	**8,760.56**	**8,843.31**	**7,548.11**	**14,081.97**	**15,274.43**	**18,484.68**	**156,247.10**

Source: Ministry of New and Renewable Energy.

Table 2.2 The targets and achievements in Phase I, Phase II, and Phase III

Phase	*Sanctioned*	*Installed*
Phase I (2010-13)	252.5 MW	117 MW
Phase I (2014-17)	713 MW	345.5 MW
Phase I (2018-21)	20.09 MW	15.67 MW

Table 2.3 Application-wise status of the installations under the off-grid and decentralized solar PV applications programme till 31.12.2022

Applications	*No. of units/capacity installed*
Solar lamps/lanterns	8,459,119
Solar pumps	500,499
Solar street lights	944,802
Solar home lighting systems	1,723,479
Solar power plants/packs	217 MWp

Source: Ministry of New and Renewable Energy.

Table 2.4 Potential of wind energy in India

Sl. no.	*State*	*Wind potential at 120 m (GW)*	*Wind potential at 150 m (GW)*
1	Andhra Pradesh	74.90	123.3
2	Gujarat	142.56	180.8
3	Karnataka	124.15	169.3
4	Madhya Pradesh	15.40	55.4
5	Maharashtra	98.21	173.9
6	Rajasthan	127.75	284.2
7	Tamil Nadu	68.75	95.1
8	Telangana	24.83	54.7
	Total eight windy states	676.55	1,136.7
9	Others	18.95	27.1
	Total	695.50	1,163.9

Source: Ministry of New and Renewable Energy.

photosynthesis process of production of energy. Table 2.5 shows the biomass power with category and resources.

2.3.2 Rural electrification and urban demand

India has been very successful in the last decade in electrifying rural areas (Dey et al., 2024), bringing electricity to nearly everyone's doorstep. With government schemes like the Deen Dayal Upadhyaya Gram Jyoti Yojana

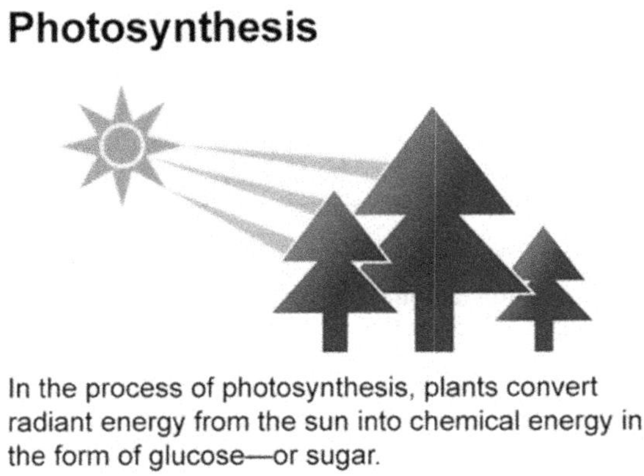

Figure 2.1 Photosynthesis process.

Table 2.5 Biomass power with category and resources

Sl. no.	*Category*	*Resources*
1	Crop residue	Bagasse (sugarcane residue), corn stalk, rice hulls & rice-straw, wheat straw, nutshells
2	Animal waste	Animal excrement (manure). Animal by-product from processing
3	Food processing waste	Potato waste, cheese whey waste, fruit pits, leftover sludge and other energy-rich FPW
4	Wood waste	Wood waste includes any type of wood or wood-based product

and the Saubhagya Yojana, millions of rural households have come under the ambit. Almost all villages in India have an electricity supply by 2024 (Kundu et al., 2024), though some places still face problems in terms of consistent and reliable power supply.

With industrialization, population increase, and enhanced living standards, electricity demand is accelerating in urban areas. Cities like Delhi, Mumbai, and Bengaluru are facing increased electrification with the use of air conditioners, business, and the expansion of transport networks. More appliances, EVs, and modern houses will demand more power supply from India as the middle class increases in this country (Dhankhar et al., 2024).

2.3.3 Power generation companies

The generation of power in India is done by both public and private companies (Dhingra et al., 2024). However, the public sector has dominated for a long time, but the private companies have made their mark as well in the past few years.

NTPC Limited: NTPC stands for the former National Thermal Power Corporation. It is India's largest power generation company and a big name in the public sector. It operates several coal-based power plants and is investing in renewable energy projects. NTPC plans to cut carbon emissions in an effort to align with the clean energy goals of India (Garg et al., 2024).

Adani Power: With operations across several coal-based power plants, the company is one of the largest private power companies in India and is now also a major investor in renewable energy, mainly focusing on solar energy. The Adani Group aims to be one of the world's major renewable energy companies (Bajaj et al., 2024).

Tata Power: Tata Power is a large private power generation company in India, utilizing coal, hydropower, solar, and wind energy. Tata Power has been a leader in the concept of renewable energy and has been focusing much on solar and wind power, which are seeing rapid growth in its renewable energy sector (Charles and Majid, 2024).

Besides these major players, several state-owned companies and smaller private players are helping add to India's power generation capacity. It is a very competitive industry where both the public and the private sectors collaborate to augment the generation capacity, enhance efficiency, and introduce renewable energy (Ministry of Power, Government of India, 2022e).

Traditionally, India's energy landscape depended on fossil fuels, but today, it is changing towards renewable energy (Chawla et al., 2024). Industrialization and urbanization are driving a surge in energy demand in the country. Balancing the rise in such a demand needs more investment in clean energy technologies, support to government policies for renewable energy, and a focus on better energy consumption across all sectors (Ministry of Power, Government of India, 2022).

2.4 FUTURE OF RENEWABLE ENERGY IN INDIA

India has reached a significant milestone in its renewable energy journey, with the country's total renewable energy capacity crossing the 200 GW (gigawatt) mark as of October 10, 2024 (Kamble et al., 2024). According to the Central Electricity Authority, the total renewable energy-based electricity generation capacity now stands at 201.45 GW. This achievement underscores India's growing commitment to clean energy and its progress in building a greener future (Dhayal et al., 2024).

This milestone reflects the result of years of dedicated efforts to harness India's natural resources (Mohapatra and Pohit, 2024). From sprawling solar parks to wind farms and hydroelectric projects, the country has steadily built a diverse renewable energy base. These initiatives have not only reduced reliance on fossil fuels but also strengthened the nation's energy security. When factoring in the 8,180 MW (megawatt) of nuclear capacity, the total non-fossil fuel-based power now accounts for almost half of the country's installed electricity generation capacity, signalling a strong move toward clean energy leadership on the global stage (Gayen et al., 2024).

Biopower, including biomass and biogas energy, adds another 11.32 GW to the renewable energy mix (Tiwari et al., 2024). These bioenergy projects are vital for utilizing agricultural waste and other organic materials to generate power, further diversifying India's clean energy sources (Niyogi et al., 2024). Together, these renewable resources are helping the country reduce its dependence on traditional fossil fuels, while driving progress towards a more sustainable and resilient energy future.

Several states in India have emerged as leaders in renewable energy capacity, contributing significantly to the nation's progress (Behera et al., 2024).

The Government of India has implemented a range of measures and initiatives aimed at promoting and accelerating renewable energy capacity across the nation, with an ambitious target of achieving 500 GW of installed electric capacity from non-fossil sources by 2030 (Abhyankar et al., 2023). Key programs include the National Green Hydrogen Mission, PM-KUSUM, PM Surya Ghar, and PLI schemes for solar PV modules. These efforts reflect the government's commitment to fostering a sustainable energy future while addressing the challenges posed by climate change and energy security (Ministry of New and Renewable Energy (MNRE), 2023).

2.5 CONCLUSION

India's quest for sustainable and renewable energy is ambitious and urgent. Given that it is one of the fastest-growing economies in the world, India faces a challenge in meeting this requirement with a reduction in carbon emissions. That 500 GW of renewable energy by 2030 and being visualized to achieve net-zero emissions by 2070 is an ambitious commitment from India towards the decarbonization of fossil fuels and ushering clean, green energy into the nation. Three crucial factors will determine India's renewable future: proper implementation of supportive policies, overcoming technical challenges that include the modernization of the grid and energy storage, and an increase in private investment. However, it is also important to ensure equal access to energy for all groups, especially vulnerable sections and rural and underprivileged communities, so that everyone will benefit. While India does terrific work in the solar, wind, and other renewable energy avenues, much is left to be done. The scope for innovation and leading initiatives in renewable energy can keep India ahead in the global arena as it further improves international collaboration with emphasis on this being coupled with counter challenges placed on finances and the environment. Renewable energy in the future will weigh heavily on the development of the country but also globally for efforts against climate change.

REFERENCES

Abhyankar, N., Deorah, S.M. and Phadke, A.A., 2023. Least-Cost Pathway for India's Power System Investments through 2030.

Adelekan, O.A., Ilugbusi, B.S., Adisa, O., Obi, O.C., Awonuga, K.F., Asuzu, O.F. and Ndubuisi, N.L., 2024. Energy transition policies: a global review of shifts towards renewable sources. *Engineering Science & Technology Journal*, *5*(2), pp.272–287.

Ahmed, A. and Chakraborty, D., 2024. Influence of liberalization on the Indian manufacturing sector: empirical evidence over 1987–2018. In Rajib Bhattacharyya and Debashis Mazumdar (Eds.), *Contemporary Issues in International Trade* (pp. 113–125). Leeds, United Kingdom: Emerald Publishing Limited.

Al Mubarak, F., Rezaee, R. and Wood, D.A., 2024. Economic, societal, and environmental impacts of available energy sources: a review. *Eng*, *5*(3), pp.1232–1265.

Algarni, S., Tirth, V., Alqahtani, T., Alshehery, S. and Kshirsagar, P., 2023. Contribution of renewable energy sources to the environmental impacts and economic benefits for sustainable development. *Sustainable Energy Technologies and Assessments*, *56*, p.103098.

Ali, F., Dawood, A., Hussain, A., Alnasir, M.H., Khan, M.A., Butt, T.M., Janjua, N.K. and Hamid, A., 2024. Fueling the future: biomass applications for green and sustainable energy. *Discover Sustainability*, *5*(1), p.156.

Anjum, S., Zafar, M.M. and Kumari, A., 2024. A review of diseases attributed to air pollution and associated health issues: a case study of Indian metropolitan cities. *Diseases and Health Consequences of Air Pollution*, pp. 145–169.

Asghar, M., Ali, S., Hanif, M. and Ullah, S., 2024. Energy transition in newly industrialized countries: a policy paradigm in the perspective of technological innovation and urbanization. *Sustainable Futures*, *7*, p.100163.

Bajaj, D., Patil, D.R. and Madhavi, P., 2024. A study on evolution of Adani Group of Industries-it's resilience in Indian industries. *Involvement International Journal of Business*, *1*(2), pp.129–141.

Batra, G., 2023. Renewable energy economics: achieving harmony between environmental protection and economic goals. *Social Science Chronicle*, *2*(2), pp.1–32.

Behera, P., Sethi, L. and Sethi, N., 2024. Balancing India's energy trilemma: assessing the role of renewable energy and green technology innovation for sustainable development. *Energy*, *308*, p.132842

Binyamin, S.S., Slama, S.A.B. and Zafar, B., 2024. Artificial intelligence-powered energy community management for developing renewable energy systems in smart homes. *Energy Strategy Reviews*, *51*, p.101288.

Central Electricity Authority (CEA), 2022a. *Coal in India's Energy Mix: Contribution to Electricity Generation Capacity*.

Central Electricity Authority (CEA), 2022b. *Electricity Consumption Patterns in India: Residential Sector and Household Energy Use*.

Central Electricity Authority (CEA), 2022c. *Electricity Consumption in Commercial Sector: Offices, Shops, Hotels, and Hospitals in India*.

Charles, J. and Majid, M.A., 2024. Advances and development of wind–solar hybrid renewable energy technologies for energy transition and sustainable future in India. *Energy & Environment*, *35*(5), pp.2517–2565.

Chawla, U., Mohnot, R., Fadahunsi, A. and Mulchandani, D., 2024. The bright revolution: accelerating adoption of solar energy in India's dynamic landscape. *International Journal of Energy Economics and Policy*, *14*(4), pp.226–233.

Dé, R., 2024. *Economic Policy in Independent India: Crisis, Coalitions, and Contingency*. Cambridge University Press.

Dey, P., Gupta, M.P., Sushil and Perez-Arriaga, I.J., 2024. Rural electrification in India: a wicked problem. *Systems Research and Behavioral Science*, *42*, pp.875–894.

Dhankhar, S., Sandhu, V. and Muradi, T., 2024. E-Mobility Revolution: Examining the Types, Evolution, Government Policies and Future Perspective of Electric Vehicles. *Current Alternative Energy*, *6*. https://doi.org/10.2174/0124054631308595240612110422

Dhayal, K.S., Agrawal, S., Agrawal, R., Kumar, A. and Giri, A.K., 2024. Green energy innovation initiatives for environmental sustainability: current state and future research directions. *Environmental Science and Pollution Research*, *31*, pp.1–19.

Dhingra, S., Raut, R., Kumar, M. and Naik, B.K.R., 2024. Factors impacting Indian healthcare supply chain performance and influence in the public and private sector: the mediating role of blockchain technology adoption. *Benchmarking: An International Journal*. https://doi.org/10.1108/BIJ-10-2023-0687

Eweade, B.S., Joof, F. and Adebayo, T.S., 2024. Analyzing India's coal, natural gas, and biomass energy consumption: evidence from a Fourier technique to promote sustainable development. *Natural Resources Forum*, *49(2)*, pp. 1238–1256.

Garg, A., Patange, O., Vishwanathan, S.S., Nag, T., Singh, U. and Avashia, V., 2024. Synchronizing energy transitions toward possible Net Zero for India: affordable and clean energy for all. *A report prepared for Office of the Principle Scientific Advisor (PSA) to Government of India and Nuclear Power Corporation of India Limited (NPCIL)*.

Gayen, D., Chatterjee, R. and Roy, S., 2024. A review on environmental impacts of renewable energy for sustainable development. *International Journal of Environmental Science and Technology*, *21*(5), pp.5285–5310.

Girgibo, N., Hiltunen, E., Lü, X., Mäkiranta, A. and Tuomi, V., 2024. Risks of climate change effects on renewable energy resources and the effects of their utilisation on the environment. *Energy Reports*, *11*, pp.1517–1534.

Giri, P., Paul, S. and Debnath, B.K., 2024. A fuzzy Graph Theory and Matrix Approach (fuzzy GTMA) to select the best renewable energy alternative in India. *Applied Energy*, *358*, p.122582.

Gupta, S., Gupta, A., Bherwani, H. and Kumar, R., 2024. Green hydrogen: potential master key for combating climate change. In Malti Goel and Gautam Sen (Eds.), *Climate Action and Hydrogen Economy: Technologies Shaping the Energy Transition* (pp. 39–60). Singapore: Springer Nature Singapore.

Hassan, Q., Viktor, P., Al-Musawi, T.J., Ali, B.M., Algburi, S., Alzoubi, H.M., Al-Jiboory, A.K., Sameen, A.Z., Salman, H.M. and Jaszczur, M., 2024. The renewable energy role in the global energy transformations. *Renewable Energy Focus*, *48*, p.100545.

Hoyos-Seba, J.J., Arias, N.P., Salcedo-Mendoza, J. and Aristizabal-Marulanda, V., 2024. Animal manure in the context of renewable energy and value-added products: a review. *Chemical Engineering and Processing-Process Intensification*, *196*, p.109660.

Husain, A.M., Hasan, M.M., Khan, Z.A. and Asjad, M., 2024. A robust decision-making approach for the selection of an optimal renewable energy source in India. *Energy Conversion and Management*, *301*, p.117989.

International Energy Agency (IEA), 2022. *India Energy Outlook 2022*.

Kamble, A.D., Das, S., Vijaya, Das, B., Bordoloi, U., Hazarika, P. and Kalita, P., 2024. Role of solar energy in the development of the Indian Economy. In Sudipta De, Avinash Kumar Agarwal, and Pankaj Kalita (Eds.), *Challenges and Opportunities of Distributed Renewable Power* (pp. 489–535). Singapore: Springer Nature Singapore.

Kaur, D. and Rehman, M., 2024. Renewable energy made in India: navigating geopolitics in achieving sustainability. *Journal of Chinese Economic and Business Studies*, *22*, pp.1–22.

Kumar, C.M.S., Singh, S., Gupta, M.K., Nimdeo, Y.M., Raushan, R., Deorankar, A.V., Kumar, T.A., Rout, P.K., Chanotiya, C.S., Pakhale, V.D. and Nannaware, A.D., 2023. Solar energy: a promising renewable source for meeting energy demand in Indian agriculture applications. *Sustainable Energy Technologies and Assessments*, *55*, p.102905.

Kumar, M.S., Kumari, P., Lad, Y.A. and Maria, S., 2024. Scenario of solar energy and policies in India. *Clean Energy*, *8*(5), pp.117–128.

Kundu, A., Rahaman, M. and Khan, K., 2024. Access of urban poor to basic services: concerns of sustainability and equity. *Social Change*, *54*(1), pp.7–35.

Majeed, Y., Khan, M.U., Waseem, M., Zahid, U., Mahmood, F., Majeed, F., Sultan, M. and Raza, A., 2023. Renewable energy as an alternative source for energy management in agriculture. *Energy Reports*, *10*, pp.344–359.

Mehrotra, A., Agarwal, A. and Styczynski, A.B., 2024. Gas-based economy: a road towards a prosperous and cleaner India. In Annika Bose Styczynski (Ed.), *India's Energy Revolution* (pp. 42–65). London, United Kingdom: Routledge.

Ministry of Coal, Government of India, 2022a. *India's Coal Reserves and its Position as the Second Largest Producer and Consumer of Coal*.

Ministry of Coal, Government of India, 2022b. *The Role of Coal in India's Industrial and Transport Sectors*.

Ministry of Coal, Government of India, 2023. *Historical Role of Coal in India's Energy System: Railways, Steam Engines, and Early Industries*.

Ministry of New and Renewable Energy (MNRE), 1995. *Assessment of Renewable Energy Potential in India*. Government of India.

Ministry of New and Renewable Energy (MNRE), 2010. *Jawaharlal Nehru National Solar Mission: Framework for the Implementation of Solar Energy in India*.

Ministry of New and Renewable Energy (MNRE), 2018. *Annual Report 2017–2018: Assessment of Renewable Energy Potential in India*. Government of India.

Ministry of New and Renewable Energy (MNRE), 2022c. *India's Renewable Energy Ambitions and the Road to 500 GW by 2030*.

Ministry of New and Renewable Energy (MNRE), 2022d. *The Role of Renewable Energy in Enhancing India's Energy Security*. Government of India.

Ministry of New and Renewable Energy (MNRE), 2023. *India's Commitment to Renewable Energy: Sustainable Development, Energy Security, and Environmental Responsibility.*

Ministry of Power, Government of India, 2022a. *Challenges to Energy Security and Air Pollution in India: The Role of Electric Vehicles and Oil Reduction.*

Ministry of Power, Government of India, 2022b. *Coal-Based Power Plants and Hydropower Development in India: A Historical Overview.*

Ministry of Power, Government of India, 2022c. *Collaboration Between Public and Private Sectors in India's Power Generation Capacity.*

Ministry of Power, Government of India, 2022d. *Electricity Reforms and the Growth of Private Investment in Energy Projects Post-1991.*

Ministry of Power, Government of India, 2022e. *Energy Consumption Patterns in Rural and Urban India.*

Ministry of Power, Government of India, 2022f. *The Impact of 1991 Reforms on India's Energy Sector: Privatization, Deregulation, and Power Shortages.*

Ministry of Power, Government of India, 2022g. *The Rise of Natural Gas in Power Generation and Challenges in India's Energy Sector.*

Ministry of Power, Government of India, 2023. *Annual Report on Energy Consumption in India.*

Mishra, S., Saini, G., Chauhan, A., Upadhyay, S. and Balakrishnan, D., 2023. Optimal sizing and assessment of grid-tied hybrid renewable energy system for electrification of rural site. *Renewable Energy Focus*, *44*, pp.259–276.

Mohapatra, S. and Pohit, S., 2024. Charting the path to a developed India: Viksit Bharat 2047. *National Council of Applied Economic Research.*

Mondal, S., Mondal, A., Rathore, J., Koundinya, K.K. and Sharma, A.K., 2024. India's shift toward sustainable energy: a comprehensive approach to renewable energy integration and environmental sustainability. *Journal of Renewable and Sustainable Energy*, *16*(4), p.042701.

Nassar, Y. and Khaleel, M., 2024. Sustainable development and the surge in electricity demand across emerging economies. *International Journal of Electrical Engineering and Sustainability*, 2, pp.51–60.

Niyogi, A., Sarkar, P., Bhattacharyya, S., Pal, S. and Mukherjee, S., 2024. Harnessing the potential of agriculture biomass: reuse, transformation and applications in energy and environment. *Environmental Science and Pollution Research*, *32*, pp.1–24.

Osman, M., Saad, M.M., Ouf, M. and Eicker, U., 2024. From buildings to cities: how household demographics shape demand response and energy consumption. *Applied Energy*, *356*, p.122359.

Prajapati, H., Patel, C., Gupta, S., Singh, G.K. and Gupta, D.K., 2024, January. Challenges and barriers of rooftop solar adaption-a study from rural part of Gujarat India using weighted average method and Garrett ranking method. In *AIP Conference Proceedings* (Vol. 2960, No. 1). AIP Publishing. https://doi.org/10.1063/5.0183781

Raihan, A., Voumik, L.C., Zimon, G., Sadowska, B., Rashid, M. and Akter, S., 2024c. Prioritising sustainability: how economic growth, energy use, forest area, and globalization impact on greenhouse gas emissions and load capacity in Poland? *International Journal of Sustainable Energy*, *43*(1), p.2361410.

Ramesh, S., 2024. An Independent India: 1947AD to 2020AD. In *The Political Economy of India's Economic Development: 5000BC to 2024AD, Volume II: The Gupta Empire to an Independent India and Beyond* (pp. 265–301). Cham: Springer Nature Switzerland.

Sarkar, P.K., 2024. 3 Power Sector Transformation: An Indian Perspective.

Senawiang, P., Piangmal, C. and Sripaurya, N., 2024. Empowering through education: Gandhi's Swadeshi principle and the path to self-reliance. *Journal of Education and Learning Reviews*, *1*(5), pp.23–34.

Siddik, A.B., Khan, S., Khan, U., Yong, L. and Murshed, M., 2023. The role of renewable energy finance in achieving low-carbon growth: contextual evidence from leading renewable energy-investing countries. *Energy*, *270*, p.126864.

Sikarwar, A.S., Ansari, N.A. and Singh, P., 2024. Current scenario and future perspectives of renewable and sustainable energy in India: progress and its challenges. *Journal of Environmental Nanotechnology*, *13*(1), pp.172–181.

Singh, M.K., Malek, J., Sharma, H.K. and Kumar, R., 2024. Converting the threats of fossil fuel-based energy generation into opportunities for renewable energy development in India. *Renewable Energy*, *224*, p.120153.

Singh, V., 2023. *The Dispossession Regime of India's Ultra Mega Power Projects (UMPPs): Case of a Captive Coal Based UMPP in Jharkhand* (Doctoral dissertation, Northeastern University).

Singh, V., 2024. Global warming and climate change. In *Textbook of Environment and Ecology* (pp. 283–295). Singapore: Springer Nature Singapore.

Srivastava, A., Jain, S. and Selvaraju, S.R., 2024. The Regulatory and Market Landscape for Climate Finance into India's Renewable Energy Sector.

Tiwari, T., Kaur, G.A., Singh, P.K., Balayan, S., Mishra, A. and Tiwari, A., 2024. Emerging bio-capture strategies for greenhouse gas reduction: navigating challenges towards carbon neutrality. *Science of the Total Environment*, *929*, p.172433.

Utturkar, P.S., Prajapat, V.R., Jadhav, S.V. and Manyar, H.G., 2024. Exploring water's role in sustainable electricity generation for power in future. In Akhilesh Kumar Yadav, Kanchan Yadav, and Vijay P. Singh (Eds.), *Integrated Management of Water Resources in India: A Computational Approach: Optimizing for Sustainability and Planning* (pp. 449–471). Cham: Springer Nature Switzerland.

Vardhan, B.S., Swain, A., Khedkar, M., Srivastava, I. and Bokde, N.D., 2024. An overview of Indian power sector and its energy management. *Renewable Energy Focus*, *50*, p.100597.

Vijayalaxmi, J., Tarun, C.S. and Shremitha, N., 2024. Solar energy scenario in India. In Ali Sayigh (Ed.), *Reducing the Effects of Climate Change Using Building-Integrated and Building-Applied Photovoltaics in the Power Supply* (pp. 259–274). Cham: Springer Nature Switzerland.

Wang, X., Lu, Y., Chen, C., Yi, X. and Cui, H., 2024. Total-factor energy efficiency of ten major global energy-consuming countries. *Journal of Environmental Sciences*, *137*, pp.41–52.

Yawale, S.K., Hanaoka, T., Kapshe, M. and Pandey, R., 2023. End-use energy projections: future regional disparity and energy poverty at the household level in rural and urban areas of India. *Energy Policy*, *182*, p.113772.

Chapter 3

Hydrogen production, storage, and transportation

Future prospects

Deoashish Panjiara, Neeraj Kumar Yadav, Sudeep Yadav, and Abhay Kumar Choudhary

3.1 INTRODUCTION

Energy is essential for daily life, social civilization, and economic progress. Population growth, smarter living, and the industrial revolution are all contributing to the exponential increase in global energy demand. Presently, most of the world's energy consumption (more than 80%) is met by using fossil fuel resources such as coal, crude oil products, and natural gas, which are non-renewable and unfriendly to the environment (Rahman et al., 2024). The ever-increasing utilization of these fuels for manufacturing, conversion, and energy generation has resulted in the release of vast volumes of greenhouse gases (GHGs), with 77% being primarily carbon dioxide (CO_2), which is responsible for global warming and climate change (Lavanya et al., 2024). Fossil fuel combustion also emits air pollutants such as nitrogen oxides (NO_x), hydrocarbons (HCs), carbon monoxide (CO), sulfur oxides (SO_x), and particulate matter (PM), resulting in negative impacts on air quality and public health (Asghar et al., 2021; Hassan et al., 2021). Furthermore, the uneven global distribution of fossil fuel reserve resources has resulted in geopolitical pressure and energy security concerns, and occasionally, it has erupted into violent conflicts (Zhang et al., 2024; Taiwo et al., 2024). Therefore, to address these issues, researchers and scientists have concentrated their efforts on switching from fossil fuels to sustainable and renewable energy resources. A viable alternative to dependence on fossil fuels is the utilization of renewable energy sources, such as hydroelectric, biomass, geothermal, nuclear, solar, and wind energy, as well as biofuels (Rahman et al., 2024). Embracing renewable energy provides several benefits, including reduced pollutants and GHG emissions particularly CO_2, supporting ecological sustainability, and the potential for improved air quality (Sadeq et al., 2024; Rahman et al., 2024). However, there are certain shortcomings to renewable energy sources, such as their dependency on weather patterns and lack of storage capacity, which can reduce their dependability for steady energy output (Tackie-Otoo and Haq, 2024; Raza et al., 2024). In this context, hydrogen (H_2) has recently garnered

DOI: 10.1201/9781003581536-3

considerable interest because of its carbon-neutral energy carrier, industrial feedstocks, environmental compatibility, diversity in its applications, the potential answer for long-term energy storage, and efforts to limit carbon emissions from industrial and economic sectors (Tackie-Otoo and Haq; 2024; Taiwo et al., 2024).

3.2 PROPERTIES/CHARACTERISTICS OF HYDROGEN

Hydrogen is a highly flammable, colorless, odorless, tasteless, non-metallic, non-toxic, and diatomic gas (H_2) molecule (Onwuemezie, 2024). Henry Cavendish discovered hydrogen (H_2) in 1766. Some of the significant physico-chemical properties of hydrogen are provided in Table 3.1. Hydrogen has a very small molecular size (2.89 Å) and a higher diffusion rate (0.63 cm^2/s) in air than other fuels (Ahmad et al., 2024). With a molecular mass of 2.016 kg/kmol and a relatively low density of 0.089 kg/m^3 at 25°C and 1 atm, it is the simplest and lightest element on Earth (Taiwo et al., 2024). Hydrogen is most abundant (making up about 91% of the universe). It is found in a combined state with other elements (i.e., O, C, and N) in natural sources such as water, biomass, and fossil fuels that are utilized as feedstock for its production (Chaubey et al., 2013; Muhammed et al., 2023). However, it is not readily accessible in the free state in nature due to its highly reactive nature. As a result, there are no hydrogen reserves on Earth (Chaubey et al., 2013). It is also mentioned in the literature that the quantity of energy released from the generation of a unit of hydrogen is less than the energy consumed in its manufacturing; hence, hydrogen is regarded as an energy carrier rather than a source (Muhammed et al., 2022). Furthermore, its ability to convert easily into heat or electricity validates its usefulness as an energy carrier due to its energy transportation and storage features (Mazloomi and Gomes, 2012). Hydrogen has a nearly threefold higher gravimetric energy content and lower heating value than liquid fuels like diesel or gasoline (Ghorbani et al., 2023a). However, the volume occupied per unit mass or per unit energy of hydrogen is much higher than that of other fuels, so a vehicle running on hydrogen fuel must have a considerably bigger fuel tank than a diesel or gasoline-powered vehicle (Table 3.1). Because hydrogen has a very low volumetric energy density (0.0107 MJ/L), it is commonly compressed or liquified to increase its density for safe storage and transportation in liquid form. Hydrogen liquefaction occurs at about 20 K or –253°C with a density of 70.8 kg/m^3, which is very costly to perform, maintain, and raises safety concerns (Taiwo et al., 2024). Its extended flammability range of 4%–75% allows for operational flexibility in terms of variable fuel-air ratios for the optimal combustion conditions in internal combustion (IC) engines or turbo-generator designs, resulting in low fuel consumption and emissions (Ghorbani et al., 2023a; Raza et al., 2024). Moreover, hydrogen has a higher research octane number

(RON) than fossil-based fuels, indicating better anti-knock characteristics (Mazloomi and Gomes, 2012). It can generate energy-enriched mixtures when combined with other fuels and can be used as an alternative fuel in engines designed for different fuel types (Raza et al., 2024). Hydrogen has the highest auto-ignition temperature, making it best suited for spark ignition engines. Additionally, it is challenging to ignite hydrogen in diesel engines through the compression process (Table 3.1). Furthermore, hydrogen holds remarkable properties as a fuel, such as high flame velocity, high diffusivity, low flash point, and minimum ignition energy. Because hydrogen has a

Table 3.1 Comparison of physicochemical characteristics of H_2 with other conventional fuels

Sl. No.	*Properties*	*Diesel (average $C_{14}H_{30}$)*	*Petrol/gasoline (average C_8H_{15})*	*Methane ($CH_{4(g)}$)*	*Hydrogen ($H_{2(g)}$)*
1	Molecular weight	198.4	111.19	16.043	2.016
2	Density (kg/m^3) at 25°C and 1 atm	875–959	715–780	0.657	0.089
3	Gravimetric energy density (MJ/kg)	45.6	46.4	55.60	143
4	Volumetric energy density (MJ/L)	34.6	34.2	0.0378	0.0107
5	Low to high-heating value (MJ/kg)	42.6–45.6	43.4–46.4	50–55.5	120–141.9
6	Boiling point	187–343	27–225	–162	–253
7	Flash point	62	–45	–188	–253
8	Volume occupied by a unit mass (L/kg)	1.0–1.1	1.3–1.4	1,522.1	1,123.6
9	Volume occupied by unit energy (L/MJ)	0.026	0.03	25	100
10	Diffusivity in the air (cm^2/s)	0.0008	0.002	0.20	0.63
11	Auto ignition temperature in air (°C)	210	280	534	585
12	Flammability limits in air (vol%)	0.6–7.5	1–8	5–15	4–75
13	Research octane number (RON)	30	87	125	>130
14	Flame velocity (cm/s)	40	47	34	270
15	Viscosity (Pa s) at 25°C and 1 atm	3.45×10^{-3}	—	1.1×10^{-5}	0.89×10^{-5}
16	Specific gravity at 20°C and 1 atm	0.81–0.89	0.72–0.78	0.55	0.0696
17	Minimum ignition energy (MJ)	20	0.8	0.3	0.02
18	Stoichiometric air/fuel ratio on mass basis	14.95	14.58	17.2	34.3

Source: Muhammed et al. (2023), Taiwo et al. (2024), Ghorbani et al. (2023a), Sarathy et al. (2014), and Mazloomi and Gomes (2012).

high flame speed, the flame spreads rapidly and consistently in the cylinder mixture, enhancing the complete burning of fuel and making the engine powered by hydrogen more like the perfect thermodynamic cycle, which contributes to increased engine thermal efficiency with lower exhaust losses (Tripathi et al., 2022). Compared to other conventional fuels, hydrogen has the lowest flash point (–253°C), indicating ease of fuel combustion and demonstrating the combustion system's simplicity (Ghorbani et al., 2023a). It is also expected to utilize less sophisticated starting and ignition devices (Mazloomi and Gomes, 2012). The high diffusivity of hydrogen in the air (0.63 cm^2/s) allows for a more uniform air fuel charge in the cylinder, resulting in increased combustion efficiency. Hydrogen also has the lowest minimum ignition energy, which enables external sources like spark plugs to ignite it quickly (Tripathi et al., 2022). Furthermore, it possesses remarkable mobility due to its low viscosity and density, as demonstrated in Table 3.1, enabling it to be filled or emptied from containers for storage promptly (Thiyagarajan et al., 2022). Furthermore, this allows it to disperse more readily than other fuels in the event of an outdoor accident. However, high hydrogen mobility makes it difficult to effectively contain and store gas, implying that, unlike other fuels, efficiently storing hydrogen will require a larger sealed system (Taiwo et al., 2024).

3.3 HYDROGEN PRODUCTION METHODS

Hydrogen is one of the essential chemical substances for a variety of applications, including industrial activities and transportation, due to its adaptability and environmentally friendly nature (Jeje et al., 2024). Thermal electrolysis of water produces purple hydrogen, which splits water using heat and nuclear power. Pink hydrogen is made through water electrolysis using electricity from a nuclear power plant, whereas red hydrogen is generated by high-temperature catalytic splitting of water using energy from a nuclear power plant (AlHumaidan et al., 2023; Jeje et al., 2024). White hydrogen is the type of hydrogen that occurs naturally. By analysing the hydrogen colour code, customers gain a better understanding of how their energy consumption impacts the environment. Figure 3.1 shows an overview of the various hydrogen-generating technologies.

3.3.1 Hydrogen production from fossil fuels

Hydrogen can be generated from various feedstocks (fossil fuels, biomass, water, etc.) using a wide range of methods. Hydrogen feedstock can be categorized as renewable sources (biomass and water) and non-renewable sources (fossil fuels such as coal, natural gas, and oil). Currently, globally 96% of hydrogen is extracted from fossil fuels, which accounts for 6% and 2% of the total coal and natural gas used annually (Sharma et al., 2024).

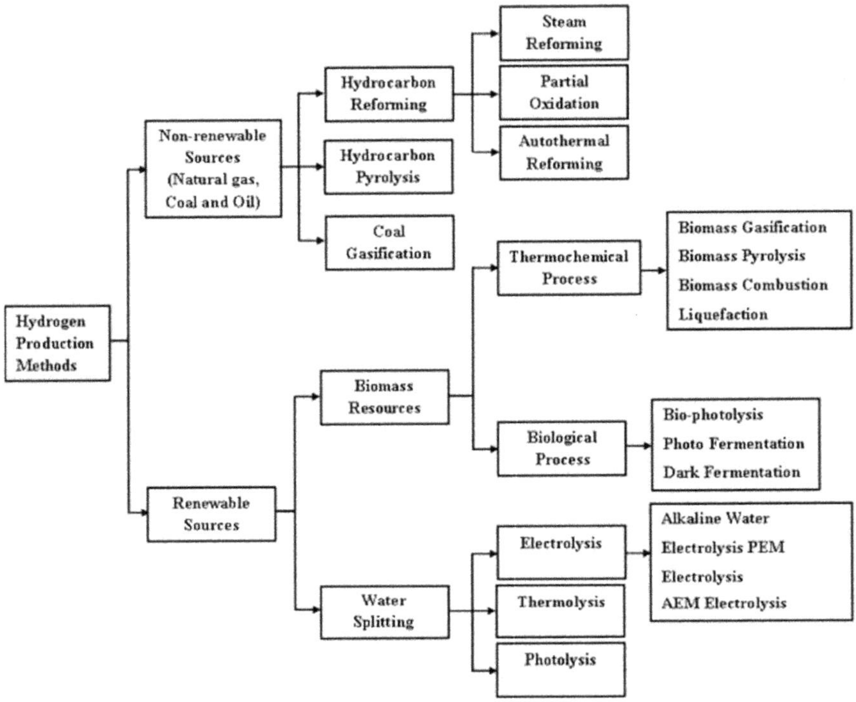

Figure 3.1 Overview of various H_2 production routes.

3.3.1.1 Hydrocarbon reforming processes

Hydrogen is produced from hydrocarbon reforming using three primary methods: steam reforming, partial oxidation, and autothermal reforming.

The different types of hydrocarbon reforming are the following:

Steam reforming

Steam reforming (SR) is a highly endothermic catalytic process that includes the interaction of steam with hydrocarbon feedstock (natural gas, methane, ethane, naphtha, diesel, etc.) at high temperatures and pressures to produce synthesis gas (syngas: mainly $CO + H_2$) (Ahmad et al., 2024). The reaction is carried out in a catalytic tube reactor containing a nickel-based or precious metal catalyst at a temperature range of 500°C–1,000°C depending upon the feedstock used. The process consists of four main steps: desulfurization, generation of syngas by steam reforming (Eq. 3.1), water gas shift (WGS) reaction (Eq. 3.2), and gas purification or methanation reaction (Eq. 3.3) (Ahmad et al., 2024). The reforming reaction is carried out at a high temperature, steam/carbon ratio of 3.5, and pressure up to 3.5 MPa to prevent coke formation on the catalyst surface and produce the desired purity of H_2 gas (Nikolaidis and Poullikkas, 2017). As a preliminary step, feedstock gas is preheated to 350°C and desulfurization is performed to remove the poisonous sulfur from the feedstock, which could contaminate and retard the H_2 production and catalyst activity (Younas et al., 2022). Figure 3.2 shows the block diagram for the steam reforming process. The general reactions involved in the process are:

$$\text{SR step}: C_nH_m + nH_2O \rightarrow nCO + (n + 0.5\text{ m})H_2 \tag{3.1}$$

$$\text{WGS step}: CO + H_2O \rightarrow CO_2 + H_2 \tag{3.2}$$

$$\text{Methanation step}: CO + 3H_2 \rightarrow CH_4 + H_2O \tag{3.3}$$

where n and m signify the number of C and H atoms in the hydrocarbon feedstock being used. The SR of methane (CH_4) can be described by applying $n = 1$ and $m = 4$ in Eq. (3.1).

Steam methane reforming (SMR) is one of the most widely used, well-established, and least expensive methods for producing industrial H_2 with a conversion efficiency of 75%–80% (Ahmad et al., 2024). The syngas formation is followed by the WGS reaction (Eq. 2.2) in which CO of the syngas is converted to CO_2 and maximizes the H_2 production by 10%–15%. Finally, the gas stream is sent via a CO_2 removal system, followed by a methanation or gas purification process such as a membrane or pressure swing adsorption (PSA), resulting in very pure H_2 gas (99.99%) (Nikolaidis and Poullikkas, 2017). Excessive CO_2 emission from the SMR is a great

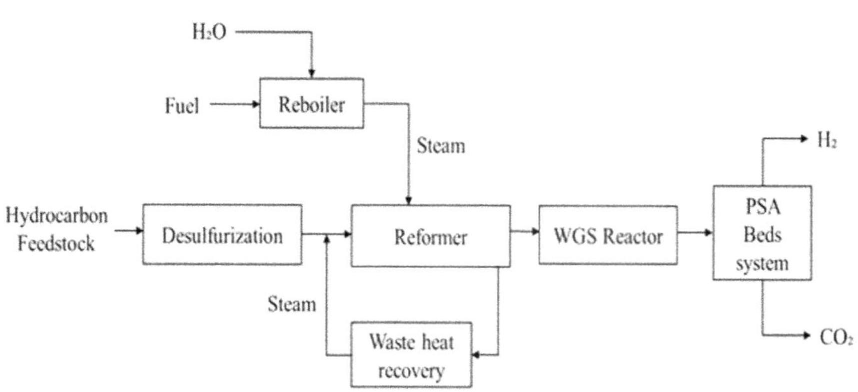

Figure 3.2 Schematic diagram of the steam reforming process.

challenge, which is the major cause of global warming. In this regard, the adoption of CCS technology can reduce carbon emissions while improving environmental sustainability.

MSR is endothermic in nature and subject to thermodynamic equilibrium. Several parameters influence the performance of the MSR process, including feedstock type, catalyst and fuel choices, steam/carbon ratio, temperature, pressure, and reactor design (Jeje et al., 2024). MSR faces various obstacles, including the capital costs associated with heat transfer and heat-resistant equipment, the high energy costs of operating under highly endothermic SMR conditions, and the short catalyst life span due to sulfur poisoning, coke deposition, and the sintering of catalytic sites at high temperatures (Muhammed et al., 2023). To overcome these drawbacks, extensive interdisciplinary research and material development are needed to improve catalyst performance, and increase efficiency, and lower greenhouse gas emissions, making the SMR process a viable and sustainable choice for producing H_2.

Partial oxidation

Partial oxidation (POX) is an exothermic process in which hydrocarbon feedstock is heated with a limited quantity of pure O_2 at high temperatures to produce hydrogen-rich syngas in a reactor/reformer (Chaubey et al., 2013). Depending on the feedstock quality, POX can be performed with or without catalysts and is classified into two types: thermal partial oxidation (TPOX) and catalytic partial oxidation (CPOX) (AlHumaidan et al., 2023).

The reaction conditions for TPOX are comparatively higher, with temperatures ranging from 1,200°C to 1,500°C and pressures (3–10 MPa) employing feedstocks such as natural gas, oil, and coal (Ahmad et al., 2024). In this process, the feedstock is first desulfurized to remove sulfur impurities, and then pure O_2 is mixed with the desulfurized feedstock to produce syngas through partial oxidation. The downstream process stages, such as WGS units (Eq. 6), syngas separation, and purification (Eq. 7), are carried out in the same manner as steam reforming. The syngas (H_2/CO ratio of ~2) generated by the TPOX method utilizing natural gas as feedstock can be used in the Fischer-Tropsch synthesis process (Ahmad et al., 2024).

The general reactions in the POX process are as follows:

$$\text{POX step}: C_nH_m + 0.5nO_2 \rightarrow n\text{CO} + 0.5m\text{H}_2\,(\text{catalytic}) \tag{3.4}$$

$$C_nH_m + nH_2O \rightarrow n\text{CO} + (n + 0.5m)\text{H}_2\,(\text{non}-\text{catalytic}) \tag{3.5}$$

$$\text{WGS step}: CO + H_2O \rightarrow CO_2 + H_2 \tag{3.6}$$

$$\text{Methanation step}: CO + 3H_2 \rightarrow CH_4 + H_2O \tag{3.7}$$

TPOX of the residual oil can be shown using Eq. (3.5) with n = 1 and m = 1.3 at 6 MPa. Similarly, TPOX may be described for the coal feed by applying Eq. (3.5), with n = 1 and m = 0 at 5.5 MPa m (approx.), and the technique is called coal gasification to obtain H_2.

The CPOX process produces syngas by reacting O_2, steam, and hydrocarbon feedstocks (methane, natural gas, gasoline, diesel, alcohol, and biodiesel) over a Ni-based or noble metal catalyst (Eq. 3.4) (AlHumaidan et al., 2023). The CPOX process operates at a lower temperature range of 700°C–1,000°C and pressure (0.1 MPa) than the TPOX process (Ahmad et al., 2024). The CPOX process has various advantages, including increasing the conversion rate, limiting the generation of toxic substances (NH_3 and HCN), reducing soot formation, and enhancing product selectivity. However, the primary concern is catalyst deactivation caused by coke deposition, sulfur poisoning, support degradation, metal vaporization, and high-temperature sintering (AlHumaidan et al., 2023).

POX method offers several advantages, such as feedstock flexibility, faster dynamic response, and compact size over the other reforming technologies (Chaubey et al., 2013). However, the process is extremely capital-intensive due to the cost of the pure O_2 plant and a desulfurization unit (Nikolaidis and Poullikkas, 2017). Recently, significant efforts have been made to develop multi-catalysts, design the CPOX reactor, and use plasma fields and radio frequency to prevent coke production.

Autothermal reforming

Autothermal reforming (ATR) is a catalytic hybrid process that combines the concepts of two processes, POX followed by SR, in a single reactor, to increase syngas production yield with a high H_2/CO_2 ratio (Ahmad et al., 2024). The ATR method utilizes the heat released by the exothermic POX process to drive the endothermic SR reaction (Nikolaidis and Poullikkas, 2017). The ATR process involves introducing O_2 and steam into an ATR reactor containing hydrocarbon feedstock (natural gas), where the POX reaction operates in the combustion zone and steam reforming occurs in the catalytic zone to produce syngas (Taipabu et al., 2022). Further, the steps of the downstream process are identical to those of steam reforming (SR). The simultaneous occurrence of both processes is demonstrated by the following Eq. (3.8).

$$C_nH_m + \left(\frac{n}{2}\right)H_2O + \left(\frac{n}{4}\right)O_2 \rightarrow nCO + (n/2 + m/2)H_2 \tag{3.8}$$

By setting n = 1 and m = 4, the Eq. (3.8) can be expressed for methane feedstock. ATR operates at moderate pressures (80–100 bar) and temperatures ranging from 900°C to 1,100°C (AlHumaidan et al., 2023). Compared to

SMR, the ATR process has several advantages, including a smaller size, lower costs, and faster dynamic response. With advancements in reactor design and catalysis, the ATR technique is expected to be a prospective method for syngas generation for the fuel sector due to its high molar H_2 synthesis, lower capital costs, and compact design compared to the SR and POX processes (Ahmad et al., 2024).

3.3.1.2 Hydrocarbon pyrolysis

Hydrocarbon pyrolysis (HCP) is a hydrogen generation technology in which a hydrocarbon is thermally decomposed into H_2 and carbon (C) in a pyrolyser with no oxidants (air or steam) (AlHumaidan et al., 2023). The pyrolysis temperature for coal ranges between 350°C and 400°C, whereas for hydrocarbons like methane, it reaches 1,400°C or more (El-Shafie et al., 2019). Turquoise hydrogen refers to hydrogen produced from methane pyrolysis with an energy efficiency of 58%.

The general reaction is shown as follows:

$$C_nH_m \rightarrow nC + 0.5mH_2 \tag{3.9}$$

Pyrolysis is typically classified into two types: rapid pyrolysis in the absence of a catalyst and catalytic pyrolysis with a catalyst. The utilization of transition metal (Ni, Fe, and Co) catalysts in a pyrolyser can dramatically reduce reaction temperature and heat input during the pyrolysis of heavy hydrocarbon feedstocks (Onwuemezie, 2024). The process has also been utilized for producing valuable carbon products (fullerenes and CNTs) for a range of uses (Ahmad et al., 2024). Burning carbon by-products, hydrogen, methane, or electricity can all provide the energy required for the process. Due to the absence of air or steam, the pyrolysis Eq. (3.9) does not yield any gaseous carbon oxides, such as CO and CO_2, as by-products. This process requires less energy (37.7 kJ/mol for methane pyrolysis) per mole H_2 generation than the MSR process (63.4 kJ/mol for MSR process), and the thermal energy needs can be covered by H_2 burning (~15%–20%), resulting in a significant drop in GHG emissions during the operation (Sánchez-Bastardo et al., 2021). Furthermore, the pyrolysis process doesn't need a WGS unit, CO_2 separation, or energy-intensive CCS technology (Ahmad et al., 2024). In addition, the pyrolysis process has various advantages, including feedstock flexibility, process compactness, and GHG-free by-products (El-Shafie et al., 2019). The major challenge associated with this process includes H_2 separation due to its low partial pressure in the reaction mixture and membrane durability at elevated temperatures required for the de-carbonization equilibrium (Nikolaidis and Poullikkas, 2017). Despite its numerous technological and financial benefits, the pyrolysis process has not been widely commercialized because of issues with managing solid carbon particles in the gas phase, carbon fouling inside the reactor, conversion rate constraints, and high operating temperatures (AlHumaidan et al., 2023).

3.3.1.3 Coal gasification

Coal gasification is a thermochemical process to produce syngas by reacting coal with gasifying agents (air, O_2, steam, and CO_2) at high temperatures (900°C–1,100°C) and moderate pressure (5–10 bar) in a gasifier (El-Shafie et al., 2019). It is one of the feasible technologies for large-scale H_2 generation cost-effectively due to global coal reserves and plays a key role in the various energy sectors including (i) syngas to synthesize fertilizer, methanol, and liquid hydrocarbon; (ii) H_2 production; (iii) synthetic natural gas; and (iv) production of industrial grade fuel gas (Younas et al., 2022). The main difficulty with coal gasification in H_2 production is the substantial CO_2 emissions (29.33 kg CO_2/kg H_2) and other harmful pollutants (Younas et al., 2022; Ahmad et al., 2024). In this regard, carbon capture, storage, and utilization (CCSU) is a technique established to capture CO_2 and utilize it; nevertheless, it makes the process more energy-intensive, raising production costs. The gasifier design configuration, as well as the type or characteristics of coal, have a considerable impact on the gasification process performance (Muhammed et al., 2023). The various types of gasifiers employed for this process include co-current/counter-current fixed bed, entrained flow, fluidized bed, free radical, and plasma (AlHumaidan et al., 2023). According to El-Shafie et al. (2019), up to 97% pure hydrogen can be produced during the gasification process using the Koppers Totzek entrained flow gasifier. Plasma gasification, which runs at 1,700°C, has gained appeal due to its benefits, including higher hydrogen generation, increased efficiency, sustainability, and lower solid waste compared to traditional gasification systems (Midilli et al., 2021). The integrated gasification combined cycle (IGCC) concept, which combines in situ CO_2 capture and storage with a high H_2 generation rate, is another development in the gasification process (Sahraei et al., 2014).

3.3.2 Hydrogen production from biomass

Biomass is recognized as a novel sustainable feedstock for hydrogen synthesis, and it can be generated from plant and animal resources such as energy crops, agricultural residue, lignocellulosic biomass, and bio-waste (Taipabu et al., 2022). The H_2 production from biomass-based substrate promotes carbon neutrality while emitting low levels of SO_x and NO_x (Vuppaladadiyam et al., 2022). Biological and thermochemical processes are the two primary routes for producing hydrogen from biomass-based resources (Nikolaidis and Poullikkas, 2017). Thermochemical techniques provide a higher rate of H_2 production than biological processes due to faster kinetics. Biological H_2 production techniques utilize living organisms that produce hydrogen via metabolic processes in mild settings. These pathways are sustainable and environmentally beneficial; however, they are slow and have a low yield (mol H_2 per mol substrate) of H_2 that depends on the substrate composition (Ahmad et al., 2024).

3.3.2.1 Thermochemical process

Biomass can be converted into H_2-rich syngas via thermochemical methods. The two most popular thermochemical methods for converting biomass into biochar, liquid fuel, and syngas are pyrolysis and gasification (Taipabu et al., 2022). Both techniques yield CH_4 and CO with other gaseous products, which can be used to generate more H_2 via stream reforming and the WGS reaction. Combustion and liquefaction are the other processes, which are considered less preferred because they generate little H_2 while emitting pollutants and necessitate high operating pressures (5–20 MPa) in the absence of air (Nikolaidis and Poullikkas, 2017). Thermochemical techniques have advantages over biological processes in that a pretreatment phase of biomass is not required.

Biomass gasification

Biomass gasification is the most affordable and well-established method of producing H_2. It is a thermochemical process that converts biomass into gaseous fuel or hydrogen-rich syngas by using a gasifying agent (air/oxygen or steam) at high temperatures (500°C–1,400°C) and pressures (1–33 bar), depending on the reactor type and end product plan (Morya et al., 2022). The fluidized bed gasifier is the best and most extensively used due to its excellent performance in terms of high stability, improved temperature control, high mass transfer, and heat transfer capabilities between liquid and solid phases (Younas et al., 2022). In gasification, the biomass feedstock is fed into a gasifier at elevated pressure and temperature in a reactive oxidation environment (air and/or steam), resulting in partial oxidation of biomass (Sarmah et al., 2023). Equations (3.10) and (3.11) depict the general biomass transformation reactions involved in the gasification process.

$$\text{Biomass} + \text{air} \rightarrow H_2 + CO + CO_2 + N_2 + CH_4 + \text{Other CHs} + \text{tar} + H_2O + \text{char} + \text{ash} \tag{3.10}$$

$$\text{Biomass} + \text{steam} \rightarrow H_2 + CO + CO_2 + CH_4 + \text{Other CHs} + \text{tar} + \text{char} + \text{ash} \tag{3.11}$$

The resulting syngas (Eqs. 3.10 and 3.11) is primarily a mixture of CO and H_2, with additional components such as CO_2, H_2S, and SO_2, which are separated and purified in subsequent steps (desulfurization, WGS unit, and reactor beds). The use of superheated steam as the most effective gasifying agent instead of air enhances the H_2 concentration of the produced syngas while eliminating impurities such as N_2 gas (Ahmad et al., 2024). The H_2 yield in the gasification process is affected by several parameters, including catalyst type, steam temperature, particle size, biomass type, and steam/biomass ratio (Nikolaidis and Poullikkas, 2017). The utilization of a catalyst in the gasification process can considerably boost H_2 yield, reduce tar formation, and improve the gasification rate, and the most commonly used catalysts are nickel or cobalt-based (Younas et al., 2022). Biomass gasification is associated with various challenges, including char and tar disposal issues, ash-like concerns, catalyst deactivation, and the cost of biomass collection and transportation, all of which hinder commercialization for H_2 production (Ahmad et al., 2024; Yadav et al., 2024; Younas et al., 2022). Recently, co-gasification technology of coal and biomass has been reported by some authors (Mallick et al., 2017; Seçer et al., 2018). Biomass has lower energy density, higher volatile content, and is more reactive than coal. Co-gasification of coal and biomass (up to 10 wt.%) increases H_2 and CO production while lowering GHG emissions (Younas et al., 2022).

Biomass pyrolysis

Biomass pyrolysis is similar to hydrocarbon pyrolysis; however, it uses biomass as the feedstock rather than hydrocarbons. Pyrolysis is the process of heating biomass at high temperatures (500°C–900°C) and pressure (0.1–0.7 MPa) in the absence of air or oxygen to produce liquid fuel, solid char, and syngas (Eq. 3.12) (Ahmad et al., 2024). The produced CH_4 and other hydrocarbon (C_nH_m) gases are converted into syngas via a catalytic steam reforming reaction (Eq. 3.13). Moreover, the product stream is followed by a WGS reaction (Eq. 3.14) to synthesize some additional H_2. Finally, the product stream is directed through PSA beds to produce purified H_2 gas. Temperature, biomass type, reaction time, heating rate, and type of catalyst used are the essential factors that significantly affect the H_2 generation yield (Taipabu et al., 2022). The step-by-step reactions that occur during biomass pyrolysis are depicted in Eqs. (3.12–3.14).

$$\text{Biomass} + \text{heat} \rightarrow H_2 + CO + CO_2 + CH_4 + C_nH_m \text{ gases} + \text{tar} + \text{char} \tag{3.12}$$

$$C_nH_m + nH_2O \rightarrow nCO + (n + 0.5m)H_2 \tag{3.13}$$

$$CO + H_2O \rightarrow CO_2 + H_2 \tag{3.14}$$

Pyrolysis is categorized into four types: slow or traditional, fast, flash, and vacuum based on operating conditions. Amongst those, slow and fast pyrolysis are typically used for hydrogen generation. Slow pyrolysis utilizes mild process conditions, such as a low heating rate and a longer holding time (>5 minutes), resulting in charcoal as the major product due to incomplete combustion. Fast pyrolysis includes heating biomass at a high temperature for only a short time (0.5–10 seconds) in the absence of air, resulting in a liquid product as the primary product, as well as syngas and char (Vuppaladadiyam et al., 2022). Fast pyrolysis at a high temperature is the more suitable of these two processes for producing H_2. Small particle sizes, high moisture content, and a prolonged holding time all contribute to increased H_2 production (Younas et al., 2022). Literature studies revealed that catalytic pyrolysis of biomass feedstock using zeolite Socony Mobil–5 (ZSM-5), Ni-based catalyst, and metal oxides (TiO_2, ZrO_2, SiO_2, and Al_2O_3) as catalysts can significantly improve the quality and quantity of H_2 yield (Hosseini et al., 2015; Taipabu et al., 2022; Younas et al., 2022). Microwave-assisted, steam-assisted, hydro, and co-pyrolysis are new pyrolysis technologies that provide higher H_2 yields with high specific surface area biochar as a solid byproduct (Vuppaladadiyam et al., 2022).

3.3.2.2 Biological processes

The hydrogen production from the biological conversion processes is considered pollution-free, carbon-neutral, sustainable, and eco-friendly compared to traditional methods (Sarmah et al., 2023; Taipabu et al., 2022). This method uses microorganisms (microalgae, bacteria, and cyanobacteria) or enzymes as catalysts to convert various organic substrates (wastewater and biomass) into hydrogen. The process commences generally at a low-temperature range of 30°C–60°C and atmospheric pressure, making it less energy-intensive (Taipabu et al., 2022). Biological processes are further subcategorized into three main methods: bio-photolysis, dark fermentation, and photo fermentation.

Fermentation is the microbial conversion of organic wastes into small amounts of H_2, alcohol, acetic acid, and CO_2. Organic acid, starch, glucose, and cellulose are the most commonly used fermentation substrates (Onwuemezie, 2024). The procedure is carried out in a bioreactor with or without oxygen, utilizing a broad range of microbial bacteria through physical, chemical, and enzymatic activities. These technologies are especially promising because they use waste items as feedstock for generating H_2 at a low cost while treating bio-waste (Kumar and Fiori, 2024). The two primary categories of fermentation processes are dark fermentation and photofermentation.

Bio-photolysis

The bio-photolysis process mimics the natural photosynthetic process that produces H_2 gas (Onwuemezie, 2024). In this process, algal plants bearing chlorophyll generate an H_2 synthetization enzyme (Fe-Fe hydrogenase enzyme), which is used to split water molecules into H_2 and O_2 by direct and indirect bio-photolysis utilizing sunlight (Ahmad et al., 2024).

Direct bio-photolysis is a two-step process (Sharma et al., 2024). Microalgae (green algae) use photosynthesis to split water into protons (H^+), electrons (e^-), and oxygen (O_2) (Nikolaidis and Poullikkas, 2017). The hydrogenase enzymes then catalyse the generated protons, converting them into H_2 gas (Sharma et al., 2024; Ahmad et al., 2024). The reactions involved in direct bio-photolysis are presented in Eqs. (3.15) and (3.16) (Sharma et al., 2024).

$$\text{Solar energy} + 2H_2O \rightarrow 4H^+ + 4e^- + O_2\,(\text{Photosynthesis reaction}) \tag{3.15}$$

$$4H^+ + 4e^- \rightarrow 2H_2\left(H_2 \text{ production reaction}\right) \tag{3.16}$$

The hydrogenase enzyme's catalytic activity is irreversibly influenced by oxygen content, hence the oxygen content is kept low (≤1%) throughout the entire process (Kumar and Fiori., 2024). Some microalgae mutants with low chlorophyll pigment content and high O_2 tolerance have been identified, allowing for a high rate of H_2 generation (Ahmad et al., 2024).

The indirect bio-photolysis process consists of two sequential steps to generate H_2 and O_2. The first step is the photosynthesis reaction in the presence of sunlight to form glucose ($C_6H_{12}O_6$) and O_2 (Eq. 3.17). In the second step, $C_6H_{12}O_6$ and H_2O are transformed to H_2 and CO_2 via fermentation reaction (Eq. 3.18). The reactions involved in indirect bio-photolysis to produce H_2 from H_2O utilizing blue-green algae or cyanobacteria are explained in Eqs. (3.17) and (3.18) (Kumar and Fiori, 2024).

$$6H_2O + 6CO_2 + \text{light energy} \rightarrow C_6H_{12}O_6 + 6O_2\,(\text{Photosynthesis reaction}) \tag{3.17}$$

$$C_6H_{12}O_6 + 6H_2O + \text{light energy} \rightarrow 12H_2 + +6CO_2\,(\text{Fermentation reaction}) \tag{3.18}$$

In general, algal H_2 generation is a renewable and sustainable approach to using H_2O and CO_2. However, key challenges associated with this approach to H_2 generation include poor H_2 yield and the large requirement to capture sufficient light for photosynthesis (Ahmad et al., 2024).

Dark fermentation

Dark fermentation exploits anaerobic bacteria on organic matter like lignocellulosic biomass, sugars, municipal solid waste, and industrial wastewater under dark and anoxic conditions to generate biohydrogen (bio-H_2). The dark fermentation process employs facultative bacteria (*Citrobacter*, *Bacillus*, *Enterobacter*) and obligate anaerobes (*Clostridium*, *Desulfovibrio* species, and *Ethanoligenens*) to produce bio-H_2 (Ahmad et al., 2024). The process requires a pretreatment step to break down the complex molecules of the substrates containing fermentable components (sugars and organic acids) into smaller, fermentable forms (Sadeq et al., 2024). A bioreactor containing substrates is infected with an anaerobic bacterial culture, which degrades the organic substrate into bio-H_2 and other byproducts such as organic acids (butyric acid and acetic acid), CO_2, and alcohols (Arun et al., 2022). The chemical reactions involved in the dark fermentation of glucose as a substrate are presented in Eqs. (3.19) and (3.20).

$$C_6H_{12}O_6 + 2H_2O \rightarrow 2CH_3COOH + 4H_2 + 2CO_2 \text{ (acetic acid)} \quad (3.19)$$

$$C_6H_{12}O_6 \rightarrow CH_3CH_2CH_2COOH + 2H_2 + 2CO_2 \text{ (butyric acid)} \quad (3.20)$$

Ideally, the acetic (Eq. 3.19) and butyric acid (Eq. 3.20) metabolic routes yield 4 and 2 mol of H_2 from one mole of glucose, respectively. More than 80% of the total end products are acetic and butyric acids generated by Eqs. (3.19) and (3.20). Furthermore, dark fermentation produces mixed biogas that is mainly composed of H_2 and CO_2 with trace amounts of CO, H_2S, and/or CH_4 (Morya et al., 2022). To achieve the requisite purity of H_2 gas, gas separation and purification units are required. The type of substrate material, its potency level, temperature, pH, the type of biological reactor and its operating parameters, gas partial pressure, and hydraulic holding time are some of the variables that determine the H_2 yield in the dark fermentation process (Morya et al., 2022; Sharma et al., 2024).

This method has several advantages over photo fermentation, including a simpler reactor design and a higher H_2 generation rate, making it the most favourable (Nikolaidis and Poullikkas, 2017). Furthermore, since it doesn't require light, this process utilizes anaerobic bacteria to break down organic substrates, and it may constantly produce H_2 from a variety of potentially useful substrates, such as waste and refuse products (Kumar and Fiori, 2024)

Photofermentation

Photofermentation is the fermentative process that utilizes sunlight to degrade the organic acid substrates (acetic, lactic, and butyric) and carbohydrates to produce H_2 and CO_2 using the photosynthetic microorganisms, such as *Rhodospirillum*, *Rhodobacter*, and *Rhodobium*, by the action of nitrogenase enzyme (Nikolaidis and Poullikkas, 2017; Sarmah et al., 2023). Under anaerobic conditions, a bioreactor containing organic substrate and photosynthetic bacteria is exposed to sunlight or artificial light, converting it into H_2 and byproducts such as CO_2. Several techniques, including gas sparging, adsorption, and membrane separation, are utilized to separate the generated H_2 from the other constituents (Sadeq et al., 2024). The overall chemical reactions in the photo fermentation conversion of organic acids (acetic, lactic, and butyric) into H_2 and CO_2 are presented in Eqs. (3.21)–(3.23) (Kumar and Fiori, 2024).

$$CH_3COOH + 2H_2O + \text{light energy} \rightarrow 4H_2 + 2CO_2 \tag{3.21}$$

$$C_3H_6O_2 + 4H_2O + \text{light energy} \rightarrow 7H_2 + 3CO_2 \tag{3.22}$$

$$C_4H_8O_2 + 6H_2O + \text{light energy} \rightarrow 10H_2 + 4CO_2 \tag{3.23}$$

H_2 production from the photo fermentation process employing bacteria is influenced by factors such as light intensity, illumination pattern, and wavelength (Ahmad et al., 2024). Using industrial wastes as feedstock for this process is a huge challenge because of the presence of colors and heavy metal ions, which may decrease the penetration of light and require pre-treatment before its use (Kumar and Fiori, 2024).

Combined fermentation is a hybrid technique that uses dark fermentation followed by photo fermentation to increase H_2 output (Nikolaidis and Poullikkas, 2017). This integrated technique employs anaerobic bacteria during dark fermentation without light to digest carbohydrates and produce H_2 together with organic acids, which are then used as feedstock for photosynthetic bacteria to generate further H_2 (Kumar and Fiori, 2024). The hybrid fermentation methodology is more adaptable, cost-effective, and feasible than the other fermentation methods (Chaudhary et al., 2024).

3.3.3 Hydrogen production from water splitting

Water is readily available in nature as an extremely stable molecule (enthalpy −285.83 kJ/mol, STP), and it is commonly recognized as a massive hydrogen source (Ahmad et al., 2024). Water splitting is a method of producing hydrogen at low temperatures that employs a catalyst and membrane for the dissociation reaction and gas separation, as well as chemical reactants and/or electrical, photonic, and thermal energy (Taipabu et al., 2022). Hydrogen can be generated from water splitting techniques, which are mainly three types: electrolysis, thermolysis, and photolysis (Sarmah et al., 2023).

3.3.3.1 Electrolysis

Electrolysis of water is the most established technology in clean hydrogen production with a high purity of ~99.9% without carbon emissions (Taipabu et al., 2022). Water electrolysis is an endothermic process that uses direct current (DC) to split water electrochemically, generating hydrogen at the cathode and oxygen at the anode (Eq. 3.24) (Nikolaidis and Poullikkas, 2017). The process occurs in an electrolyser, which comprises two electrodes (anode and cathode), an electrolyte, and a separator (Sadeq et al., 2024). The water electrolysis reaction (Eq. 3.24) is

$$H_2O + \text{Electricity} + \text{Heat} \rightarrow H_2 + \frac{1}{2}O_2 \tag{3.24}$$

The theoretical cell voltage for water dissociation is 1.23 V. However, experimental data show a significantly higher value (1.48 V) due to reaction kinetics and ohmic resistances of the electrolytes and cell components (Sharma et al., 2024). Electrolysis technologies can be categorized as alkaline, Proton Exchange Membrane (PEM), or Solid Oxide Electrolysis (SOE) based on the electrolyte, ionic species (H^+, OH^-, O^{2-}), and working temperature (Kumar and Himabindu, 2019).

Alkaline water electrolysis (AWE) uses NaOH or KOH (conc. 20%–30%) as electrolytes and non-noble metal catalysts like nickel in electrodes and operates at a lower temperature (30–80°C) (Taipabu et al., 2022). Asbestos is employed as a diaphragm between the electrodes, or anode and cathode, to separate gas products such as H2 and O2, ensuring safety and purity (Jeje et al., 2024). The reactions of AEM are presented below (Eqs. 3.25–3.27).

$$\text{Anode}: 2OH^- \rightarrow H_2O + \frac{1}{2}O_2 + 2e^- \tag{3.25}$$

$$\text{Cathode}: 2H_2O + 2e^- \rightarrow H_2 + 2OH^- \tag{3.26}$$

$$\text{Overall}: H_2O \rightarrow H_2 + \frac{1}{2}O_2 \tag{3.27}$$

AWE is a well-established and dominant technology that generates hydrogen with an efficiency of 50%–60% and a purity of 99.5%–99.9%

(Sarmah et al., 2023). AWE has limitations such as low electrical efficiency, low operating pressure, and a limited current density (Taipabu et al., 2022). The low efficiency of AWE is caused by high polarization losses and slow electrochemical kinetics (Ahmad et al., 2024). Furthermore, problems with pure water, electrode deterioration, and catalyst poisoning could be addressed through optimized system operation and material development. Anion Exchange Membrane (AEM) electrolysis is a new approach currently under development that uses an anion exchange membrane instead of an asbestos diaphragm (Jeje et al., 2024).

The PEM electrolysis is very similar to the PEM fuel cell that uses a proton-conducting membrane as an electrolyte for the exchange of protons (H^+ species) between electrodes and operates at low temperature (≤80°C) (Kumar and Himabindu, 2019; Nasser et al., 2022). In PEM electrolysis noble metal Pt/Pd is employed as the HER catalyst and IrO_2/RuO_2 as the OER catalyst (Kumar and Himabindu, 2019). In a PEM electrolyser, water is fed at the anode, split into H^+ ions, O_2 gas, and electrons (e^-) (Eq. 3.28). H^+ species migrate through the PEM membrane towards the cathode, where they recombine with electrons to generate H_2 gas (Eq. 3.29). The reactions of PEM electrolysis are shown below.

$$\text{Anode}: H_2O \rightarrow 2H^+ + \frac{1}{2}O_2 + 2e^- \tag{3.28}$$

$$\text{Cathode}: 2H^+ + 2e^- \rightarrow H_2 \tag{3.29}$$

$$\text{Overall}: H_2O \rightarrow H_2 + \frac{1}{2}O_2 \tag{3.30}$$

The PEM electrolyser offers many advantages, such as compact and simple design, high energy efficiency, high operation pressure, fast response to dynamic load variation, high H_2 purity (99.999), and low-temperature operation (20°C–80°C) (Kumar and Himabindu, 2019; Nasser et al., 2022). However, high water purity, catalyst durability, membrane stability, and the high cost of electrode materials are the major setbacks to the PEM electrolysis technology (Jeje et al., 2024; Sarmah et al., 2023).

The SOE is an emerging technology and is being researched on laboratory scales. SOE uses yttria-stabilized zirconia (YSZ) as a solid electrolyte and operates at high temperatures (500°C–950°C) (Jeje et al., 2024). Water is introduced in the steam (990°C–1,000°C) form at the cathode, where it splits into H_2 gas and oxide species (O^{2-}) in contrast to PEM and alkaline electrolyser (Ahmad et al., 2024). The O^{2-} species pass through the solid electrolyte to the cathode side, oxidizing to form O_2 gas. The reactions of SOE are given by Eqs. (3.31–3.33).

$$\text{Anode}: O^{2-} \rightarrow \frac{1}{2}O_2 + 2e^- \tag{3.31}$$

$$\text{Cathode}: H_2O + 2e^- \rightarrow H_2 + O^{2-} \tag{3.32}$$

$$\text{Overall}: H_2O \rightarrow H_2 + \frac{1}{2}O_2 \tag{3.33}$$

SOE demonstrates increased energy efficiency (85%–90%), owing to the use of inexpensive thermal energy or waste heat from nuclear and thermal power plants as heat sources (Nasser et al., 2022). SOE has several severe limitations, including electrode instability, safety and sealing issues, thermal management, and a high operating temperature, which shortens its operational life (Jeje et al., 2024).

AWE is the most cost-effective method of producing electrolyser-grade H_2 with high purity (Jeje et al., 2024). AWE and PEM electrolysis are currently being utilized on a large scale, while SOE is still in the experimental stages (Younas et al., 2022). By using electrolysis technology, surplus energy from renewable sources such as solar and wind can be utilized to produce renewable H_2, which can then be stored and used as fuel in fuel cells to generate electricity (Ahmad et al., 2024). However, electrolysis technology has substantial drawbacks such as high operational and capital costs caused by the use of expensive electrode and electrolyte materials, as well as energy-intensive operations (Sadeq et al., 2024). Water electrolysis currently supplies only 4% of global H_2 consumption due to cost constraints, while the majority of it is generated as a by-product of the chlor-alkali sector (Sharma et al., 2024; Kumar and Himabindu, 2019).

3.3.3.2 Thermolysis

Thermolysis, also known as thermochemical water splitting (TWS), is a high-temperature process that splits H_2O into H_2 and O_2 in a single step at elevated temperatures (≥2,500°C) through the following reaction: H_2O + heat energy → H_2 + O_2 (Nikolaidis and Poullikkas, 2017; Ahmad et al., 2024). Without the intervention of any chemicals, the reaction requires a considerable thermodynamic energy input of 286 kJ/mol to dissociate water into its constituent elements, H_2, and O_2 (Ahmad et al., 2024).

Several technological limitations of direct thermolysis exist, including the need for high temperatures, gas separation issues, the durability of construction materials, and relatively low conversions, necessitating the development of various TWS cycles (Ghorbani et al., 2023b). A thermochemical cycle is a repeating set of redox reactions that occur at different temperatures and represent one of the most promising routes to dissociate H_2O using thermal energy to produce chemical energy (H_2) at a temperature range of 500°C–2,000°C usually in two to six steps (Safari and Dincer, 2020; Ghorbani et al., 2023b). The number of steps in TWS cycles controls process parameters such as operating temperature and the chemistry of the materials used. Increasing the number of steps in TWS cycles for H_2 generation lowers the operating temperature; nevertheless, more steps in the TWS cycle increase the process complexity (Muhammed et al., 2023; Ghorbani et al., 2023b).

The two-step TWS cycle involves thermally reducing a metallic oxide (MO_x) as a reactive intermediate at high temperatures, resulting in the release of O_2 (Eq. 3.34). The reduced oxide then combines with the steam and is reoxidized to produce H_2 and pristine oxide (Eq. 3.35), which is then transferred to the first step.

$$\text{Reduction}: MO_x \rightarrow MO_{x-1} + 1/2O_2 \,(\text{Endothermic step}) \tag{3.34}$$

$$\text{Oxidation}: MO_{x-1} + H_2O \rightarrow MO_x + H_2 \,(\text{Exothermic step}) \tag{3.35}$$

$$\text{Overall}: H_2O \rightarrow H_2 + 1/2O_2 \tag{3.36}$$

More than 300 TWS cycles have been proposed to date, although only a small number of them are practical for large-scale H_2 generation. The process's challenges include the need for sophisticated materials resistant to corrosion and heat, the toxicity of the materials involved, complex reaction handling and scaling, and efficiency, which depends on proper integration with heat sources (Nikolaidis and Poullikkas, 2017; Ahmad et al., 2024).

3.3.3.3 Photolysis (photo-electrolysis)

The photolysis technique for H_2 production uses the energy of visible light or solar photons absorbed by a suitable photocatalyst to split H_2O into H_2 and O_2 (Nikolaidis and Poullikkas, 2017; Ahmad et al., 2024). Photolysis is recognized as an essential component of the photo-electrolysis or photoelectrochemical (PEC) water splitting process, which is used to generate hydrogen from H_2O molecules (Sarmah et al., 2023). The process's performance is determined by the semiconductor photocatalysts used, which include TiO_2, CdS, sulfides, and other metal oxides (Sadeq et al., 2024). The electrons travel to the cathode via the external circuit, where they interact with protons to generate H_2 gas. The reactions involved in the PEC water splitting process are outlined in Eqs. (3.38)–(3.40).

$$\text{Light or photon excitation}: 2hv \rightarrow 2h^{\bullet} + 2e^{-} \tag{3.37}$$

$$\text{Anode}: 2h^{\bullet} + H_2O \rightarrow 1/2O_2 + 2H^{+}\,(\text{Oxidation}) \tag{3.38}$$

$$\text{Cathode}: 2H^{+} + 2e^{-} \rightarrow H_2\,(\text{Reduction}) \tag{3.39}$$

$$\text{Overall cell}: hv + H_2O \rightarrow H_2 + 1/2O_2 \tag{3.40}$$

Photocatalyst materials should have the following desirable properties: high stability with pH variations, adequate bandgap energy, minimal toxicity, and acceptable photocatalytic efficiency (Sadeq et al., 2024). The benefits of this technology include utilizing free, abundant, and renewable energy, resulting in a sustainable way to produce hydrogen with low greenhouse gas emissions (Sarmah et al., 2023). Photolysis of H_2O has some limitations, including low H_2O dissociation efficiency, instability, and auto-corrosion of electrode materials. These issues must be addressed to make the approach more economically viable and scalable for large-scale H_2 generation (Younas et al., 2022).

The solar photovoltaic (PV) cells integrated with an electrolyser (PV-E) are utilized to produce H_2 (Taipabu et al., 2022). Sunlight is absorbed by semiconductor-based PV cells to generate electricity, which is then used to split water by traditional electrolysers (AWE or PEM electrolysers). Despite being commercialized and developed, PV-E technology is still expensive (\$5.78–23.27/kg H_2) (Ahmad et al., 2024).

Table 3.2 outlines the various H_2 production methods, including their feedstock, energy sources, advantages and drawbacks, technology readiness levels (TRL), efficiency, produced H_2 color, and hydrogen cost (\$/kg). Among the discussed H_2 production technologies, H_2 generation from hydrocarbon reforming processes such as SR, POX, and ATR is the most favourable and mature (TRL: 9), achieving up to 85% efficiency and providing 97% of all global H_2 demand. Hydrocarbon pyrolysis creates solid carbon as a byproduct, whereas coal gasification is the most significant CO_2 emitter (20 kg CO_2/kg H_2), and its future market potential is falling (Sadeq et al., 2024). H_2 generation from biomass-based feedstock via thermochemical processes such as gasification and pyrolysis is a green H2 production route and an emerging technology, but their efficiency (35%–50%) is slightly lower. Another disadvantage is that biomass feedstocks vary seasonally, resulting in variable H_2 concentrations in output. Biological H_2-generating methods (TRL:1-3) such as bio-photolysis, dark fermentation, and photo-fermentation may be preferable H_2 production methods for achieving environmentally friendly practices. However, there are issues about low efficiency, limited scalability, and the fact that they are laboratory-sized and

Table 3.2 Comparison of various hydrogen production routes

Production method and TRL	*Feedstock and energy source*	*Efficiency (%) and produced H_2 color*	*Advantages*	*Drawbacks*	*H_2 cost (\$/kg)*
Steam reforming (SR); TRL: 9	Feedstock: natural gas; Energy source: standard fossil fuels	Eff. (%): 74–85; H_2 color: grey, blue with CCS	Commercialized, cost-effective technology, existing infrastructure, high yield efficiency	CO_2 emission, complex catalytic process, higher temperature requirement, unsuitable soot	2.08
Partial oxidation (POX); TRL: 9	Feedstock: natural gas, coal, and heavy oil; Energy source: standard fossil fuels	Eff. (%): 60–75; H_2 color: brown, grey, blue	Established technology, existing infrastructure	CO_2 emission, produced heavy oils and petroleum coke	1.48
Autothermal reforming (ATR); TRL: 9	Feedstock: natural gas; Energy source: standard fossil fuels	Eff. (%): 60–75; H_2 color: brown, grey, blue	Proven technology, existing infrastructure	CO_2 emission, dependence on fossil fuels	1.48
Hydrocarbon pyrolysis; TRL: 4–5	Feedstock: natural gas, coal and oil; Energy source: internally generated steam	Eff. (%): 60; H_2 color: turquoise	Reduced process steps, emission-free	Carbon by-product, fossil fuels depended	1.59–1.7
Coal gasification; TRL: 8	Feedstock: coal; Energy source: coal itself (authothermal gasification)	Eff. (%): 35–45; H_2 color: Brown	H_2 with low-cost synthetic fuel, affordable	CO_2 emission, needs CCS technology to capture CO_2, high reactor costs	1.34
Biomass pyrolysis; TRL: 3–5	Feedstock: crop residue, bagasse and municipal waste; Energy source: internally generated steam	Eff. (%): 35–50; H_2 color: Green	Abundant and cheap feedstock, CO_2 neutral	Char and tar deposits on catalytic sites, high temperature requirement, and seasonal availability, high handling cost, biomass	1.25–2.2
Biomass gasification; TRL: 6–7	Feedstock: crop residue, bagasse, and municipal waste; Energy source: internally generated steam	Eff. (%): 40–50; H_2 color: Green	Abundant and cheap feedstock, CO_2 neutral	Tar formation, higher temperature and pressure requirement, and global warming potential	1.77–2.05

(continued)

Table 3.2 (*Cont.*)

Production method and TRL	*Feedstock and energy source*	*Efficiency (%) and produced H_2 color*	*Advantages*	*Drawbacks*	*H_2 cost ($/kg)*
Bio-photolysis; TRL: 1–3	Feedstock: water + algae; Energy source: Solar	Eff. (%): 10; H_2 color: Green	O_2 by-product, operation under mild conditions, CO_2 consumed	Low H_2 yields and production rates, sunlight, and larger reactor requirements	2.13
Dark fermentation; TRL: 1–3	Feedstock: organic biomass	Eff. (%): 60–80; H_2 color: Green	Simple and O_2 free process, H_2 production without light, waste recycling contribution, CO_2 neutral	Low H_2 yields and production rates, removal of fatty acids, and a large reactor volume are needed	2.57
Photo-fermentation; TRL: 1–3	Feedstock: organic biomass; Energy source: Solar	Eff. (%): 0.1; H_2 color: Green	Use of different wastewater and organic wastes, waste recycling contribution, and CO_2 neutral	Low H_2 yields and production rates, sunlight, and a large reactor volume are needed	2.83
Electrolysis; TRL: 9	Feedstock: water; Energy source: grid, solar, wind, and nuclear	Eff. (%): 60–80; H_2 color: green, pink	Commercial technology, easily available and cheap feedstock, O_2 by-product with zero emission	High capital and maintenance costs, energy-intensive process	10.30
Thermolysis; TRL: 1–3	Feedstock: water; Energy source: solar and nuclear	Eff. (%): 20–45; H_2 color: green, red	O_2 by-product only, clean and reliable process, ample feedstock	Corrosion issues, high capital cost, and toxicity issues of elements	7.98–8.4
Photolysis; TRL: 1–3	Feedstock: water; Energy source: solar	Eff. (%): 0.06; H_2 color: green	O_2 by-product only, emission-free, ample feedstock	Non-effective photolytic materials, low conversion efficiency, and sunlight requirement	8–10

Source: Nikolaidis and Poullikkas (2017), Younas et al. (2022), Sharma et al. (2024), Nasser et al. (2022), Chaudhary et al. (2024), and AlHumaidan et al. (2023).

Eff., efficiency; TRL, technology readiness levels.

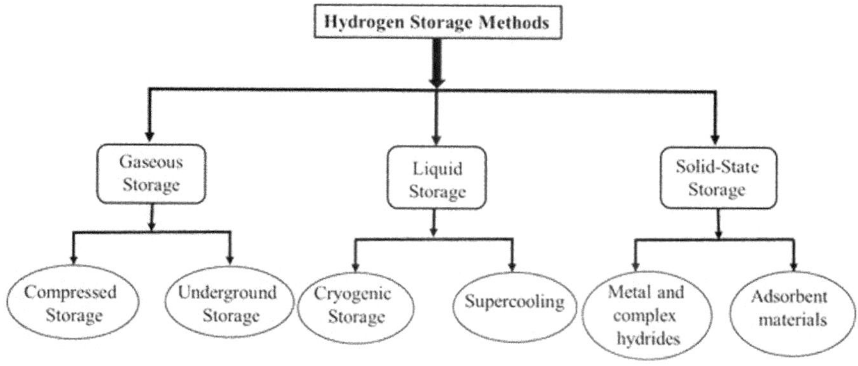

Figure 3.3 Overview of hydrogen storage technologies.

under development. Water electrolysis is a significantly developed (TRL: 9) method for producing H_2 without emissions.

3.4 HYDROGEN (H_2) STORAGE METHODS

The storage of hydrogen is one of the primary obstacles that must be overcome for its diverse applications. Hydrogen energy storage aims to be both safe and efficient, allowing for convenient use. Hydrogen has a low volumetric energy density but a high gravimetric energy density (Zhang et al., 2016). Various hydrogen storage systems, each with unique features and capacities, could be beneficial for the future hydrogen economy. There are three ways hydrogen can be stored, i.e., in the gaseous form, liquid form, and solid form. An overview of hydrogen storage systems is shown in Figure 3.3.

3.4.1 Gaseous H_2 storage system

Compressed hydrogen at high pressure is the most often used storage method (Tarhan and Çil, 2021; Durbin and Malardier-Jugroot, 2013). Hydrogen compression is a well-developed technology that provides high rates of hydrogen filling and release (Rivard et al., 2019; Abe et al., 2019). Hydrogen can be kept as a gas in cylinders, containers, or underground holes at high pressure up to 700 bar (Zhang et al., 2016). Cylindrical vessels are easy to store hydrogen, while it is difficult to transport hydrogen in spherical vessels. According to Zheng et al. (2012), about 80% of worldwide refueling stations of hydrogen used compressed gas storage in 2010. The storage of hydrogen in compressed form requires a specially designed vessel that can withstand high temperatures and pressure. The materials used for storing hydrogen in compressed form are built from aluminum, steel, and carbon fiber-reinforced plastic composites (Zheng et al., 2012). The metallic material made for hydrogen storage is of low cost and can withstand pressure up to 30 bar (Barthélémy et al., 2017; Moradi and Groth, 2019). The containers possess an exceptionally low gravimetric energy density since only about 1 wt.% of H_2 is efficiently kept there (Barthélémy et al., 2017). Steel and aluminum alloys are the most commonly used construction materials (Barthélémy et al., 2017). In another storage vessel, the material consists of a metallic wall with a fiber resin composite wrapped around the cylindrical portion (Abdalla et al., 2018). The main advantage of this composite vessel is that it has 30%–40% less weight than the pure metallic vessel; however, with the expense of a high cost (approx. 50%) (Moradi and Groth, 2019). Additionally, the storage vessels are constructed entirely of composite material, lined with metal and polymer, often high-density polyethylene (HDPE) (Barthélémy et al., 2017; Abdalla et al., 2018). The composite material storage vessel is used to store high-pressure hydrogen up to 700 bars. The energy density of hydrogen at 700 pressures is 5.7 MJ/L (Rivard et al., 2019). Onboard pressurized vessels are not widely accepted and pose a higher danger of explosions from abrupt shocks. Practical obstacles with refueling at 700 bar exist, as do safety concerns at filling stations (Prabhukhot et al., 2016). Infrastructure for hydrogen fuel is also required, including dispensing stations and transportation (Ogden et al., 2018). Compressed hydrogen gas has generally short-term applications.

The long-term storage of hydrogen is generally done underground. Underground storage is basically the in-situ conversion of a large quantity of hydrocarbon into hydrogen and storing it within the ground. Underground structures like salt caverns, aquifers, and depleted natural gas/oil reserves have sufficient storage capacity to store massive volumes of hydrogen (Tarkowski, 2019; Zivar et al., 2021). The underground storage of hydrogen has various advantages, such as it can store large amounts of hydrogen for a longer duration, it is safer than other methods due to the absence of oxygen, and it is also economical (Zivar et al., 2021). However, it has some limitations of location suitability, leakage, water management during hydrogen production, contamination, and environmental issues (Zivar et al., 2021; Heinemann et al., 2018).

3.4.2 Liquid H_2 storage

Hydrogen stored in liquid form has a higher density than compressed gas, allowing for more energy storage per volume (Zhang et al., 2016). Liquid hydrogen has a density of around 71 g/L at –253°C (Abe et al., 2019) and an energy density of 8 MJ/L H_2 (Durbin and Malardier-Jugroot, 2013). Systems for storing liquid hydrogen at low pressure are affordable. Liquid hydrogen can be stored at –253°C, whereas pressurized hydrogen gas can be kept at room temperature (Zheng et al., 2012). Liquid hydrogen storage requires a low boiling point. To maintain hydrogen efficiency, specific containers with insulation techniques are needed (Abe et al., 2019). If H_2 is mistakenly permitted to wash through the opening, nothing bad will happen because of the low adiabatic expansion energy (Tang et al., 2023). It is well known that liquid hydrogen is less corrosive, and the storage vessel can be fabricated from materials like stainless steel or aluminum alloy (Usman, 2022). Apart from the various advantages of a liquid hydrogen storage system, it is accompanied by various disadvantages. The conversion of hydrogen into liquid form is highly costly and power-consuming. It is estimated that about 30%–40% of the total heating value of hydrogen is lost during the conversion of liquid hydrogen (Barthélémy et al., 2017; Usman, 2022). Another problem associated with the liquid hydrogen storage system is the boil-off phenomenon. In the boil-off phenomenon, a small amount of hydrogen is continuously lost (Bosu and Rajamohan, 2024). A unique vessel design and costly insulation are needed to lessen the boil-off phenomenon, which raises the vessel's overall weight and lowers its gravimetric energy density (Tang et al., 2023).

3.4.3 Solid state H_2 storage

The solid-state hydrogen storage is predominantly done in a metal hydride. Metal hydride storage of hydrogen is highly recommended because it offers high volumetric energy densities and provides better safety than other storage methods. The hydrogen is chemically bound with metal at low pressure, which provides better stability of hydrogen (Von Colbe et al., 2019). Additionally, there are numerous potential uses for various kinds of metal hydrides, spanning from short-term to long-term energy storage and small- to large-scale applications. Metal hydrides are chemicals generated when metals or alloys react with hydrogen under certain conditions. These materials absorb hydrogen, forming a solid solution or intermetallic complex that may be stored at low pressures and ambient temperatures (Klopčič et al., 2023). The process of absorbing hydrogen is known as chemisorption, which releases energy. This reaction is also reversible and is known as desorption. To release hydrogen, an equal amount of energy from outside is required (Usman, 2022). The general reaction for the formation of a metal hydride is given by Eq. (3.41):

$$\mathrm{M} + 0.5n\mathrm{H}_2 \leftrightarrow \mathrm{MH}_n \tag{3.41}$$

where M is the metal or alloy, n is any number, and H is the hydrogen.

Metal hydrides are better than compression and liquefaction in terms of hydrogen storage capacity (Tarhan and Çil, 2021). Metal hydride is a safer alternative to gas compression and liquefaction due to its milder operating conditions (Klopčič et al., 2023). Furthermore, hydrogen can be loaded and released as often as needed, as long as no impurities obstruct the process (Tarhan and Çil, 2021). Additionally, a lot of metal hydrides often need a substantial amount of energy (20–55 kJ/mol H_2). However, metal hydride has some disadvantages like high weight, high material cost, high desorption temperature, and production of unwanted gases when the hydrogen is being released (Ren et al., 2017). There are various types of metal hydrides available, such as light metal hydrides, intermetallic hydrides, and transition metal hydrides (Usman, 2022). The MgH_2 metal hydrides can absorb and release hydrogen at a high rate at lower temperatures. It is reported that the absorption and desorption kinetics of MgH_2 is very promising at very low temperatures of 150°C (Klopčič et al., 2023). The Mg metal is available in sufficient amounts; it is of low cost, and has a higher gravimetric energy density than liquid hydrogen (Møller et al., 2017). Ball milling significantly reduces the time required to load hydrogen in MgH_2 and also decreases the size of the crystal grains. Reule et al. (2000) stated that ball milling MgH_2 produces lower desorption temperatures and higher desorption rates. The researchers hypothesized that longer milling times lead to a decrease in crystal size. AlH_3 is the other metal hydride that has high hydrogen content (10.1 wt.%) and high gravimetric energy density (Gandhi et al., 2010; Usman, 2022). The metal hydride AlH_3 releases hydrogen at a very low temperature of 100°C. However, AlH_3 has a significant disadvantage: reversible reactions with AlH_3 require extremely high hydrogen pressure (Pukazhselvan et al., 2012).

The metal hydride in the form of a complex metal hydride is also widely used for hydrogen storage. There are various types of complex hydrides, such as borohydride (BH_4^-), alanates (AlH_4^-), and amides (NH_2^-) (Rusman and Dahari, 2016). A variety of borohydrides are available for the storage of hydrogen, such as sodium borohydride, magnesium borohydride, lithium borohydride, and calcium borohydride. Among the borohydrides, lithium borohydride may be the most important, with a high hydrogen content of 18.5 wt.% (Rusman and Dahari, 2016). However, the temperature of hydrogen release is extremely high, and the kinetics of hydrogen release are slow (Liu et al., 2018). However, it has the disadvantage of slow hydrogen release. The practical application of amide hydrogen storage is achievable by using only a catalyst for hydrogen release (Usman, 2022).

The other important method used to store hydrogen in a solid state is the physical adsorption method. In the physical adsorption method of storing hydrogen, adsorbent materials store hydrogen through physisorption, which occurs when hydrogen molecules stick to the surface of a solid material due to van der Waals forces (Sadeq et al., 2024). Physical adsorption of hydrogen allows for quick loading and unloading, eliminating the need for long refuelling times. The process has an adsorption/desorption enthalpy of

Table 3.3 Comparison of different hydrogen storage methods

H_2 storage method	*Conditions*	*Hydrogen capacity (Wt.% H_2)*	*Volumetric density (g/L)*	*Advantages*	*Disadvantages*
Compressed gas (gaseous state)	T = RT P = 1–700 bar	5.7–100	0.0814–40.8	Well-developed technology, fast refilling and discharge, high purity H_2, low energy demand	Risk of explosion, losses (15%), low volumetric density, high cost of a large storage tank
Underground	—	—	—	High capacity, durability, safety, and minimum environmental impact	Location suitability, leakage, contamination, requirement of maintenance, and monitoring
Liquid state	T = −253°C P = 1 bar	14–100	51–70.8	High volumetric density, well-developed technology, suitable for transport, and high rate of hydrogen release	Higher losses (30%), high energy usage, liquefaction, necessitate a cooling system, boil off, and evaporation losses
Solid state (metal and complex hydrides like MgH_2, $FeTiH_2$, and $NaAlH_4$)	—	1.89–7.6	80–114	Safer storage method with high purity, high storage capacity, and easy to operate	Slow kinetics, high hydrogen release temperature, and high weight of storage materials

Source: Bosu and Rajamohan (2024), Usman (2022), Zivar et al. (2021), and Tarhan and Çil (2021).

*P, pressure; RT, room temperature; T, temperature.

1–10 kJ/mol, making it relatively energy efficient (Usman, 2022). Materials with a large surface area, such as activated carbon, allow for considerable hydrogen uptake, especially under high-pressure and low-temperature circumstances (Sadeq et al., 2024). Emerging materials, such as porous aromatic frameworks (PAFs), porous Boron Nitride (p-BN), and nanostructured graphene, are being studied for their potential characteristics and capacity to store hydrogen efficiently (Sadeq et al., 2024). Table 3.3 shows the comparison of the different storage methods of hydrogen with advantages and disadvantages.

3.5 HYDROGEN (H_2) TRANSPORTATION ROUTES

Hydrogen is a dynamic energy source, used in power generation, transportation, and industry. The hydrogen economy depends on the transport of hydrogen from the manufacturing location to the point of end-users or storage (Faye et al., 2022). There are several approaches for transporting hydrogen. Hydrogen transportation routes can be broadly classified into three main types: gaseous transport, liquid transport, and the use of hydrogen carrier materials (Noh et al., 2023). With this method of transport, compressed hydrogen is stored and transported using high-pressure gas cylinders or specifically built tube trailers that are strong enough for handling such high pressures (Sadeq et al., 2024). In the context of liquid hydrogen transport technology, hydrogen gas gets transformed into a liquid state via a cooling process at very low temperatures (–253°C) (Zhang et al., 2024). This method offers the handling of substantial volumes of hydrogen in a constrained space and is regarded as perfect for very long-distance travel. The following method is for conveying H_2 by using hydrogen carrier materials, in which hydrogen is chemically linked to the carrier molecules (Muhammed et al., 2023).

3.5.1 Gaseous transport

The technique of gaseous state hydrogen transport involves compressing hydrogen gas at high pressure and transporting it to a designated container while adhering to all safety rules and guidelines. This methodology leverages typical means such as high-pressure tube trailers and pipeline systems for gaseous H_2 transportation. Initially, compressors are used to reduce the volume of hydrogen in tube trailers, enhancing the effectiveness of transportation (Sdanghi et al., 2019). Depending on the conditions and particular requirements, compressed hydrogen can be transported by road, rail, or ship using tube trailers, which are made up of several cylinders fastened to a chassis, to storage facilities or end customers (Sadeq et al., 2024).

Pipelines are commonly utilized to transport hydrogen over short, medium, and even long distances, particularly in regions with developed hydrogen economies. For the pipeline transportation system, hydrogen gas must be compressed before being injected into the pipelines (Zhang et al., 2024). Depending on the distance and pipeline construction, compression pressures can vary. Materials compatible with hydrogen must be used to prevent issues like hydrogen embrittlement, which can damage pipeline materials due to hydrogen absorption (Sun and Cheng, 2022). Recently, more sophisticated techniques have been introduced, including ionic liquid piston compression and electrochemical compression (Sdanghi et al., 2019). Electrochemical compression operates on principles analogous to those used in PEM electrolysis (Sadeq et al., 2024; Sdanghi et al., 2019).

Hydrogen embrittlement is a substantial obstacle facing the infrastructure for transporting gaseous hydrogen, as it refers to the degradation of the structural integrity of materials due to the diffusion and absorption of hydrogen atoms (Zhang et al., 2024). Current research efforts are focusing on the development of innovative materials, coatings, and pipeline topologies to mitigate the adverse consequences of hydrogen embrittlement and improve the overall functionality and durability of hydrogen pipelines (Luo et al., 2020). Furthermore, protective coatings, including those that are ceramic or metallic in nature, may function as barriers to the ingress of hydrogen, thereby mitigating the potential for embrittlement within the substrate material (Laadel et al., 2022). Using advanced hydrogen purification techniques such as PSA, membrane separation, or cryogenic distillation is effective in lowering impurity concentrations, which in turn reduces the risk of embrittlement (Kim et al., 2022).

Transportation of hydrogen in gaseous form allows for high-purity applications such as fuel cells. However, gaseous hydrogen is characterized by its flammable nature and remarkably low ignition energy, raising considerable safety issues in its management, storage, and transport (Sadeq et al., 2024).

3.5.2 Liquid transport

Liquid hydrogen transport involves using a cryogenic approach to convert gaseous hydrogen into liquid form through liquefaction at extremely low temperatures. In the hydrogen production site, it passes through a liquefaction process, which turns it into a liquid by cooling it to an extremely low temperature of about –253°C (Sadeq et al., 2024). Although the process of liquidation has high energy consumption, it enhances transportation efficiency. Tankers deliver liquid hydrogen to their destination by rail, road, or ship, based on available resources and customer needs (Ahluwalia et al., 2023). The execution of protective precautions is critical throughout the whole shipping process. Monitoring liquid hydrogen's temperature and pressure, using leak detection technology, and adhering to pertinent laws and standards are all necessary to guarantee safety (Klebanoff et al., 2017).

3.5.3 Hydrogen carrier materials

Hydrogen carriers are a feasible alternative to transporting liquid or gaseous hydrogen because they overcome the associated problems. Ammonia, liquid organic hydrogen carriers (LOHCs), and metal hydrides are some of the most common hydrogen carriers used in transportation (Zhang et al., 2024). Compared to gaseous or liquid hydrogen, these carriers are less flammable and explosive, guaranteeing safety during storage and transportation. At the production site, hydrogen is paired with metal hydrides to make alloys based on Mg or Na via an exothermic reaction (Chen et al., 2021). The closed-loop transporting system is being developed by recycling the depleted metal hydride material back to the production facility for re-hydrogenation (Muhammed et al., 2023). The Haber-Bosch process produces ammonia by combining nitrogen and hydrogen gas at high pressure in the presence of iron as a catalyst. To produce LOHC, hydrogen is coupled with other liquid organic compounds that contain active hydrogens, such as *N*-ethylcarbazole (NEC) or dibenzyltoluene (Niermann et al., 2019).

3.6 APPLICATION OF HYDROGEN IN DIFFERENT SECTORS

Hydrogen, the most ubiquitous element in the universe, has a diverse array of applications across various sectors and holds significant potential as a versatile fuel source or energy carrier. To reduce pollution, it can be blended with natural gas or utilized in gas turbines or Internal Combustion Engines (ICEs), in which hydrogen is directly combusted to provide both thermal energy and mechanical work (Sadeq et al., 2024). Furthermore, hydrogen plays a critical role in the decarbonization of heavy industries such as refining, steel manufacturing, and chemical production (Zhang et al., 2024). Figure 3.4 depicts an overview of major H_2 application sectors. Hydrogen has various useful applications, some of which are described below.

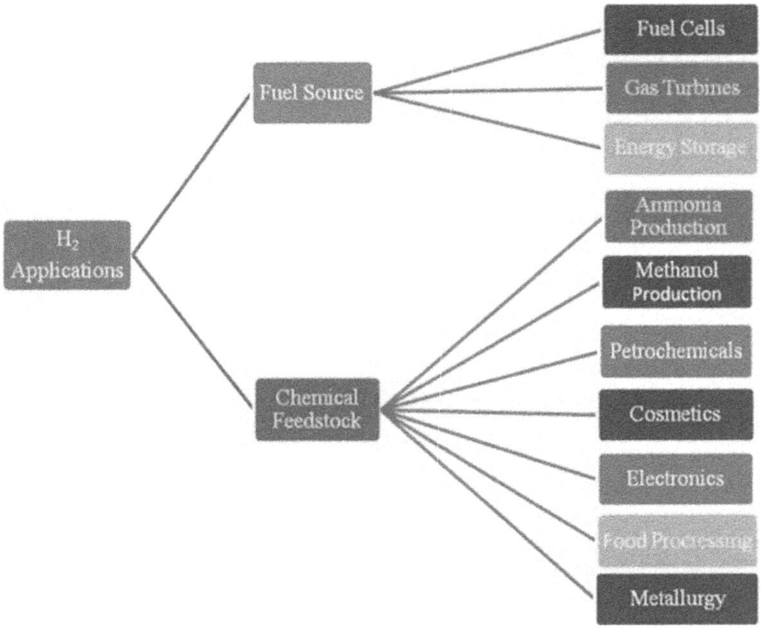

Figure 3.4 Overview of various application sectors of H_2.

3.6.1 Fuel source

3.6.1.1 *Fuel cells*

Fuel cells are combustion-free electrochemical devices that convert the chemical energy associated with hydrogen or other hydrogen-rich fuels into electrical energy through oxidation and reduction reactions (Panjiara and Pramanik, 2022). The electrolytes, cathode, and anode are the main components of a fuel cell (Sajid et al., 2022). These include direct alcohol fuel cells (DAFCs), solid oxide fuel cells (SOFCs), phosphoric acid fuel cells (PAFCs), alkali fuel cells (AFCs), molten carbonate fuel cells (MCFCs), and proton exchange membrane fuel cells (PEMFCs) (Muhammed et al., 2023; Choudhary and Pramanik, 2020). Fuel cells convert chemical energy directly into electricity, providing higher energy efficiency (>60%) than traditional combustion-based power generation and greater reliability than batteries in conditions of extreme weather (Muhammed et al., 2023). Hydrogen may now be widely used in transportation, electricity generation, and other fields thanks to fuel cells. These fuel cells can be tailored in different capacities and sizes to fulfil a diversity of usage, from personal electronics to large-scale energy generation (Zakaria et al., 2021).

3.6.1.2 *Gas turbines*

Gas turbines (GT) are widely used for power generation due to their characteristics, such as the ability to generate high power, fast start-up, and the ability to be integrated into a compact space with other power generation devices, such as the Rankine cycle. Hydrogen is utilized in the GT for combustion to generate electricity or mechanical power due to its high diffusivity and low ignition energy (Escamilla et al., 2022). However, these features of hydrogen also provide challenges, such as a higher chance of kickback and higher NO_x emissions (Ahmad et al., 2024; Zhang et al., 2024). To address these issues, researchers are working to develop advanced combustion technologies such as flameless hydrogen combustion, which can improve safety features, increase efficiency, and drastically reduce NOx emissions (Sadeq et al., 2024). For example, when using hydrogen as fuel, computational studies have been done to optimize combustor designs with the goal of lowering NO_x emissions and preventing flashback.

3.6.1.3 *Energy storage*

HydESS is gaining traction as a comparatively low-cost method of storing renewable energy for commerce and transportation (Arsad et al., 2022). Electrolysis can be used in the energy system to create hydrogen from excess power created from renewable sources. Using electricity, the electrolysis process splits water (H_2O) into hydrogen (H_2) and oxygen (O_2). Once created, hydrogen can be stored for long periods and then burned or converted back into energy using fuel cells (Sadeq et al., 2024).

3.7 CONCLUSIONS

This chapter provides an in-depth assessment of several H_2-generating techniques using diverse feedstocks. It also discusses the technical aspects of various H_2 storage routes, followed by an explanation of different transportation techniques. Finally, it examines numerous applications of hydrogen as a fuel or energy carrier in different sectors. Amongst the various H_2 generation methods, the SMR process using natural gas as feedstock is commercially utilized to generate H_2 gas, with 48% global production, conversion efficiency of 74%–85%, and the lowest cost. Nevertheless, the process generates a significant amount of CO_2, such as 7 kg of CO_2/kg H_2, which accounts for 3% of CO_2 emissions worldwide.

REFERENCES

Abdalla, A.M., Hossain, S., Nisfindy, O.B., Azad, A.T., Dawood, M. and Azad, A.K., 2018. Hydrogen production, storage, transportation and key challenges with applications: A review. *Energy Conversion and Management*, 165, pp.602–627.

Abe, J.O., Popoola, A.P.I., Ajenifuja, E. and Popoola, O.M., 2019. Hydrogen energy, economy and storage: Review and recommendation. *International Journal of Hydrogen Energy*, 44(29), pp.15072–15086.

Ahluwalia, R.K., Roh, H.-S., Peng, J.-K., Papadias, D., Baird, A.R., Hecht, E.S., Ehrhart, B.D., Muna, A., Ronevich, J.A., Houchins, C., Killingsworth, N.J. and Aceves, S.M., 2023. Liquid hydrogen storage system for heavy duty trucks: Configuration, performance, cost, and safety. *International Journal of Hydrogen Energy*, 48(35), pp.13308–13323.

Ahmad, S., Ullah, A., Samreen, A., Qasim, M., Nawaz, K., Ahmad, W., Alnaser, A., Kannan, A.M. and Egilmez, M., 2024. Hydrogen production, storage, transportation and utilization for energy sector: A current status review. *Journal of Energy Storage*, 101, p.113733.

AlHumaidan, F.S., Halabi, M.A., Rana, M.S. and Vinoba, M., 2023. Blue hydrogen: Current status and future technologies. *Energy Conversion and Management*, 283, p.116840.

Arsad, A.Z., Hannan, M.A., Al-Shetwi, A.Q., Mansur, M., Muttaqi, K.M., Dong, Z.Y. and Blaabjerg, F., 2022. Hydrogen energy storage integrated hybrid renewable energy systems: A review analysis for future research directions. *International Journal of Hydrogen Energy*, 47(39), pp.17285–17312.

Asghar, U., Rafiq, S., Anwar, A., Iqbal, T., Ahmed, A., Jamil, F., Khurram, M.S., Akbar, M.M., Farooq, A., Shah, N.S. and Park, Y.K., 2021. Review on the progress in emission control technologies for the abatement of CO_2, SOx and NOx from fuel combustion. *Journal of Environmental Chemical Engineering*, 9(5), p.106064.

Barthélémy, H., Weber, M. and Barbier, F., 2017. Hydrogen storage: Recent improvements and industrial perspectives. *International Journal of Hydrogen Energy*, 42(11), pp.7254–7262.

Bosu, S. and Rajamohan, N., 2024. Recent advancements in hydrogen storage-comparative review on methods, operating conditions and challenges. *International Journal of Hydrogen Energy*, 52, pp.352–370.

Chaubey, R., Sahu, S., James, O.O. and Maity, S., 2013. A review on development of industrial processes and emerging techniques for production of hydrogen from renewable and sustainable sources. *Renewable and Sustainable Energy Reviews*, 23, pp.443–462.

Chaudhary, K., Bhardvaj, K. and Chaudhary, A., 2024. A qualitative assessment of hydrogen generation techniques for fuel cell applications. *Fuel*, *358*, p.130090.

Chen, Z., Ma, Z., Zheng, J., Li, X., Akiba, E. and Li, H.-W., 2021. Perspectives and challenges of hydrogen storage in solid-state hydrides. *Chinese Journal of Chemical Engineering*, 29, pp.1–12.

Choudhary, A.K. and Pramanik, H., 2020. Addition of rhenium (Re) to Pt-Ru/f-MWCNT anode electrocatalysts for enhancement of ethanol electrooxidation in half-cell and single direct ethanol fuel cell. *International Journal of Hydrogen Energy*, 45(24), pp.13300–13321.

Durbin, D.J. and Malardier-Jugroot, C., 2013. Review of hydrogen storage techniques for on board vehicle applications. *International Journal of Hydrogen Energy*, 38(34), pp.14595–14617.

El-Shafie, M., Kambara, S. and Hayakawa, Y., 2019. Hydrogen production technologies overview. *Journal of Power Energy Engineering*, 7, pp.107–154.

Escamilla, A., Sánchez, D. and García-Rodríguez, L., 2022. Assessment of power-to-power renewable energy storage based on the smart integration of hydrogen and micro gas turbine technologies. *International Journal of Hydrogen Energy*, 47(40), pp.17505–17525.

Faye, O., Szpunar, J. and Eduok, U., 2022. A critical review on the current technologies for the generation, storage, and transportation of hydrogen. *International Journal of Hydrogen Energy*, 47(29), pp.13771–13802.

Gandhi, K., Dixit, D.K. and Dixit, B.K., 2010. Hydrogen desorption energies of aluminum hydride (Al_nH_{3n}) clusters. *Physica B: Condensed Matter*, 405(15), pp.3075–3081.

Ghorbani, B., Zendehboudi, S., Saady, N.M.C. and Dusseault, M.B., 2023a. Hydrogen storage in North America: Status, prospects, and challenges. *Journal of Environmental Chemical Engineering*, 11(3), p.109957.

Ghorbani, B., Zendehboudi, S., Zhang, Y., Zarrin, H. and Chatzis, I., 2023b. Thermochemical water-splitting structures for hydrogen production: Thermodynamic, economic, and environmental impacts. *Energy Conversion and Management*, 297, p.117599.

Hassan, A., Ilyas, S.Z., Jalil, A. and Ullah, Z., 2021. Monetization of the environmental damage caused by fossil fuels. *Environmental Science and Pollution Research*, 28, pp.21204–21211.

Heinemann, N., Booth, M.G., Haszeldine, R.S., Wilkinson, M., Scafidi, J. and Edlmann, K., 2018. Hydrogen storage in porous geological formations–onshore play opportunities in the midland valley (Scotland, UK). *International Journal of Hydrogen Energy*, 43(45), pp.20861–20874.

Hosseini, S.E., Abdul Wahid, M., Jamil, M.M., Azli, A.A. and Misbah, M.F., 2015. A review on biomass-based hydrogen production for renewable energy supply. *International Journal of Energy Research*, 39(12), pp.1597–1615.

Jeje, S.O., Marazani, T., Obiko, J.O. and Shongwe, M.B., 2024. Advancing the hydrogen production economy: A comprehensive review of technologies, sustainability, and future prospects. *International Journal of Hydrogen Energy*, 78, pp.642–661.

Kim, T., Song, Y., Kang, J., Kim, S.K. and Kim, S., 2022. A review of recent advances in hydrogen purification for selective removal of oxygen: Deoxo catalysts and reactor systems. *International Journal of Hydrogen Energy*, 47(59), pp.24817–24834.

Klebanoff, L.E., Pratt, J.W. and LaFleur, C.B., 2017. Comparison of the safety-related physical and combustion properties of liquid hydrogen and liquid natural gas in the context of the SF-BREEZE high-speed fuel-cell ferry. *International Journal of Hydrogen Energy*, 42(1), pp.757–774.

Klopčič, N., Grimmer, I., Winkler, F., Sartory, M. and Trattner, A., 2023. A review on metal hydride materials for hydrogen storage. *Journal of Energy Storage*, 72, p.108456.

Kumar, P. and Fiori, L., 2024. Thermochemical and biological routes for biohydrogen production: A review. *Energy Conversion and Management*, X, p.100659.

Kumar, S.S. and Himabindu, V., 2019. Hydrogen production by PEM water electrolysis: A review. *Materials Science for Energy Technologies*, 2(3), pp.442–454.

Laadel, N.-E., El Mansori, M., Kang, N., Marlin, S. and Boussant-Roux, Y., 2022. Permeation barriers for hydrogen embrittlement prevention in metals – A review on mechanisms, materials suitability and efficiency. *International Journal of Hydrogen Energy*, 47(76), pp.32707–32731.

Lavanya, M., Shrivastava, S., Lakshmi, T., Sandadi, E.R., Gour, S., Thomas, N.A., Priya, S.S. and Sudhakar, K., 2024. An overview of hydrogen storage technologies: Key challenges and opportunities. *Materials Chemistry and Physics*, 325, p.129710.

Liu, Y., Ren, Z., Zhang, X., Jian, N., Yang, Y., Gao, M. and Pan, H., 2018. Development of catalyst-enhanced sodium alanate as an advanced hydrogen-storage material for mobile applications. *Energy Technology*, 6(3), pp.487–500.

Luo, H., Sohn, S.S., Lu, W., Li, L., Li, X., Soundararajan, C.K., Krieger, W., Li, Z. and Raabe, D., 2020. A strong and ductile medium-entropy alloy resists hydrogen embrittlement and corrosion. *Nature Communications*, 11(1), p.3081.

Mallick, D., Mahanta, P. and Moholkar, V.S., 2017. Co-gasification of coal and biomass blends: Chemistry and engineering. *Fuel*, 204, pp.106–128.

Mazloomi, K. and Gomes, C., 2012. Hydrogen as an energy carrier: Prospects and challenges. *Renewable and Sustainable Energy Reviews*, 16(5), pp.3024–3033.

Midilli, A., Kucuk, H., Topal, M.E., Akbulut, U. and Dincer, I., 2021. A comprehensive review on hydrogen production from coal gasification: Challenges and opportunities. *International Journal of Hydrogen Energy*, 46(50), pp.25385–25412.

Møller, K.T., Jensen, T.R., Akiba, E. and Li, H.W., 2017. Hydrogen: A sustainable energy carrier. *Progress in Natural Science: Materials International*, 27(1), pp.34–40.

Moradi, R. and Groth, K.M., 2019. Hydrogen storage and delivery: Review of the state of the art technologies and risk and reliability analysis. *International Journal of Hydrogen Energy*, 44(23), pp.12254–12269.

Morya, R., Raj, T., Lee, Y., Pandey, A.K., Kumar, D., Singhania, R.R., Singh, S., Verma, J.P. and Kim, S.H., 2022. Recent updates in biohydrogen production strategies and life–cycle assessment for sustainable future. *Bioresource Technology*, 366, p.128159.

Muhammed, N.S., Gbadamosi, A.O., Epelle, E.I., Abdulrasheed, A.A., Haq, B., Patil, S., Al-Shehri, D. and Kamal, M.S., 2023. Hydrogen production, transportation, utilization, and storage: Recent advances towards sustainable energy. *Journal of Energy Storage*, 73, p.109207.

Muhammed, N.S., Haq, B., Al Shehri, D., Al-Ahmed, A., Rahman, M.M. and Zaman, E., 2022. A review on underground hydrogen storage: Insight into geological sites, influencing factors and future outlook. *Energy Reports*, 8, pp.461–499.

Nasser, M., Megahed, T.F., Ookawara, S. and Hassan, H., 2022. A review of water electrolysis–based systems for hydrogen production using hybrid/solar/wind energy systems. *Environmental Science and Pollution Research*, 29(58), pp.86994–87018.

Niermann, M., Beckendorff, A., Kaltschmitt, M. and Bonhoff, K., 2019. Liquid organic hydrogen carrier (LOHC): Assessment based on chemical and economic properties. *International Journal of Hydrogen Energy*, 44(13), pp.6631–6654.

Nikolaidis, P. and Poullikkas, A., 2017. A comparative overview of hydrogen production processes. *Renewable and Sustainable Energy Reviews*, 67, pp.597–611.

Noh, H., Kang, K. and Seo, Y., 2023. Environmental and energy efficiency assessments of offshore hydrogen supply chains utilizing compressed gaseous hydrogen, liquefied hydrogen, liquid organic hydrogen carriers and ammonia. *International Journal of Hydrogen Energy*, 48(20), pp.7515–7532.

Ogden, J., Jaffe, A.M., Scheitrum, D., McDonald, Z. and Miller, M., 2018. Natural gas as a bridge to hydrogen transportation fuel: Insights from the literature. *Energy Policy*, 115, pp.317–329.

Onwuemezie, L., 2024. Overview and prospects of low-emissions hydrogen (H_2) energy systems: Roadmap for a sustainable H_2 economy. *Energy*, 360, p.100008.

Panjiara, D. and Pramanik, H., 2022. Study on the effect of calcium hypochlorite and air as mixed oxidant and a synthesized low-cost Pd-Ni/C anode electrocatalyst for electrooxidation of glycerol in an air-breathing microfluidic fuel cell. *The Canadian Journal of Chemical Engineering*, 100, pp.S94–S111.

Prabhukhot Prachi, R., Wagh Mahesh, M. and Gangal Aneesh, C., 2016. A review on solid state hydrogen storage material. *Advances in Energy Power*, 4(11), pp.11–22.

Pukazhselvan, D., Kumar, V. and Singh, S.K., 2012. High capacity hydrogen storage: Basic aspects, new developments and milestones. *Nano Energy*, 1(4), pp.566–589.

Rahman, A., Murad, S.W., Mohsin, A.K.M. and Wang, X., 2024. Does renewable energy proactively contribute to mitigating carbon emissions in major fossil fuels consuming countries? *Journal of Cleaner Production*, 452, p.142113.

Raza, A., Mahmoud, M., Arif, M. and Alafnan, S., 2024. Underground hydrogen storage prospects in the Kingdom of Saudi Arabia. *Fuel*, 357, p.129665.

Ren, J., Musyoka, N.M., Langmi, H.W., Mathe, M. and Liao, S., 2017. Current research trends and perspectives on materials-based hydrogen storage solutions: A critical review. *International Journal of Hydrogen Energy*, 42(1), pp.289–311.

Reule, H., Hirscher, M., Weißhardt, A. and Kronmüller, H., 2000. Hydrogen desorption properties of mechanically alloyed MgH_2 composite materials. *Journal of Alloys and Compounds*, 305(1–2), pp.246–252.

Rivard, E., Trudeau, M. and Zaghib, K., 2019. Hydrogen storage for mobility: A review. *Materials*, 12(12), p.1973.

Rusman, N.A.A. and Dahari, M., 2016. A review on the current progress of metal hydrides material for solid-state hydrogen storage applications. *International Journal of Hydrogen Energy*, 41(28), pp.12108–12126.

Sadeq, A.M., Homod, R.Z., Hussein, A.K., Togun, H., Mahmoodi, A., Isleem, H.F., Patil, A.R. and Moghaddam, A.H., 2024. Hydrogen energy systems: Technologies, trends, and future prospects. *Science of the Total Environment*, 939, p.173622.

Safari, F. and Dincer, I., 2020. A review and comparative evaluation of thermochemical water splitting cycles for hydrogen production. *Energy Conversion and Management*, 205, p.112182.

Sahraei, M.H., McCalden, D., Hughes, R. and Ricardez-Sandoval, L.A., 2014. A survey on current advanced IGCC power plant technologies, sensors and control systems. *Fuel*, 137, pp.245–259.

Sajid, A., Pervaiz, E., Ali, H., Noor, T. and Baig, M.M., 2022. A perspective on development of fuel cell materials: Electrodes and electrolyte. *International Journal of Energy Research*, 46(6), pp.6953–6988.

Sánchez-Bastardo, N., Schlögl, R. and Ruland, H., 2021. Methane pyrolysis for zero-emission hydrogen production: A potential bridge technology from fossil fuels to a renewable and sustainable hydrogen economy. *Industrial & Engineering Chemistry Research*, 60(32), pp.11855–11881.

Sarathy, S.M., Oßwald, P., Hansen, N. and Kohse-Höinghaus, K., 2014. Alcohol combustion chemistry. *Progress in Energy and Combustion Science*, 44, pp.40–102.

Sarmah, M.K., Singh, T.P., Kalita, P. and Dewan, A., 2023. Sustainable hydrogen generation and storage: A review. *RSC Advances*, 13(36), pp.25253–25275.

Sdanghi, G., Maranzana, G., Celzard, A. and Fierro, V., 2019. Review of the current technologies and performances of hydrogen compression for stationary and automotive applications. *Renewable and Sustainable Energy Reviews*, 102, pp.150–170.

Seçer, A., Küçet, N., Fakı, E. and Hasanoğlu, A., 2018. Comparison of co–gasification efficiencies of coal, lignocellulosic biomass and biomass hydrolysate for high yield hydrogen production. *International Journal of Hydrogen Energy*, 43(46), pp.21269–21278.

Sharma, G., Dewangan, A.K., Yadav, A.K. and Ahmad, A., 2024. Current status of research on hydrogen generation, storage and transportation technologies: A state-of-the-art review towards sustainable energy. *Process Safety and Environmental Protection*, 191(A), pp.1445–1460.

Sun, Y. and Cheng, Y.F., 2022. Hydrogen-induced degradation of high-strength steel pipeline welds: A critical review. *Engineering Failure Analysis*, 133, p.105985.

Tackie-Otoo, B.N. and Haq, M.B., 2024. A comprehensive review on geo-storage of H2 in salt caverns: Prospect and research advances. *Fuel*, 356, p.129609.

Taipabu, M.I., Viswanathan, K., Wu, W., Hattu, N. and Atabani, A.E., 2022. A critical review of the hydrogen production from biomass-based feedstocks: Challenge, solution, and future prospect. *Process Safety and Environmental Protection*, 164, pp.384–407.

Taiwo, G.O., Tomomewo, O.S. and Oni, B.A., 2024. A comprehensive review of underground hydrogen storage: Insight into geological sites (mechanisms), economics, barriers, and future outlook. *Journal of Energy Storage*, 90, p.111844.

Tang, D., Tan, G.L., Li, G.W., Liang, J.G., Ahmad, S.M., Bahadur, A., Humayun, M., Ullah, H., Khan, A. and Bououdina, M., 2023. State-of-the-art hydrogen generation techniques and storage methods: A critical review. *Journal of Energy Storage*, 64, p.107196.

Tarhan, C. and Çil, M.A., 2021. A study on hydrogen, the clean energy of the future: Hydrogen storage methods. *Journal of Energy Storage*, 40, p.102676.

Tarkowski, R., 2019. Underground hydrogen storage: Characteristics and prospects. *Renewable and Sustainable Energy Reviews*, 105, pp.86–94.

Thiyagarajan, S.R., Emadi, H., Hussain, A., Patange, P. and Watson, M., 2022. A comprehensive review of the mechanisms and efficiency of underground hydrogen storage. *Journal of Energy Storage*, 51, p.104490.

Tripathi, A., Park, S., Agarwal, A.K. and Valera, H., 2022. Prospects of Dual-Fuel Injection System in Compression Ignition (CI) Engines Using Di-Methyl Ether (DME). In Avinash Kumar Agarwal, Hardikk Valera (Eds.), *Greener and Scalable E-Fuels for Decarbonization of Transport*. Singapore: Springer Nature, PP. 223–259.

Usman, M.R., 2022. Hydrogen storage methods: Review and current status. *Renewable and Sustainable Energy Reviews*, 167, p.112743.

Von Colbe, J.B., Ares, J.R., Barale, J., Baricco, M., Buckley, C., Capurso, G., Gallandat, N., Grant, D.M., Guzik, M.N., Jacob, I. and Jensen, E.H., 2019. Application of hydrides in hydrogen storage and compression: Achievements, outlook and perspectives. *International Journal of Hydrogen Energy*, 44(15), pp.7780–7808.

Vuppaladadiyam, A.K., Vuppaladadiyam, S.S.V., Awasthi, A., Sahoo, A., Rehman, S., Pant, K.K., Murugavelh, S., Huang, Q., Anthony, E., Fennel, P. and Bhattacharya, S., 2022. Biomass pyrolysis: A review on recent advancements and green hydrogen production. *Bioresource Technology*, 364, p.128087.

Yadav, S., Choudhary, A.K., Yadav, P. and Pal, D.B., 2024. Lignocellulosic biomass gasification for bio-circular economy sustainability: A multiple criteria analysis framework. *Biomass Conversion and Biorefinery*, pp.1–11. DOI:10.1007/s13399-024-06387-3

Younas, M., Shafique, S., Hafeez, A., Javed, F. and Rehman, F., 2022. An overview of hydrogen production: Current status, potential, and challenges. *Fuel*, 316, p.123317.

Zakaria, Z., Kamarudin, S.K., Abd Wahid, K.A. and Abu Hassan, S.H., 2021. The progress of fuel cell for Malaysian residential consumption: Energy status and prospects to introduction as a renewable power generation system. *Renewable and Sustainable Energy Reviews*, 144, p.110984.

Zhang, F., Zhao, P., Niu, M. and Maddy, J., 2016. The survey of key technologies in hydrogen energy storage. *International Journal of Hydrogen Energy*, 41(33), pp.14535–14552.

Zhang, L., Jia, C., Bai, F., Wang, W., An, S., Zhao, K., Li, Z., Li, J. and Sun, H., 2024. A comprehensive review of the promising clean energy carrier: Hydrogen production, transportation, storage, and utilization (HPTSU) technologies. *Fuel*, 355, p.129455.

Zheng, J., Liu, X., Xu, P., Liu, P., Zhao, Y. and Yang, J., 2012. Development of high pressure gaseous hydrogen storage technologies. *International Journal of Hydrogen Energy*, 37(1), pp.1048–1057.

Zivar, D., Kumar, S. and Foroozesh, J., 2021. Underground hydrogen storage: A comprehensive review. *International Journal of Hydrogen Energy*, 46(45), pp.23436–23462.

Chapter 4

A case study on renewable energy sources, power demand, and policies in Indian prospective

Ambneesh Mishra, Dan Bahadur Pal, and Shraddha Awasthi

4.1 INTRODUCTION

The necessity of renewable energy in building a sustainable future has been highlighted by the growing urgency of tackling climate change. Renewable energy can keep global warming well below 2°C over pre-industrial levels, underscoring its importance in mitigating climate change. In addition, renewable energy could supply over half of the world's energy needs, significantly reducing carbon emissions and improving air quality. It has been demonstrated that switching to entirely renewable energy is both technically and financially possible in several areas, including transportation, heating, cooling, industrial operations, and electricity (Jacobson, 2021). The expansion of renewable energy in India highlights the role of solar, wind, hydropower, bioenergy, and new technologies like green hydrogen. Policies that encourage development and adoption, like carbon credit trading and green energy subsidies, are essential. Moreover, the shift to renewable energy sources has the potential to significantly reduce emissions in addition to boosting jobs and GDP (Brouwer et al., 2014).

Acknowledging the urgent priorities of an environmental crisis and the need to achieve energy security has rendered India a prominent player in the field of renewable energy. Assimilating the country's sincere dedication to sustainable development and rich natural resources, the government is rapidly embracing the clean energy option. Solar, wind, hydropower, and biomass are all technologies being developed in the energy sector that diminish carbon emissions, promoting economic growth. This transformation also shows that India's aspirations for prosperity can be achieved alongside global calls for sustainable development through renewable energy. Renewable energy has become a major player for progression in the backdrop of shrinking fossil fuel supplies and the rapid pace of climate change. India's economic growth is, of course, greatly reliant upon expanding energy consumption, although it has also been triggered by the environmental problems associated with conventional energy sources. Under the dual impetus of domestic and international commitments, India is

DOI: 10.1201/9781003581536-4

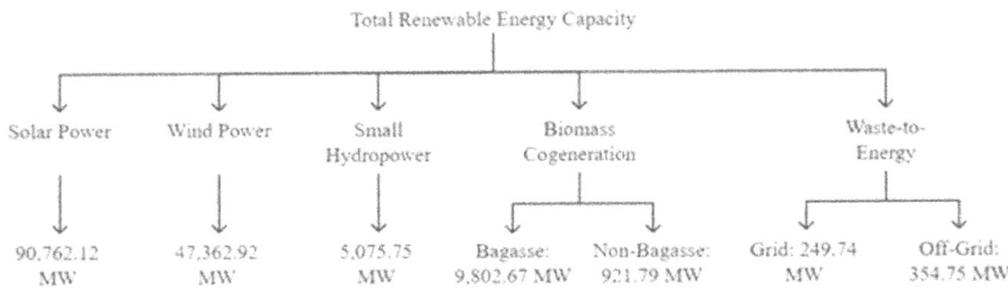

Figure 4.1 Flow diagram representing India's total renewable energy capacity.

harnessing its potential in various aspects of renewable energies: solar, wind, hydropower, biomass, and geothermal (Kumari, 2024).

India's renewable energy sector has shown significant progress as of September 2024, with cumulative installed capacities across various sources. Wind power reached a cumulative capacity of 47,362.92 MW, while solar power continues to lead with 90,762.12 MW. Small hydropower installations now total 5,075.75 MW. Biomass cogeneration, both bagasse and non-bagasse, stands at 9,802.67 MW and 921.79 MW, respectively. Waste-to-energy projects contribute an additional 249.74 MW, and off-grid waste-to-energy has reached 354.75 MW. Altogether, the cumulative renewable energy capacity in India is 154,529.74 MW, reflecting sustained growth and commitment to clean energy solutions. Figure 4.1 shows the flow Diagram representing India's Total Renewable Energy Capacity.

4.2 SOLAR ENERGY

The most plentiful renewable energy source on Earth is solar radiation, which can be used directly or indirectly through wind, biomass, hydropower, the ocean, and other sources. There are two ways to use solar energy, which we experience as heat and light: the thermal method uses the heat for cooking, drying, heating, purifying, and power generation, among other uses; the photovoltaic method turns the light from the sun into electricity, which can be used for a variety of things, including communications, lighting, pumping, and power supply in areas that are not electrified. The cumulative amount of annual solar radiation that falls on the Earth's surface is almost 7,500 times the world's yearly primary energy use (450 EJ). Solar radiation striking the Earth's surface annually (about 3,400,000 EJ) exceeds all predicted non-renewable energy resources, encompassing fossil fuels and nuclear. Currently, fossil fuels account for 80% of global energy use. India receives an average of 4–7 kWh of solar energy per square metre per day, with 250–300 bright days per year. Annual radiation energy is highest in Western Rajasthan and lowest in the northeast part of the country (Kumar et al., 2010). Around 5,000 trillion kWh of energy is obtained by India each year, with 4–7 kWh/m^2 being received daily in the majority of areas. Because solar energy is so widely available, it is the most secure energy source. Theoretically, the entire nation's electrical needs could be met by a small portion of the total amount of incoming solar energy if it were used well. According to the National Institute of Solar Energy, the nation has 748 GW of solar potential, assuming that solar PV modules cover 3% of the undeveloped land. To increase solar energy production in India, the Indian government has implemented several programs, including Solar Park Schemes, VGF Schemes, CPSU Schemes, Defence Schemes, Canal Bank and Canal Top Schemes, Bundling Schemes, Grid Connected Solar Rooftop Schemes, and others (Seikh and Chatterjee, 2024).

According to the data of MNRE (2024a), India's total solar power capacity has reached an amazing 90.76 GW, demonstrating the country's dedication to renewable energy development. This achievement consists of numerous main segments, each of which contributes uniquely to the overall capacity:

- Ground-mounted solar plants, with a capacity of 70.05 GW, dominate India's solar energy landscape and account for most of the solar infrastructure.
- Grid-Connected Solar Rooftops: Solar rooftops connected to the grid contribute 14.30 GW to the total, demonstrating the integration of smaller, distributed solar systems in the residential, commercial, and industrial sectors.
- Hybrid Projects (Solar Component): Hybrid projects that mix solar with other renewable sources add 2.63 GW, demonstrating flexibility in meeting electricity demands while optimizing renewable sources.
- Off-Grid Solar: With 3.78 GW, off-grid solar systems—which are frequently used in rural and isolated locations—highlight initiatives to bring sustainable energy access to areas outside the service area of traditional grids. Several measures have been put in place by the Indian government to hasten the expansion of solar energy throughout the nation. Important steps include encouraging foreign investment in solar projects and permitting up to 100% FDI via the automated method. Furthermore, for projects scheduled to be put into service by June 30, 2025, the government has abolished Inter-State Transmission System (ISTS) fees for the interstate selling of wind and solar energy.

A renewable purchase obligation (RPO) trajectory has been announced through 2029–2030 to promote long-term planning and guarantee a consistent rise in the use of renewable energy. Along with the establishment of a Project Development Cell aimed at luring and facilitating investment in the solar industry, standards for the deployment of solar photovoltaic systems and devices have also been established. The government has published Standard Bidding Guidelines to expedite the competitive bidding process for purchasing electricity from grid-connected solar PV and wind projects. To promote on-time payments, orders mandate that power be provided only upon a letter of credit (LC) or advance payment. This ensures timely payments to renewable energy (RE) generators by distribution licensees. The Green Energy Open Access Rules 2022, which support renewable energy, and "The Electricity (Late Payment Surcharge and Related Matters) Rules 2002," which address late payments in the sector, are other initiatives. By establishing a platform for the sale of renewable energy, such as solar power, through energy exchanges, the Green Term Ahead Market (GTAM) has improved market dynamics and accessibility for renewable energy (MNRE 2024b).

4.3 WIND ENERGY

A record-breaking year for wind power worldwide is highlighted in the 2023 Annual Report from the World Wind Energy Association (2023), With 116 GW added in 2023 alone, the total installed capacity reached 1,047 GW, representing a growth rate of 12.5%. With the installation of roughly 75 GW, or almost two-thirds of the new capacity, China made a substantial contribution to this increase. Wind energy is now responsible for 10% of worldwide power production, highlighting its critical role in the world's shift to renewable energy sources. Given the country's ambitious targets, India is expected to remain fourth in terms of overall capacity in the coming years. The new impetus in India is likely to come from repowering and offshore wind, which is expected to start in the coming years. According to the Union Minister for New & Renewable Energy and Power, as of June 30, 2023, the nation's installed wind energy generation capacity was at 43,773 MW. In FY 2022–2023, the country produced 71,814 million units of electricity using wind energy. Andhra Pradesh, Gujarat, Karnataka, Kerala, Madhya Pradesh, Maharashtra, Rajasthan, Tamil Nadu, and Telangana were the states that produced the most wind energy in 2022–2023 (PIB, 2023). The earth's rotation, the sun's heat energy, the cooling effects of the oceans and polar ice caps, temperature differences between the land and the sea, and the physical impact of mountains and other obstructions are some of the intricate processes that produce winds.

Wind energy is a very sustainable form of electricity because it produces very little waste. Additionally, India has a significant wind potential, which could contribute to lower energy costs, with lower operational and upkeep costs as well. Wind power is expected to be a significant source of meeting India's energy requests (Thapar et al., 2018).

4.4 HYDRO ENERGY

Seikh and Chatterjee (2024) state that India has a substantial energy poverty problem even though it has the second-largest population and the fifth-largest economy in the world. Furthermore, to meet electricity demands, a move toward renewable energy sources is required due to rising pollution levels. India's numerous rivers and streams make hydropower a feasible substitute and an important source of energy. This energy source could help the nation meet its low-carbon growth goals and close the energy deficit. Despite its growth over time, hydropower's proportion of India's total energy mix has declined in recent decades. However, the expansion of hydropower in India is still controversial since projects frequently encounter resistance citing issues including ecological effects, population displacement, heightened natural catastrophe risk, and socioeconomic difficulties. India divides its hydropower projects into two categories: major hydropower facilities and smaller hydropower projects with capacities up to 25 MW. For small and tiny hydropower projects, the Department of Hydro and Renewable Energy (HRED) has calculated a potential capacity of 21,133 MW from 7,133 locations nationwide. The hilly states of India, especially Arunachal Pradesh, Himachal Pradesh, Jammu and Kashmir, and Uttarakhand, contain around half of this potential. The states of Kerala, Karnataka, Chhattisgarh, and Maharashtra also have great hydropower potential. Hydropower has the potential to produce significant amounts of energy, despite its difficulties. The International Hydropower Association's (IHA) data from 2021 has outlined the worldwide significance of hydropower. In 2020, the total hydropower installed capacity globally rose 1.6%, from 1,306 GW in 2019 to 1,330 GW. Industries produced a total of 4,370 TWh of clean electricity. Twenty-one gigawatts of advanced projects, extending past 15.6 GW in 2019. China spearheaded this expansion, adding 13.8 GW of new capacity, or roughly two-thirds of the worldwide rise, followed by Turkey with 2.5 GW. Turkey's Ilisu and Lower Kaleköy projects, China's Jixi pumped storage (1.8 GW), and Angola's Lauca facility (2.1 GW) were notable completions. With 6.8 GW, the Wudongde project in China was the most significant addition to the grid. With more than 370 GW of built hydropower capacity, China is the world leader, followed by the US (102 GW), Canada (82 GW), and Brazil (109 GW). With a total hydropower capacity of 50 GW, India is ranked fifth in the world, solidifying its position as a major contributor to global renewable energy targets.

India's hydropower capacity includes contributions from both small and large hydropower projects, with small hydropower generating approximately 5,075.75 MW and large hydropower contributing 46,928.17 MW. Together, this brings India's total hydropower capacity to around 52,003.92 MW, highlighting hydropower's vital role in the nation's renewable energy mix and its potential to meet growing energy needs (MNRE, 2024c). Figure 4.2 shows India's Hydropower Capacity Distribution.

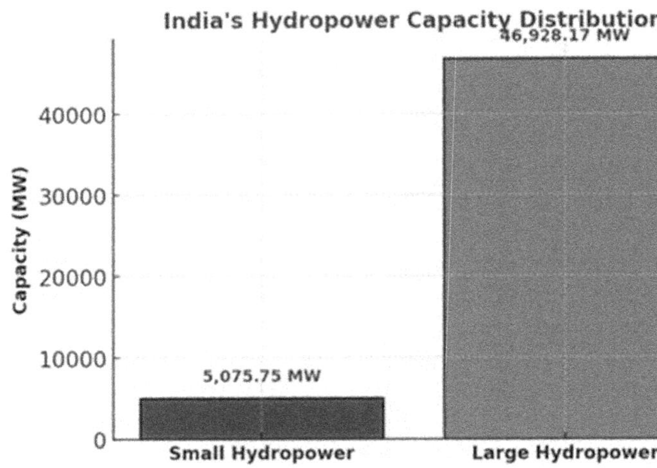

Figure 4.2 India's hydropower capacity distribution.

4.5 GEOTHERMAL

Geothermal energy is the energy that is drawn from the heat of the Earth. By employing water under pressure as a medium to transfer heat, deep extraction bore wells are bored to a depth of three kilometres or more to release the stored heat in the earth's layers. The top countries using their geothermal resources are the United States, the Philippines, Indonesia, Mexico, Italy, and Iceland. In India, research on geothermal energy was first conducted in 1862. With a projected capacity of 10,600 MW, the Geological Survey of India has investigated 340 hot springs and a few possible geothermal provinces. In actuality, though, it has yet to be utilized. Puga Valley in the Ladakh district is the most promising geothermal region among those in India that have been identified (Shanker and GSI, 1991; Jha and Puppala, 2017). An overall analysis suggests that there may be a 10 gigatonne geothermal power potential (Yadav and Sircar, 2020). Tattapani, in the Balrampur district, was selected by the Chhattisgarh government in 2013 to build the country's first geothermal power plant. An agreement to build Ladakh's first geothermal power plant was inked in 2021. 8.3 million units might be produced yearly by 1 MW geothermal installations, according to several worldwide studies. Although this is a substantial sum, the establishment of a geothermal plant in India costs roughly 30–35 crores. This is primarily due to the high cost of drilling and the unpredictability of project-specific exploration. In India, geothermal energy is therefore still in its infancy.

4.6 BIOENERGY

The latest update from MNRE (2024d) aims to achieve net zero by 2070 and have 50% of installed electricity capacity come from non-fossil fuel-based energy sources by 2030. India has set out on an ambitious energy transformation journey. Utilizing domestically accessible Renewable Energy alternatives to their full potential is essential to meeting the ambitious Renewable Energy targets and achieving energy sector self-reliance. Modern bioenergy is one such option. Energy recovery from biomass and other trash is a feasible alternative because of the nation's abundance of these resources. In addition to producing clean fuels, modern bioenergy offers several social and environmental advantages. The national bioenergy program has the following categories:

(a) **Waste to Energy**
In 2018, the Ministry of New and Renewable Electricity (MNRE) launched a program aimed at generating electricity from municipal solid waste, industrial, agricultural, and urban wastes and residues. The goal of this project is to use these waste sources to generate power, biogas, and bioCNG. Except for municipal solid waste (MSW) to power projects, the program provides Central Financial Assistance for the establishment of large biogas, bioCNG, and power plants under the revised standards for the 2021–2026 timeframe. In addition to manufacturing biofertilizers, India has significant potential for producing bio-CNG (Compressed Biogas-CBG) for use in cooking, industry, and car fueling. By 2023, the Ministry of Petroleum and Natural Gas (MoPNG) hopes to generate 15 million metric tons (MMT) of CBG from 5,000 facilities as part of its SATAT (Sustainable Alternative Towards Affordable Transportation) program. Ensuring the financial viability of these projects requires the assistance of oil marketing companies, which can provide long-term price agreements for CBG. The MNRE program's BioCNG segment, by complementing and enhancing the activities set out in SATAT, supports India's move to cost-effective, environmentally friendly energy choices.

(b) **Biomass**
This program, originally adapted from an emphasis on cogeneration, now focuses on producing biomass briquettes and pellets for power generation. This is in harmony with the National Mission on Co-firing Biomass in Thermal Power Plants, which aims at reducing the environmentally hazardous stubble burning, particularly in the northern region. Under the revised guidelines for 2021–2026, the program seeks to provide Central Financial Assistance for the

projects aimed at the production of pellets and briquettes, as well as for non-bagasse-based power generation projects, as part of the endeavours to promote sustainable biomass utilization for energy generation in India.

(c) **Biogas**

The Ministry of New and Renewable Energy (MNRE), as part of the National Biogas Programme, has been promoting the construction of small biogas plants since 1981–1982. In rural areas, the program provides clean gaseous fuel primarily used for lighting and cooking. To enable decentralized power generation and thermal applications from 3 to 250 kW, MNRE has sponsored biogas generation for mid-sized biogas plants with a production range of 25 m^3 to 2,500 m^3/day. The energy source can better meet the demand for rural energy. MNRE has extended the program with improved financial assistance for the establishment of biogas plants across a broader range—from 1 to 2,500 m^3 of biogas per day—under the updated plan for the years 2021–2026. This update seeks to encourage the use of biogas as a clean, sustainable energy source and further support rural energy needs.

4.7 GREEN HYDROGEN

As per MNRE (2024e), India has set goals to achieve net zero by 2070 and energy independence by 2047. It is anticipated that green hydrogen will significantly contribute to achieving these objectives. Green hydrogen is created by the electrolysis process, which uses electricity from renewable resources like solar, wind, or hydropower to split water into hydrogen and oxygen. The clean, emission-free gasoline produced by this technology has enormous potential to displace fossil fuels and lower carbon emissions. Biomass-based green hydrogen production is another technique that uses gasification to create hydrogen. The production methods used to develop green hydrogen are both clean and sustainable. Furthermore, green hydrogen is regarded as an attractive substitute for non-renewable energy sources. The demand for green hydrogen is rapidly increasing due to its ability to decarbonize industries like transportation, shipping, and steel manufacturing. Green hydrogen is cleaner and more environmentally friendly, and therefore can substitute traditional fossil fuels in transportation, which reduces greenhouse gas emissions. It can also be used in industry to manufacture steel, methanol, and ammonia, which are now fossil fuel dependent. Additionally, green hydrogen serves as a consistent energy source and can, therefore, be used as a backup for renewable energy plants. Furthermore, Green hydrogen can be used as fuel in fuel cells to make electricity, power automobiles, and even be used in fertilizer production and heating systems. Because of their high energy density and superior efficiency over conventional combustion engines, hydrogen fuel cells are a desirable alternative for vehicle power. Additionally, green hydrogen can be harnessed in microgrids, providing autonomy in energy and supplying power to remote areas. The importance of green hydrogen cannot be overemphasized due to its need in India's fight for energy independence. Energy security would be viable with the onboard generation of Green Hydrogen by variable renewable energies: solar, wind, and hydro. The production shall gradually reduce dependency on fossil fuel supply, adding to a consistent and reliable energy supply. The most vital advantages include the reduction of costs or avoiding high financial losses caused by imported crude oil. On a side note, local areas could make a living associated with generating Green Hydrogen via production off-site from all land waste biomass.

4.7.1 Power demand and energy distribution

The analysis of India's power system from April to November 2023 provides important new information about the country's energy distribution and demand. Over 1,102,887 million Units (MU) of energy were needed during this time in all states and regions. Of that amount, 1,099,907 MU were successfully provided, leaving an unfulfilled need of about 2,980 MU, or a 99.7% supply efficiency. Certain states—like Chandigarh, Delhi, and Tamil Nadu—achieved a 100% energy supply, while others—like Haryana (0.5%) and Uttarakhand (0.8%)—saw only slight shortages. The energy requirements, supply, and deficits among states and regions are detailed in Table 4.1.

4.8 POLICIES AND REGULATORY FRAMEWORKS

4.8.1 Centre Rules MNRE (2024)

(a) **The Electricity (Amendment) Rules, 2022.**
The Electricity Rules, 2005, were modified in several ways by the Electricity (Amendment) Rules, 2022. Notably, the new regulations allow distribution companies to automatically collect monthly expenses from consumers that result from changes in fuel prices and power purchase costs. Additionally, the MoP has required the establishment of a centralized renewable energy tariff under the new regulations.

(b) **Electricity (Promoting Renewable Energy Through Green Energy Open Access) Rules, 2022.**
To further accelerate our ambitious renewable energy programs and ensure that everyone has access to affordable, dependable, sustainable, and green energy, the Ministry of Power notified the Electricity (Promoting Renewable Energy Through Green Energy Open Access) Rules, 2022 on June 6, 2022.

Table 4.1 Energy demand and energy distribution in India (PIB Delhi, 2023)

State/Region	*Energy requirement (MU)*	*Energy supplied (MU)*	*Energy not supplied (MU)*	*% Not supplied*
Chandigarh	1,289	1,289	0	0
Delhi	26,126	26,123	3	0
Haryana	45,855	45,605	250	0.5
Himachal Pradesh	8,348	8,324	24	0.3
UT of J&K and Ladakh	12,770	12,577	193	1.5
Punjab	51,458	51,453	5	0
Rajasthan	69,851	69,352	499	0.7
Uttar Pradesh	107,129	106,875	254	0.2
Uttarakhand	10,524	10,444	80	0.8
Northern Region	334,239	332,930	1,309	0.4
Chhattisgarh	26,253	26,202	51	0.2
Gujarat	98,536	98,509	27	0
Madhya Pradesh	64,255	64,192	63	0.1
Maharashtra	138,065	137,891	174	0.1
Dadra & Nagar Haveli	6,719	6,719	0	0
Goa	3,413	3413	0	0
Western Region	343,422	343,107	315	0.1
Andhra Pradesh	54,616	54,561	55	0.1
Telangana	54,096	54,089	7	0
Karnataka	59,533	59,382	151	0.3
Kerala	20,203	20,203	4	0
Tamil Nadu	84,230	84,221	9	0
Puducherry	2,371	2,371	0	0
Southern Region	275,083	274,856	227	0.1
Bihar	30,448	29,963	485	1.6
Jharkhand	9,721	9,401	320	3.3
Odisha	28,999	28,984	15	0.1
West Bengal	48,465	48,403	63	0.1
Eastern Region	136,056	135,170	887	0.7
Arunachal Pradesh	646	646	0	0
Assam	9,062	8,983	78	0.9
Meghalaya	1,464	1,301	164	11.2
Nagaland	640	640	0	0
North-Eastern Region	14,086	13,844	242	1.7
All India	1,102,887	1,099,907	2,980	0.3

(c) **Electricity (Promoting Renewable Energy Through Green Energy Open Access) Amendment Rules, 2023.**
According to the Amendment Rules, any consumer may choose to buy green energy for all or a specific percentage of their usage. They may submit a demand to their distribution licensee for this reason. The distribution licensee is responsible for obtaining and supplying this amount of green energy. The customer will have the option to submit separate requests for solar and non-solar energy. Additionally,

the Amendment Rules stipulate that the banked energy credit must be changed during the same banking cycle and cannot be carried over to the following banking cycles: After each banking cycle, any unused excess banked energy will be deemed to have expired, and the Renewable Energy generating station will be eligible to receive Renewable Energy Certificates.

(d) **Green Hydrogen Standard for India**
For hydrogen generated from renewable sources to be categorized as "green," the government has specified the emission threshold that must be fulfilled. Both biomass-based and electrolysis-based hydrogen production techniques are covered under the definition.

(e) **Renewable Generation Obligation as per Revised Tariff Policy, 2016**
The Renewable Generation Obligation (RGO) requires power producers who develop coal or lignite-based thermal generating stations with a project's Commercial Operation Date (COD) on or after April 1, 2023, to either establish renewable energy generating capacity (measured in MW) of at least 40% of the thermal generating station's capacity or acquire and supply renewable energy equal to the thermal generating station's capacity.

4.9 POLICIES

(a) **Carbon Credit Trading Scheme, 2023**
The Carbon Credit Trading Scheme will, therefore, build a robust infrastructure in carbon credit trading. Each of these credits is 1 metric ton of carbon dioxide equivalent (tCO2e) in value and represents defined amounts of emissions reduction, removal, or avoidance. These would probably be exchanged across businesses and organizations to reduce GHG emissions in the country. In this way, the program aims to encourage needed entities to decrease their carbon footprint through emission reduction.

(b) **National Framework for promoting energy storage**
The Indian Ministry of Power has adopted a broad-based National Framework to incentivize the growth and adoption of Energy Storage Systems as an integral part of its power infrastructure. This initiative aligns with India's goal of switching to renewable energy sources to reduce greenhouse gas emissions. This framework aims at the reduction of energy prices, a steady supply of renewable energy sources (Renewable Energy-Round the Clock), and the reduction of emissions by increasing the use of ESS and reducing reliance on fossil fuel power plants.

(c) **National Green Hydrogen Mission**
Making India a global leader in the production and supply of green hydrogen. Creating export potential for green hydrogen and its derivatives.

(d) **PM-Gati Shakti National Master Plan related to Renewable Energy** According to the MNRE on renewable energy inclusions in Gati Shakti, the ministry has established a cell where all communications on PM Gati Shakti-related tasks may be completed. Moreover, the PM-Gati Shakti platform would map projects with a capacity greater than 50 MW. The projects undertaken by the renewable energy implementing agencies (REIA) will be mapped on the portal by the Gati Shakti cell in the interim. MNRE has adopted a standard format for mapping data. The resource maps of wind speed and sun radiation will be uploaded to the portal by the Gati Shakti cell in collaboration with BISAG-N and the National Institute of Wind Energy (NIWE).

4.9.1 Challenges for renewable energy in India

India's renewable energy sector is rapidly becoming a critical component of the nation's energy production. However, several barriers—financial, technical, infrastructural, and political—continue to hinder its growth and adoption (Dey et al., 2022).

4.10 FINANCIAL AND MARKET LIMITATIONS

The high cost of production is one of the main barriers to the widespread use of renewable energy. Renewable energy sources like solar and wind have comparatively cheap operating costs as compared to fossil fuels. But the initial expenses of installation, like installing wind turbines or solar panels, are far greater. Because of this, financial institutions view renewable energy projects as high-risk, which raises borrowing rates. The challenge of finding reasonably priced funding deters investment in renewable energy, especially from private investors. On the other hand, customers, who typically view price fluctuations as inevitable, can be impacted by fossil fuel plants passing on growing fuel costs (Güney and Kaygusuz, 2010; Hilton and Duddy, 2009). Renewable energy also faces additional challenges due to access constraints. The fossil fuel industry dominates the energy market because of decades of financial strength, regulatory support, and infrastructural investment. Startups in the renewable energy sector are thus at a disadvantage since they find it difficult to compete with well-established market participants. Startups frequently encounter investor scepticisms, particularly when they are unable to provide significant energy outputs or show scalability. To level the playing field for renewables, governments must take the initiative and offer financial support, tax breaks, and subsidies (Abbasi et al., 2010; Asumadu-Sarkodie and Owusu, 2016; Dey et al., 2022).

4.10.1 Infrastructure and long-distance transmission

There are a great deal of hurdles to be surmounted for renewable energy to be incorporated into India's electrical infrastructure. For one, the decentralized character of renewable sources does not agree with the power transmission infrastructure of India, which was primarily built for fossil fuel plants. Long-range energy transmission at higher costs does not allow getting the highest amount but reduces the process's efficiency in general. Moreover, geographic phenomena like the climate or the weather are obstacles to the development of renewable energies. For example, poor sunlight or wind speed in a region cannot support the projects of either solar or wind power (Ayoub and Abdullah, 2012; Dey et al., 2022; Koh and Ghazoul, 2008).

Technical and Operational Limitations: Concerns about electricity supply arise due to the reliance of renewable energy on natural factors. Renewable sources of energy are interfered with by weather changes, whereas fossil fuels provide a continuous supply of energy. Wind turbines must receive wind at a speed sufficient to operate them, while solar panels only generate electricity during the daytime. This intermittency complicates maintaining a reliable supply of electricity when demand is high.

Besides, concerns about energy quality are increasing, too. A uniform level of energy quality ensures the electricity supply is always as reliable as it must be. Such intermittent fluctuating renewable sources also cause failures of equipment, affecting industrial production. This raises bills for consumers and decreases overall effectiveness within the general energy network. Addressing such problems needs continued funding in the RD&D of cutting-edge technologies for these alternatives (Dey et al., 2022).

4.10.2 Other challenges

More especially over the past couple of years, the sector has gone into shock over the oversupply of renewable components, particularly solar panels Lastly, there are remarkable obstacles to renewables due to the strong political lobby representing non-renewable industries. Subsidies, tax exemptions, and other forms of incentives are granted to the fossil fuel industry at the expense of such resources that would hasten the transition toward renewable energy. Overcoming these political obstacles will require coordination in efforts toward the realignment of policies and resources in favour of sustainable energy development (Patel and Kumar, 2017; Dey et al., 2022).

4.11 CONCLUSION

The case study thus underlined remarkable progress by the country to transition to renewable energy, and it, therefore, means that India will surely play a very important role in tackling global energy challenges for sustainable growth. The Indian economy and growing population are increasing at a very fast rate, meaning that the need for more energy is one of the most vital reasons for the required transition coming from more sustainable sources and moving away from non-renewable sources. It has also become a key alternative in reducing energy poverty and environmental degradation. India has made great leaps forward in the area of renewable energy. Solar energy has been one of the major drivers for changing this reality, and its rates of adoption have been nothing short of phenomenal. With abundant sunlight and a raft of successful government initiatives, solar energy has been the giant in India's renewable energy portfolio, ranging from huge solar parks to rooftop installations and hybrid projects that harness solar energy in conjunction with other renewable resources. The fact that solar energy projects can also power remote and metropolitan locations and industries, in general, is indicative of the wide scope they cover. These equally have attracted massive investments further increasing the capacity of the sector. A further key addition to India's renewable energy capacity is the development of onshore wind in concert with solar. This growth is seen to be both because of strong policy support and natural winds. Some of the reasons include fiscal incentives, competitive tariffs, and technical support for wind projects. Wind energy is scalable and has very low operational costs; hence, it is vital in reaching India's ambitious clean energy targets. The other vital factor is hydroelectricity, utilizing the vast river and stream network of India. Small-scale and large-scale hydropower projects have contributed a great deal to an improved mix of renewable energy. In the development of hydropower, various obstacles are more often than not associated with environmental issues, with socioeconomic effects related to migration being key among them. These problems must be solved for full usage of hydropower with guaranteed sustainable development. Equally relevant are the initiatives in waste-to-energy and biofuel projects serving as an innovative way of energy production while solving certain problems in solid waste management using municipal refuse, industrial wastes, and agricultural residues. Biogas and biomass-based cogeneration initiatives further outline the capability of India to make use of all kinds of resources for sustainable energy generation. Even newer technologies, like green hydrogen and geothermal energy, are also in discussion. The huge potential of geothermal energy in places like Ladakh is still unexplored. Green hydrogen-derivative extracted from renewable resources can change the face of the energy sector in cleaning the carbon footprint of such sectors as steel and transportation. India's ambition to lead the world in green hydrogen generation is one of the best examples of the country's commitment to renewable energy innovation. But despite these successes, the research also pinpoints those things standing in the way of India's full potential as far as renewable energy is concerned.

REFERENCES

Abbasi, T., Premalatha, M., & Abbasi, S. (2010). The return to renewables: Will it help in global warming control? *Renewable and Sustainable Energy Reviews*, 15(1), 891–894.

Asumadu-Sarkodie, S., & Owusu, P. A. (2016). The potential and economic viability of wind farms in Ghana. *Energy Sources Part a Recovery Utilization and Environmental Effects*, 38(5), 695–701.

Ayoub, M., & Abdullah, A. Z. (2012). Critical review on the current scenario and significance of crude glycerol resulting from biodiesel industry towards more sustainable renewable energy industry. *Renewable and Sustainable Energy Reviews*, 16(5), 2671–2686.

Brouwer, A. S., Van Den Broek, M., Seebregts, A., & Faaij, A. (2014). Impacts of large-scale intermittent renewable energy sources on electricity systems, and how these can be modeled. *Renewable and Sustainable Energy Reviews*, 33, 443–466.

Centre Rules MNRE (2024). Available online at: https://mnre.gov.in/centre-rules/ Accessed on 31st October 2024.

Dey, S., Sreenivasulu, A., Veerendra, G., Rao, K. V., & Babu, P. A. (2022). Renewable energy present status and future potentials in India: An overview. *Innovation and Green Development*, 1(1), 100006.

Güney, M. S., & Kaygusuz, K. (2010). Hydrokinetic energy conversion systems: A technology status review. *Renewable and Sustainable Energy Reviews*, 14(9), 2996–3004.

Hilton, B., & Duddy, B. (2009). The effect of E20 ethanol fuel on vehicle emissions. *Proceedings of the Institution of Mechanical Engineers, Part D: Journal of Automobile Engineering*, 223(12), 1577–1586.

International Hydropower Association (2021). Hydropower Status Report 2021. Available Online at: https://www.hydropower.org/publications/2021-hydropower-status-report Accessed on 29th October 2024.

Jacobson, M. Z. (2021). The cost of grid stability with 100% clean, renewable energy for all purposes when countries are isolated versus interconnected. *Renewable Energy*, 179, 1065–1075.

Jha, S. K., & Puppala, H. (2017). Prospects of renewable energy sources in India: Prioritization of alternative sources in terms of energy index. *Energy*, 127, 116–127.

Koh, L. P., & Ghazoul, J. (2008). Biofuels, biodiversity, and people: Understanding the conflicts and finding opportunities. *Biological Conservation*, 141(10), 2450–2460.

Kumar, A., Kumar, K., Kaushik, N., Sharma, S., & Mishra, S. (2010). Renewable energy in India: Current status and future potentials. *Renewable and Sustainable Energy Reviews*, 14(8), 2434–2442.

Kumari, A. (2024). Renewable energy for sustainable development in India. *International Journal for Multidisciplinary Research*, 6(3), 2.

Ministry of New and Renewable Energy (2024a). Available online at: https://mnre.gov.in/physical-progress/ Accessed on 27th October 2024.

Ministry of New and Renewable Energy (2024b). Solar Overview. Available online at: https://mnre.gov.in/solar-overview/ Accessed in 28th October 2024.

Ministry of New and Renewable Energy (2024c). Small Hydropower Overview. Available online at: https://mnre.gov.in/small-hydro-overview/ Accessed on 30th October 2024.

Ministry of New and Renewable Energy (2024d). Bioenergy Overview. Available Online at: https://mnre.gov.in/bio-energy-overview/ Accessed on 30th October 2024.

Ministry of New and Renewable Energy (2024e). Hydrogen Overview. Available Online at: https://mnre.gov.in/hydrogen-overview/ Accessed on 30th October 2024.

Patel, R. V. P., & Kumar, A. (2017). Experimental investigation of double slope solar still for the climatic condition of Sultanpur. *International Journal of Engineering and Technology*, 9(6), 4019–4033.

PIB Delhi (2023). Status of power generation and power supply position in the country. In PIB. Ministry of Power. https://pib.gov.in/PressReleaseIframePage.aspx?PRID=1988262

Press Information Bureau (2023). Available online at: https://www.pib.gov.in/PressReleasePage.aspx?PRID=1944759 Accessed on 29th October 2024.

Seikh, M. R., & Chatterjee, P. (2024). Determination of best renewable energy sources in India using SWARA-ARAS in confidence level-based interval-valued Fermatean fuzzy environment. *Applied Soft Computing*, 155, 111495.

Shanker, R., & Geological Survey of India (1991). Geothermal atlas of India. Geological Survey of India.

Thapar, S., Sharma, S., & Verma, A. (2018). Key determinants of wind energy growth in India: Analysis of policy and non-policy factors. *Energy Policy*, 122, 622–638.

World Wind Energy Association (2023). Annual Report 2023. Available Online at: https://wwindea.org/AnnualReport2023 Accessed on 28th October 2024.

Yadav, K., & Sircar, A. (2020). Geothermal energy provinces in India: A renewable heritage. *International Journal of Geoheritage and Parks*, 9(1), 93–107.

Chapter 5

An in-depth case study on renewable energy sources, power demand trends, and policy frameworks

The Indian perspective

Sanchayita Rajkhowa, Jyotirmoy Sarma, Hiyaman Kalita, Sankar Neogi, and Dilip Chorei

5.1 INTRODUCTION

Changing the dependency of the world's economy solely on fossil fuels, which are non-renewable resources that significantly contribute to global pollution, and moving towards one that efficiently uses renewable energy (RE) sources and encourages high levels of energy efficiency is a significant task. As one of the fastest-growing economies in the world, India must simultaneously meet its growing energy needs and transition to cleaner, more sustainable electricity sources. The country's energy needs are expected to increase exponentially over the next few decades due to its large population, fast urbanization, and industrial growth. India's energy industry, historically reliant on coal and other non-renewable resources, has significantly increased greenhouse gas emissions, ultimately fueling climate change. With a total of approximately 4,443 $MtCO_2$ energy-related emissions in 2022, India has set out on an ambitious path to diversify its energy portfolio, giving priority to renewable sources including solar, wind, hydroelectric, and biomass (Climate Trasnparency 2022). The country has recognized the need for a radical change. In the present day, India's bottom-up approach (as shown in Figure 5.1) towards a green economy is attracting the interest of researchers, industrialists, as well as policy makers. It is anticipated that inducing green and energy-efficient technologies would be able to save our limited resources significantly by 2050 (IEA 2021). In addition, this approach can reduce unemployment figures by creating job opportunities and thereby improving the living standards of low-per capita-income countries (Reddy 2016). In line with attempts to keep the average global temperature from rising more than 1.5°C over the long term, global carbon dioxide (CO_2) emissions should also be reduced to zero by 2050 (IEA 2021). Glasgow Climate Pact (COP26) aligns with this aim by regulating coal power and fossil fuel subsidies (COP26 2021). In this context, COP29 focused on the role of the Local Communities and Indigenous Peoples Platform (LCIPP) in promoting

DOI: 10.1201/9781003581536-5

gender responsiveness in climate change. The discussion targeted government representatives, policymakers, UNCTs, regional centers, civil society, private sector, youth, Indigenous Peoples, local communities, and academic institutions (UNFCCC 2024). The session also discussed the implications of COP29 decisions on gender-responsive climate action and how Indigenous youth can be engaged in global climate discussions and the implementation of the Baku Workplan (LCIPP 2024) and the enhanced Lima Work Programme. With industrialized nations leading the way, COP29 also targets to triple climate finance for developing countries to US$300 billion per year by 2035 from US$100 (UNFCCC 2024). Global primary energy consumption has increased by 2% to 620 EJ in 2023, marking the second consecutive record year. Its growth rate was more than 5% higher than its 2019 pre-COVID level and 0.6% higher than its ten-year average. However, the amount of fossil fuels consumed (in absolute terms) set a new record. It dropped to 81.5% in 2023 from around 81.9% in 2022. Energy-related greenhouse gas emissions hit a record high, surpassing 40 $GtCO_2e$ for the first time, as a result of the growing use of more carbon-intensive coal and oil. In contrast, demand for natural gas, a comparatively low-carbon-intensive fossil fuel, remained steady. The most significant source of energy-related greenhouse gas emissions, accounting for around 87% of the total, is the burning of fossil fuels, which releases CO_2 (Energy Institute 2024).

The proactive actions of multiple stakeholders and the influence of dominant drivers are more responsible for the variations between the baseline and green energy scenarios than the proliferation of different technologies. At a moderate pace of roughly 1% year, RE sources like solar, wind, small hydro, hydrogen, and geothermal expanded. As a result, non-commercial energy's proportion in the overall energy mix decreased from 63.5% to 31.6% (Planning Commission 2012). Figure 5.1 shows the bottom-up approach that includes all sectors of the energy economy and energy carriers.

India's ambitious goal of reaching net zero emissions by 2070 and meeting 50% of its electricity requirements from RE sources by 2030 is a significant step in the global fight against climate change. The country's rapid growth in economic development has lifted millions out of poverty and has added cities the size of London to its urban population each year. However, India's annual CO_2 emissions have risen to become the third highest in the world, with the average household using one-tenth as much electricity as the average household in the United States. India's clean energy transition is already underway, with 40% of its power capacity coming from non-fossil fuels, and solar and wind power plants growing at a faster rate than any other major economy. To address these challenges, India has implemented several policy measures, including subsidies for petrol and diesel, electric vehicles, and hydrogen. Achieving net zero requires support from the international community, as the IEA estimates that $160 billion per year is needed across India's energy economy between now and 2030. Green hydrogen will play a significant role in achieving net zero and decarbonizing hard-to-abate

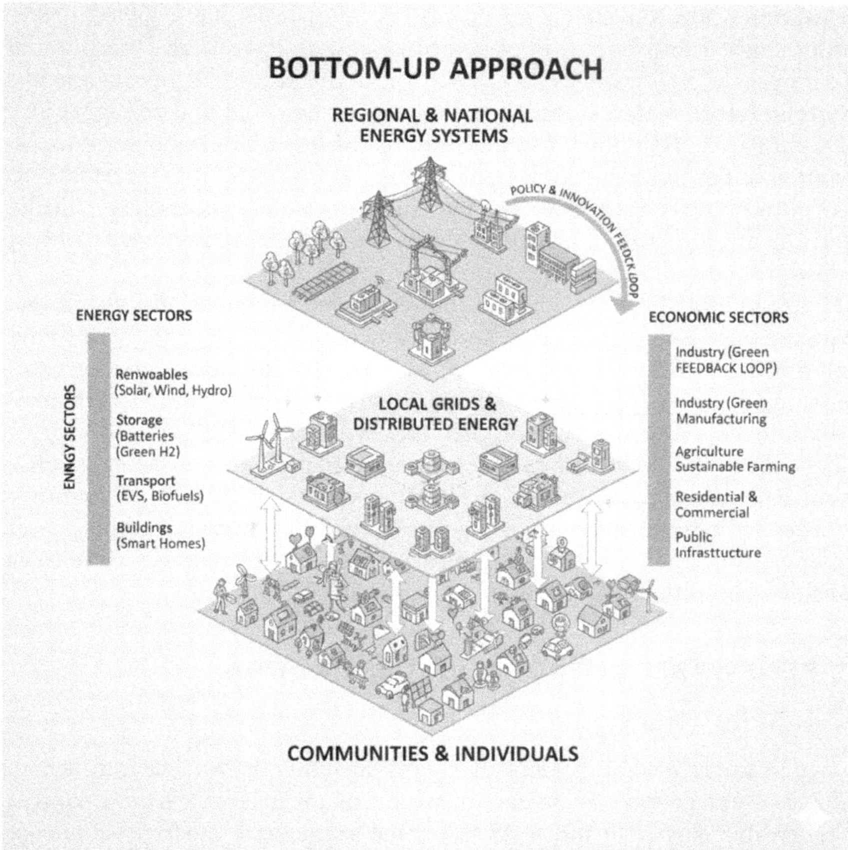

Figure 5.1 Bottom-up approach that includes all sectors of energy economy and energy carriers. (Adopted from Reddy, BS, 'India's energy system transitiond-Survival of the greenest', *Renew. Energy*, 92, 293–302, 2016. With Permission.)

sectors. India aims to become a global hub for green hydrogen production and exports (Birol & Kant 2022).

However, as fossil fuels account for 80% of India's electrical usage, this raises serious energy equity issues. But India faces a complicated energy transition issue that requires both climate finance and structural and regulatory changes. India's energy demand is predicted to increase by 25% over the next 20 years due to rising urbanization and population density, making it the third-largest energy user in the world by 2030 (Haldar & Sharma 2022).

India's energy transformation is a top-down phenomenon, with thermal power—primarily coal-fired—supplying 80% of the country's electricity needs. The costs and benefits of transition are not evenly shared, and those who lack the agency to affect the pattern or results of the energy transition—such as women, children, old people, and marginalized rural communities—are more likely to endure economic and social inequalities. To examine how the literature on energy transitions has evolved in the Indian context and how energy justice narratives have been incorporated into it over the last few decades (2008–2021), this research employs a methodical literature review technique. The study is essential and relevant since energy justice is becoming a significant area of research in the literature on sustainable development and global energy transitions. It is necessary to incorporate justice ideas into energy transition studies, both academically and in order to better align policy decisions with the requirements of just transitions.

Financial globalization is linked to greenhouse gas emissions, with India being the third-largest emitter. A study using data from 1970 to 2018 found that non-RE utilization increases emissions, RE consumption decreases emissions, financial development contributes to emissions, and growth positively impacts emissions (Akadiri & Adebayo 2022). The findings suggest policymakers should focus on implementing green economy policies, emphasizing RE sources, to prevent devastating effects on future generations. In order to determine how RE sources contribute to meeting the power demands, this study takes a methodical approach to the analysis of energy transitions literature in the Indian context.

5.2 RENEWABLE ENERGY SOURCES IN INDIA

5.2.1 Solar energy

A significant renewable source of energy around the world, and also in India, is solar energy. Solar energy is one of the fastest-growing sources of renewable energy in India. As the name suggests, it is a type of energy that is produced by the sun. Solar energy can be used in various ways to meet people's needs. In fact, various techniques are being invented to capture and use solar energy effectively. An increase in the use of this energy has shown a lowering of pollution as compared to the conventional energy sources. As International Energy Agency (2011) said, "The development of affordable, inexhaustible and clean solar energy technologies will have huge long-term benefits. It will increase countries' energy security through reliance on an indigenous, inexhaustible, and mostly import-independent resource, enhance sustainability, reduce pollution, lower the costs of mitigating climate change, and keep fossil fuel prices lower than otherwise. These advantages are global" (IEA 2011).

5.2.1.1 Current status and installed capacity

India stands as the third-largest producer of solar energy across the globe, only behind the United States and China. According to a report from the

Ministry of New and Renewable Energy posted by PIB Delhi, India's solar power installed capacity is 92.12 GW_{AC} in 2024 (MNRE 2024b). Currently, much importance is being given to increasing this capacity. As we know, India has set a goal of carbon neutrality by 2070. To achieve this goal, at least 250 GW of solar power has to be generated by 2030.

According to a working paper – 'Indian Solar Sector: Fostering Growth and Sustainable Development published by India Exim Bank', US$20.7 billion of foreign capital has been invested in India (Indian Exim Bank 2022). This is in accordance with the provision that allows 100% foreign direct investment in renewable energy. In FY2023-24, an amount of US$3.76 was received by India (Indian Exim Bank 2022).

In the present time, India has built nearly 70 solar parks. According to the newspaper 'Financial Express', Gujarat Hybrid Renewable Energy Park near Khavda in the Rann of Kutch desert in Gujarat is estimated to generate about 30 GW of power and become the world's largest renewable energy park. It covers an area of 72,600 ha of the desert (Financial Express 2020)

5.2.1.2 Regional concentrations and growth potential

According to Purohit et al. (2013), for the acceleration of decarburization in the Indian power sector, concentrating solar power (CSP) is important (Purohit et al. 2013). These CSP technologies constitute a significant part of Jawaharlal Nehru National Solar Mission (JNNSM) of the Government of India. The Northwestern regions of India have high potential due to their annual solar radiation, favourable meteorological conditions, etc.

According to information provided by Minister of State (I/C) for New and Renewable Energy and Power, Shri R. K Singh on 12 February 2019 (MNRE 2019), two annexures have been published.

Annexure-I tells us about 'Installed capacity of various renewable modes of energy'. In this annexure, detailed information on the installed capacity of hydro power, solar power, bio power, and wind power in different states and UTs of India is given, Annexure-II gives us information about 'Central Financial Assistance' (CFA) received by various states for promoting renewable energy. Table 5.1 shows the Installed capacity of various renewable modes of energy.

5.2.2 Wind energy

Another significant source of renewable energy is wind energy. It is very beneficial for the environment and mankind in general. Wind energy has a long-lasting future. Wind Energy also has several advantages. Since air and wind are natural phenomena, there is no fear of ever exhausting them. Wind energy is a kind of energy that will never die. Also, many places in India get high-speed winds.

Table 5.1 Installed capacity of various renewable modes of energy' state-wise installed capacity of grid interactive renewable power as on 31.12.2018

Sl. No.	States/UTs	*Small hydro power* (MW)	*Wind power* (MW)	*Bio power* (MW)	*Solar power* (MW)	*Total capacity* (MW)
1	Andhra Pradesh	162.11	4,076.45	500.34	2,641.76	7,380.66
2	Arunachal Pradesh	107.100			5.39	112.49
3	Assam	34.11			16.68	50.79
4	Bihar	70.70		121.20	142.45	334.35
5	Chhattisgarh	76.00		230.50	231.35	537.85
6	Goa	0.05			1.69	1.74
7	Gujarat	35.10	5,955.07	77.30	1,827.38	7,894.85
8	Haryana	73.50		205.66	217.09	496.25
9	Himachal Pradesh	860.61		7.20	3.89	871.70
10	Jammu & Kashmir	179.03			9.85	188.88
11	Jharkhand	4.05		4.30	32.41	40.76
12	Karnataka	1,230.73	4,682.80	1,799.80	5,255.11	12,968.44
13	Kerala	222.02	52.50	0.72	138.49	413.73
14	Madhya Pradesh	95.91	2,519.890	120.75	1,586.25	4,322.80
15	Maharashtra	375.570	4,788.13	2,528.69	1,607.79	9,300.18
16	Manipur	5.45			2.59	8.04
17	Meghalaya	31.03		13.80	0.08	44.91
18	Mizoram	36.47			0.20	36.67
19	Nagaland	30.67			1.00	31.67
20	Odisha	64.625		59.22	389.40	513.25
21	Punjab	173.55		326.35	905.62	1,405.52
22	Rajasthan	23.85	4,299.72	121.30	3,130.99	7,575.86
23	Sikkim	52.11			0.01	52.12
24	Tamil Nadu	123.05	8,631.19	1,003.88	2,228.44	11,986.56
25	Telangana	90.87	128.10	177.60	3,410.26	3,806.83
26	Tripura	16.01			5.09	21.10
27	Uttar Pradesh	25.10		2,117.51	891.01	3,033.62
28	Uttarakhand	214.320		130.50	303.18	648.00
29	West Bengal	98.50		319.92	42.32	460.74
30	Andaman & Nicobar	5.25			6.56	11.81
31	Chandigarh				32.40	32.40
32	Dadar & Nagar Haveli				5.46	5.46
33	Daman & Diu				13.38	13.38
34	Delhi			52.00	124.21	176.21
35	Lakshadweep				0.75	0.75
36	Pondicherry				1.73	1.73
37	Others		4.30			4.30
	Total (MW)	**4,517.445**	**35,138.150**	**9,918.54**	**25,212.26**	**74,786.40**

Source: MNRE (2019).

5.2.2.1 *Installed capacity and key regions*

At present India ranks in fourth position globally in wind power capacity, only after China, the US, and Germany. According to the Union Minister for New and Renewable Energy and Power, the installed wind generation capacity of India is 43,773 MW (MNRE 2023). During FY 2022–2023, the quantum of electricity generated by wind power was 71,814 million units (MNRE 2023). As per the Central Electricity Authority's 'Report on Optimal Generation Capacity Mix for 2029-30 Version 2.0', the likely installed capacity of wind energy by the end of the year 2029–2030 is estimated to be 99,895 MW (MNRE 2023).

5.2.2.2 *Role in India's energy mix*

Some key roles of wind power in India's energy mix are –

1. Renewable Energy Target – As discussed earlier, India aims to achieve 500 GW of renewable energy capacity by 2030. Wind energy or wind power plays a significant role in this matter. In order for India to achieve its desired target and go carbon-neutral by 2070, improvements and efficient use of wind energy are much needed.
2. Grid Integration – With the addition of wind energy in India's energy mix, the sources of energy will diversify.

5.2.3 Hydroelectric and biomass energy

Hydroelectric Energy refers to the energy generated from water, i.e., hydropower. India ranks fifth in installed hydroelectric capacity across the globe. As per information from the 'Central Electricity Authority', the country's hydroelectric capacity was around 46,000 MW (CEA 2020). The hydroelectric potential is estimated to be 148,700 MW at 60% load factor. It is to be noted that India has 733 operational hydro power plant units and 182 pipeline power plant units (NITI Aayog 2024).

Biomass Energy is another type of renewable energy. It is obtained from plants and animals, i.e., Biomass. When biomass is burnt, it leads to clean energy. Biomass, as it is readily available, if used as a source of energy, will be very profitable and affordable. Biomass energy has a lot of potential. Realizing this potential, the Ministry of New and Renewable Energy (MNRE) has notified the National Bioenergy Programme for a period of 01-04-2021 to 32-03-2026 with an outlay of Rs. 858 crores under Phase-I (MNRE 2024b).

5.2.3.1 *Contribution to the energy portfolio*

According to Pandit et al. (2023), there has been a sharp fall in India's hydropower contribution (Pandit et al. 2023). During the independence period, the energy contribution from hydropower was nearly 40%. This declined rapidly up to 25% in the early 2000s and has now further declined to 12% at the current time. However, the report also states that the growth of hydropower has gone up from 500 to 47,000 MW from 1947 to 2022 (Pandit et al. 2023).

But as mentioned previously, from the document 'Hydro Power Policy 2008' released by the Ministry of Power, Govt of India (2008), hydropower has significant potential (nearly 150,000 MW at 60% load factor) and can contribute substantially to India's energy profile. Hydropower is a green energy. It poses no harm to the environment. At the same time, it is cost-effective compared to other energy sources. These factors, combined with many others, make hydropower in India a cornerstone for achieving energy security, grid stability, and suitability.

According to Singh et al. (2014), biomass energy production in rural area is emphasized. Biomass energy production in India predominantly happens in rural areas (over 80%) (Singh et al. 2014).

According to Ghosh et al. (2021), India is home to a large biomass reserve. It is around 500 million metric tonnes per year (Ghosh et al. 2021). Out of this, only 120–150 million tonnes are utilized per year.

Their study gave us a clear view of the tremendous contribution biomass can make to the country's energy portfolio. With proper schemes, planning, and utilization of the biomass at hand, we can achieve a green, eco-friendly, inexhaustible source of energy. The use of this energy will significantly contribute to India's goal of going carbon-neutral by 2070.

5.2.3.2 *Geographic influence*

India is a land of diverse geography. Different parts of India experience different climates throughout the year. The resources available in every region are also different. So, the availability, production, and utilization of hydroelectric energy and biomass energy greatly depend on the geography.

According to the study by Singh et al. (2008), it is found that in states like Punjab, where agriculture is widely practiced, loads of agricultural biomass are produced. This biomass has immense potential for energy and power production (Singh et al. 2008)

Similarly, states like Haryana and Uttar Pradesh also produce significant crop residues that can help generate biomass energy. Regions like Maharashtra, Karnataka, and Andhra Pradesh produce sugarcane biogases, cotton stalks, etc., that can serve as a good source of biomass energy. Forest-rich regions like Central India and North-Eastern India are also good for biomass energy production.

5.3 POWER DEMAND TRENDS IN INDIA

In FY23, India's power generation grew at its fastest rate in more than 30 years. As of January 2024, India's power generation had grown by

6.80% to 1,452.43 billion kWh. In April 2023, India's power consumption was 1,503.65 BU, according to figures from the Ministry of Power. In June 2024, the nation's peak power demand was 249.85 GW. India allocated US$77.7 billion between 2015 and 2022, placing it fourth on the list of countries with the most significant investments in renewable energy. For the first nine months of FY23, the coal plants recorded a PLF of 73.7%, while for the same period in FY22, the PLF was 68.5%. India has an installed power capacity of 446.18 GW in June 2024, holding the third-ranked electricity producer and consumer in the world (IBEF 2024). A year-wise power generation chart is depicted in Figure 5.2 (IBEF 2024).

According to the International Energy Agency, of which India has been an Associate member since March 2017, Total Energy Consumption (TEC) is a crucial indicator for evaluating the effectiveness of energy consumption in economic activities and suggesting possible sustainability solutions. Although energy utilized in transformation processes and energy-producing businesses for their own operations is not included in TEC, the total amount of energy used in end-use sectors is defined under TEC. As a result, final consumption is the amount required to meet domestic consumption and primarily accounts for energy provided to consumers. Central Electricity Authority (CEA), India, has released data on installed generation capacity (Figure 5.3) and % the share of renewable sources in total power generation (Figure 5.4) (CEA 2023). Wind, solar, and similar renewable sources generated the highest power till 2023.

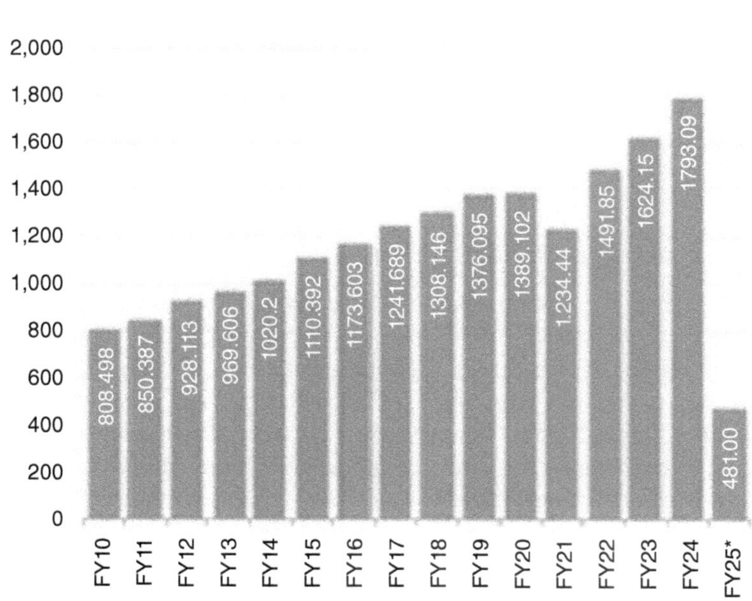

Figure 5.2 Total generation of power in India (including renewable sources) (BU).

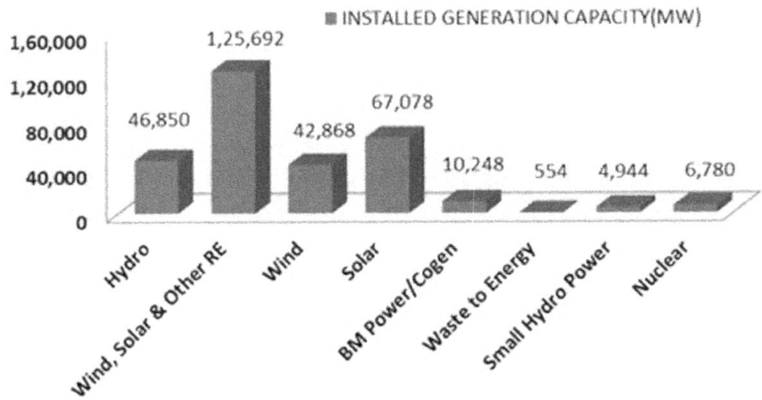

Figure 5.3 Installed generation capacity (renewable resources) as on 31.05.2023 (CEA, 2023).

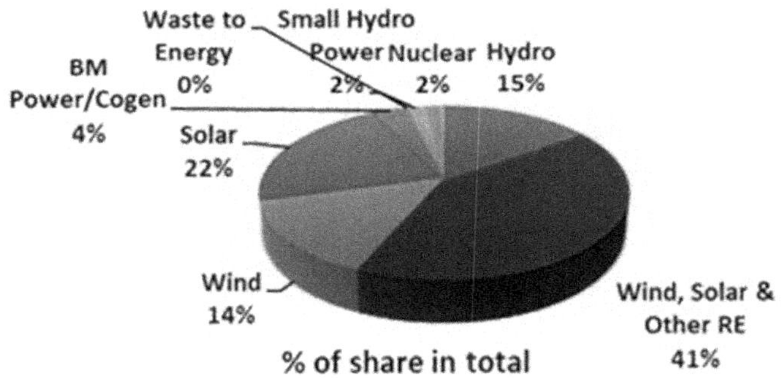

Figure 5.4 Percentage of share of renewable resources for installed generation capacity (CEA, 2023).

5.3.1 Sector-wise power consumption analysis

Energy is crucial for human and economic progress, influencing economic development. A study in India found a negative correlation between energy use and energy resource production, but a positive correlation between total primary energy consumption and GDP, population, and per capita energy consumption (Sahu 2008). Population is a significant factor in demand for additional energy resources, and per capita energy consumption positively correlates with overall energy use. As a country's economy grows, so does its energy resource demand. With the increasing power demand, there is a simultaneous hike in consumption of non-RE sources, which leads to the emission of greenhouse gases and carbon footprints. China, India, the USA, and Russia are the world's leading countries in energy consumption and emissions, accounting for 54% of global CO_2 emissions (Ahmed et al. 2023). India, a top energy-consuming country, is focusing on reducing carbon emissions through energy efficiency initiatives in recent times. However, the reduction in energy intensity is slower than expected. Continuous efforts have been made to improve their performance by lowering technical losses and theft. In order to address urgent demand needs, power utilities have been able to add more supplies to existing contracts under the open market approach. The short-term power market's share has increased over the last ten years, and in FY 2022–2023 it accounted for 12% of the overall electricity volume (Thapar & Sarangi 2024).

A study examines household energy use patterns, revealing inefficient equipment usage, rebound effects, and seasonal consumption trends. Policy recommendations include weatherizing dwellings, real-time consumption tracking, streamlining tariffs, and broadening equipment selection criteria. Residential buildings significantly impact energy usage globally. A study in India aimed to evaluate the scalability of existing research in diverse settings (Batra et al. 2013).

As of FY 2023–2024, India had 275.44 million power users, a sizable and varied population. In India, there are 70 Distribution Companies, or DISCOMs (Table 5.2). Only a small percentage of DISCOMs are profitable; the majority are public organizations. This suggests ongoing challenges in operational efficiency, cost recovery, and financial sustainability. Nevertheless, the "Sector-wise Energy Consumption" graph (Figure 5.5) shows the patterns in energy consumption in India's several industries between 2006 and 2023. Agriculture, Commercial, Industry, Residential, Transportation, Non-Energy, and Other are among the sectors covered. The horizontal axis covers the years 2006–2023, and the vertical axis shows energy consumption in million tons of oil equivalent (Mtoe). Table 5.2 shows the DISCOMs (Distribution Companies) in India.

India's economy is expanding rapidly, which has led to a sharp increase in power demand from a variety of consumer groups, each with unique load patterns.

With a steady rise in consumption due to industrial growth linked to manufacturing, production processes, and economic development, the industry sector continues to be the highest energy consumer, with a total

Table 5.2 DISCOMs (distribution companies) in India

Total number of DISCOMs	70
Number of public DISCOMs	57
Number of private DISCOMs	13
Number of profit making DISCOMs	17
Number of loss making DISCOMs	51

Source: ICED (2024).

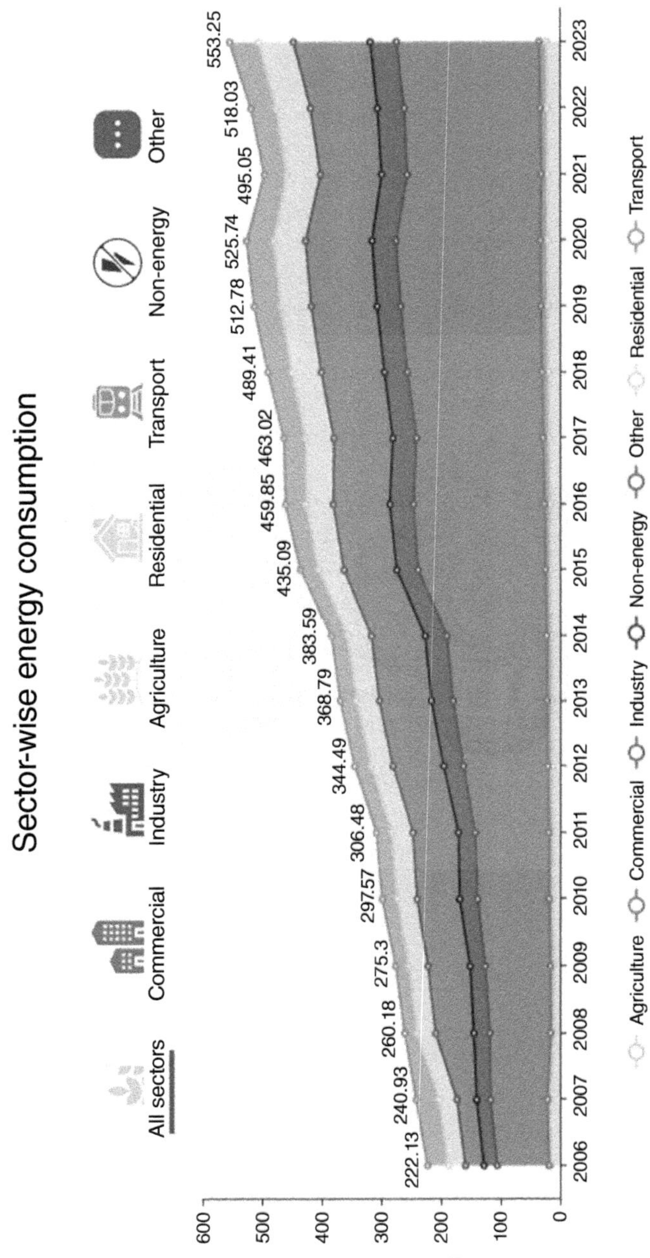

Figure 5.5 Energy consumption by various sectors from 2006 to 2023 (ICED, 2024).

of ~41%. India has adopted Demand-Side Management (DSM) to manage its energy needs efficiently. DSM can reduce peak load by up to 68.46% and save 66% of costs. Initiatives like UJALA (Unnat Jyoti by Affordable LEDs for All) have reduced CO_2 emissions by 2.3 million tons annually and saved 28,553 MUs of power (Mohit & Pawar 2024). This game-changing program, which focuses on initiatives by the government and distribution companies, is bringing stakeholders, planners, and legislators under a common umbrella to promote successful DSM adoption. For the last 16 years, the generation of RE has been 16.9% while the use of electricity has been 5.7%. Using the Simple Linear Regression Model, it was found that a single unit change in the generation of RE might have an impact of 1.43 units on the shift in home electricity consumption (Gautam et al. 2024).

5.3.2 Seasonal and regional demand variations

A report by Grid- India examined the trends in power demand in India for the duration 2009–2022, showing a notable increase in peak demand from 96 to 211 GW (CAGR of 6.26%) as well as a variety of regional patterns impacted by socioeconomic, meteorological, and cultural factors. Demand fell by 25%–30% due to events like COVID-19 lockdowns, with brief variations caused by cultural customs during solar eclipses. Different agricultural practices, climate circumstances, and industrial activity all contribute to regional variety. While governmental approaches concentrate on controlling grid flexibility and balancing supply-demand dynamics, technological improvements such as electric vehicles and renewable energy are changing demand. It also emphasized how crucial it is to comprehend past patterns in order to predict future electricity requirements and guide efficient energy planning (GRID-INDIA 2023).

Demand and supply must be continuously balanced in real time for the electrical grid, and various fuels are designed to achieve this mix. The Indian grid has always been primarily dependent on fossil fuels. Nonetheless, there are plans to significantly increase the availability of RE. At COP26 in Glasgow, the prime minister declared that 500 GW of non-fossil fuel capacity would be the goal, with the majority of that capacity coming from solar and wind.[1] This is a dramatic transition from the grid of the past, where three-quarters of the electricity originated from coal. The overall electricity demand is highest in June and July, followed by a second peak in September when the monsoon comes to an end. Throughout the year, thermal generation is essentially constant, with a minor increase from January to June and a subsequent decline.

Table 5.3 RE policies adopted in India

Policy	*Year of implementation*	*Objective*	*Focus*
National Action Plan on Climate Change (NAPCC)	2008	To address the challenges of climate change and promote sustainable development.	One of the eight missions under NAPCC is the *National Solar Mission.*
National Solar Mission (NSM)	2010	To make solar energy a key part of India's energy mix.	It focuses on promoting the use of solar energy in India. The plan aims to reduce the carbon intensity of India's economy and increase the share of clean energy. The mission includes incentives like financial support, tax exemptions, and simplified regulations to promote solar power generation.
Perform, Achieve and Trade (PAT) Scheme	2012	To improve energy efficiency in industries and reduce carbon emissions.	PAT creates a market for Energy Savings Certificates (ESCerts) which can be traded. It encourages industries to invest in RE to reduce energy consumption and emissions.
Atal Mission for Rejuvenation and Urban Transformation (AMRUT)	2015	To provide basic services in urban areas, including sustainable energy solutions.	The scheme promotes solar power installations in urban spaces, including rooftop solar systems for public buildings.
National Wind-Solar Hybrid Policy	2018	To promote the development of hybrid projects that combine wind and solar power generation, optimizing land use and balancing intermittency.	The policy encourages developers to set up projects that combine both wind and solar energy, with incentives for hybrid systems.
Pradhan Mantri Kisan Urja Suraksha Evam Utthaan Mahabhiyan (PM-KUSUM)	2019	To provide solar power solutions to farmers.	This scheme aims to install solar pumps for irrigation, promote decentralized solar power generation, and provide farmers with a supplementary income by allowing them to sell surplus power to the grid. Installation of 30.8 GW of solar power through this initiative by 2022.

Table 5.3 (Cont.)

Policy	*Year of implementation*	*Objective*	*Focus*
Ujjwal Bharat Scheme	2019	To provide 24/7 electricity to all homes and promote the use of clean energy.	The scheme focuses on expanding the electricity grid and encouraging decentralized solar installations to reach remote areas.
The Green Energy Corridor (GEC)	Ongoing	To create a dedicated transmission network for RE.	The Green Energy Corridor project aims to facilitate the smooth transmission of RE, particularly from wind and solar-rich regions to consumption centers. It includes building new transmission lines and strengthening existing infrastructure.
Faster Adoption and Manufacturing of Hybrid and Electric Vehicles (FAME) Scheme	2015 (Phase-I), 2019 (Phase-II)	To promote electric mobility and reduce dependence on fossil fuels.	The scheme encourages the use of electric vehicles (EVs) powered by RE, helping to integrate clean energy into the transportation sector.
Draft National Energy Policy	2017 (draft)	To ensure sustainable, affordable, and secure energy for all.	The policy emphasizes the need for a diverse energy mix and includes a target of 175 GW renewable energy capacity by 2022 and 500 GW by 2030. The policy supports the integration of renewable sources like solar, wind, biomass, and hydropower into the national grid.
Revised Wind-Solar Hybrid Power Policy	2021	To reduce curtailment and provide more efficient use of renewable resources.	The policy facilitates hybrid projects (wind-solar) to better manage grid demand and maximize energy generation. It also encourages the development of green hydrogen, a clean energy carrier.
Incentives for Rooftop Solar Power		To promote the installation of rooftop solar systems for residential, commercial, and industrial consumers.	The government provides subsidies for rooftop solar installations, especially for residential and small-scale applications, to reduce the cost burden.

(continued)

Table 5.3 (Cont.)

Policy	*Year of implementation*	*Objective*	*Focus*
Renewable Purchase Obligation (RPO)		To mandate the consumption of a minimum percentage of RE in the total energy consumption of utilities.	States and power utilities are obligated to purchase a set percentage of electricity from renewable sources. This has driven the demand RE and created a market for RECs (Renewable Energy Certificates).
National Biofuels Policy	2018	To promote the production and use of biofuels (ethanol, biodiesel, and bio-CNG) in India.	The policy encourages the blending of ethanol with petrol and biodiesel with diesel to reduce dependency on imported crude oil and promote cleaner fuel alternatives.
Government-Backed Auctions for RE Projects		To reduce the cost of RE.	The government has been increasingly moving towards competitive bidding for renewable energy projects. This has led to a sharp reduction in solar and wind power tariffs in India, making RE more affordable.

Source: Kumar et al. (2010), Atteridge et al. (2012), Schmid (2012), Kumar and Majid (2020), and Deshpande et al. (2023).

5.3.3 Challenges of balancing demand with intermittent supply

India's power distribution industry faces several serious challenges that reduce its dependability and efficiency. Inadequate infrastructure and electrification gaps are a significant problem, especially in rural areas where unequal distribution networks and antiquated equipment make it difficult to get electricity reliably. In order to solve this, significant expenditures are required to modernize power grids and extend coverage to outlying areas, as well as enhance transmission and distribution facilities. To supply electricity to every village and home, government programs such as the Deen Dayal Upadhyaya Gram Jyoti Yojana (DDUGJY) highlight the significance of careful planning (Kapoor 2023).

5.4 POLICY FRAMEWORKS FOR RENEWABLE ENERGY DEVELOPMENT

5.4.1 Overview of key policies and initiatives

In recent years, India has advanced significantly in the creation and marketing of renewable energy. To accelerate the shift to greener energy sources, several national policies and programs have been introduced. Most of the scientific developments in RE sector belong to the policy named National Action Plan on Climate Change (NAPCC) (Deshpande et al. 2023). As an example, the National Solar Mission, one of the eight missions under the NAPCC, is dedicated to encouraging the use of solar energy in India. Various policies related to RE in India are listed in Table 5.#.

5.4.2 National solar mission (NSM) – objectives, achievements, and challenges

As part of the National Action Plan on Climate Change (NAPCC), the National Solar Mission (NSM) was introduced in India in 2010 with the goal of advancing the country's solar energy production and utilization (Deshpande et al. 2023). By boosting the proportion of renewable energy in the nation's energy mix, it aimed to prevent climate change, lessen reliance on fossil fuels, and establish India as a global leader in solar energy.

The solar power capacity is increased by the NSM. The capacity of solar electricity generation in India has grown rapidly. India is one of the world's biggest solar markets, having deployed more than 45 GW of solar power as of December 2023 (Deshpande et al. 2023). With aggressive targets being established for 500 GW of non-fossil fuel-based energy capacity by 2030, the original aim of 20 GW of solar power capacity by 2022 was amended and exceeded (Kumar & Majid 2020). Competitive bidding procedures for solar projects have also been observed in India, which has further reduced costs. India is regarded as a leader in solar energy worldwide. With substantial foreign investment in solar energy projects, it has emerged as one of the world's leading solar marketplaces. India was a driving force behind the creation of the International Solar Alliance (ISA), a global campaign to advance solar energy (Shukl 2017). Through programs like 'Atmanirbhar Bharat' and 'PLI' (Production-Linked Incentive) for solar manufacturing, India has taken initiatives to increase its capacity for producing solar panels and cells domestically (Mubarak 2024).

Despite such notable achievements, the NSM is also associated with a few challenges. Land acquisition is one of the biggest obstacles to large-scale solar projects. Due to conflicting land use interests, it has been difficult to find adequate, non-arable land for solar parks (Amita & Soni 2011).

Another major obstacle is constructing the required infrastructure (roads, transmission lines) in isolated locations. Even while investments in India have increased, the solar industry still has financial difficulties, particularly for smaller projects or new entrepreneurs. Without appropriate funding methods, the high initial costs of solar systems may act as a disincentive. Despite price reductions, large-scale projects still face financial challenges. Solar power's intermittent nature (caused by weather and day-night cycles) is still a problem (Shukl 2017). It is technically difficult to integrate solar energy into the current grid infrastructure in an efficient manner. Batteries and other energy storage devices are necessary to mitigate variations in solar power production, although they are still pricey and not yet commonly used. There are also some policy and regulatory issues. Uncertainty has arisen in the solar energy industry as a result of inconsistent or shifting policies (Schmid 2012). Certain projects have been delayed due to unclear rules and delays in the execution of policies about land acquisition, grid connectivity, and power purchase agreements (PPAs). India imports a lot of solar cells and modules, mainly from China, raising questions about trade imbalances and the potential for dependency on external sources (Kumar & Majid 2020). Concern over the effects of large-scale solar farms on the environment is growing, especially with regard to water use, land degradation, and effects on biodiversity and local ecosystems (Atteridge et al. 2012).

5.4.3 National wind-solar hybrid policy

The Indian government launched the National Wind-Solar Hybrid Policy in 2018 with the goals of encouraging the best use of solar and wind energy resources, bolstering grid stability, and increasing power generating efficiency through hybrid projects (Kumar & Majid 2019). While addressing the challenges posed by wind and solar power's intermittent nature, the strategy aims to optimize the incorporation of renewable energy into the energy mix.

5.4.3.1 Integration of wind and solar projects

By using the complementary qualities of both resources, the policy aims to integrate wind and solar energy projects. Since solar and wind energy production frequently peaks at different times of the day or year, hybrid systems can offer a more reliable and steady power source. As long as the projects satisfy the technical and financial standards established by the government, the policy permits developers to design projects that incorporate wind and solar energy systems in different ratios (for example, 50% wind and 50% solar) (Das et al. 2020). To guarantee long-term viability and little downtime, hybrid systems must be operated and maintained in accordance with industry standards. The policy describes procedures for tracking hybrid project performance, such as assessing grid compliance and electricity generation efficiency (Das et al. 2020).

5.4.3.2 Benefits in infrastructure and cost reduction

For wind and solar hybrid projects, the policy encourages the use of shared transmission infrastructure, which decreases transmission costs and eliminates the need for separate transmission lines for each resource (Hassan et al. 2022). In order to give these plants simple access to the grid, it also offers guidelines for making sure hybrid projects are effectively integrated with the current power transmission system. Because of the shared land, combined infrastructure, and cheaper operating and maintenance expenses, hybrid projects may be less expensive to generate than individual wind and solar projects (Das et al. 2020). The quantity of energy produced but not used, commonly observed in stand-alone wind or solar installations, is reduced with the use of hybrid systems. Because extra power might be wasted during periods of low demand or when the grid is overloaded, this is especially crucial as the grid's penetration of renewable energy grows (Hassan et al. 2022). The goal of the program is to reduce dependency on fossil fuel-based backup systems and enhance grid stability by integrating solar and wind energy. Economies of scale provided by hybrid plants lower the total cost per unit of power produced. Compared to independent wind and solar plants, shared infrastructure such as land and transmission lines reduces costs (Memon & Patel 2021). By offering a more dependable, varied, and consistent power supply that lessens reliance on fossil fuels, hybrid projects help to ensure energy security. As the use of renewable energy grows, hybrid systems that combine solar and wind power produce more reliable power, which is essential for grid stability (Hassan et al. 2022). By balancing supply and demand, they lessen power generation variations and facilitate grid management. By combining wind and solar systems, the hybrid model lowers the ecological impact compared to constructing separate plants for each, hence reducing the need for land utilization. The hybrid concept reduces carbon emissions significantly because both sources are renewable (Memon & Patel 2021). This is also associated with challenges in implementation during integration with the Grid and land acquisition. A key step in boosting India's renewable energy capacity and lowering its dependency on fossil fuels is the National Wind-Solar Hybrid Policy. The strategy aims to create a more consistent, reliable, and economical power generation model that meets the nation's renewable energy goals by integrating the benefits of solar and wind energy. However, issues with financing, land acquisition, and grid interconnection must be resolved to guarantee its viability (Das et al. 2020).

5.4.4 Energy efficiency programs and climate obligations

Programs for energy efficiency are essential in the worldwide battle against climate change. These initiatives aim to reduce overall energy consumption and greenhouse gas (GHG) emissions by increasing the efficiency of energy use across a range of sectors, including residential, commercial, industrial, and transportation (Kumar & Majid 2020). Under agreements like the Paris Agreement, these initiatives are frequently incorporated into national and international climate commitments (Deshpande et al. 2023). In order for nations to fulfill their climate commitments under this agreement, which include lowering greenhouse gas emissions and containing the rise in global temperatures, energy efficiency is essential. For renewable energy to be successfully integrated into national systems, energy efficiency is essential. Energy efficiency becomes crucial for lowering total energy demand and storage load as the proportion of renewable energy in the world's energy mix rises. This reduces the need for fossil fuel-based plants to provide backup power if the renewable energy supply fluctuates. By reducing energy usage during peak periods, energy efficiency lowers the overall need for large-scale energy storage systems (Kumar et al. 2019). India's RE policies aim to accelerate the use of clean energy, with a focus on solar and wind power, while also promoting energy efficiency, decentralized energy systems, and electric mobility (Das et al. 2020). Furthermore, the government has set ambitious capacity targets, and other initiatives provide incentives for individuals, companies, and farms to transition to renewable energy. The top priorities are ensuring energy security, reducing carbon emissions, and promoting sustainable economic growth. Countries can significantly advance their climate obligations under the Paris agreement by integrating energy efficiency programs into national and international climate frameworks (Atteridge et al. 2012). However, realizing the full potential of energy efficiency in climate change mitigation will require removing obstacles pertaining to investment, technology, and policy.

5.5 RENEWABLE ENERGY FOR SUSTAINABLE DEVELOPMENT OF INDIA

Since fossil fuel-dominated energy production has substantial negative environmental externalities, the importance of producing electricity from renewable sources is growing. The importance of addressing social and environmental effects has grown in operational strategy, project development, and policymaking. Large amounts of ash and particles are produced due to the high ash concentration and the low conversion efficiency of pulverized coal plants (around 33%). Significant volumes of carbon dioxide (CO_2), sulfur dioxide (SO_2), and nitrogen oxides (NO_x) are also released by coal-fired power plants. Studies infer that almost 40% of all carbon emissions are caused by the power industry.

Table 5.4 CO_2 emissions reduction potentials from grid-connected renewable energy systems (Mahesh and Jasmin 2013)

Year	*Installed capacity (MW)*	*Installed capacity (TWh)*	*Baseline CO_2 (kg/kWh)*	*Mitigation potential million tonnes of CO_2*
1992	32	0.28032	0.86	0.27
1997	902	7.90152	0.86	7.49
2002	1,658	14.52408	0.86	13.77
2007	7,761	67.98636	0.86	64.45
2012	24,503	214.6463	0.86	203.48
2017	54,503	477.4463	0.86	451.61

India has the second-biggest biogas and improved cooking stoves (ICS) projects in the world, as well as the most extensive small gasifier program. Nevertheless, several of the biogas facilities that have been installed are no longer in operation.

To highlight the environmental advantages of investments in this industry, it is crucial to estimate the mitigation potential of renewable energy sources. These systems currently reduce 203.48 million tons of CO_2, but with anticipated investments in the renewable energy industry, that number is expected to increase to 452.61 million tonnes by 2017. The importance of RE sources in shifting to a sustainable energy system was recognized as early as the 1970s (MNRE 2011), but the progress has been below expectations. As of September 30, 2014, India had a cumulative RE installation of 33.8 Gigawatts (GW), including off-grid and captive renewable capacities (MNRE 2011). At that time, RE accounted for only just over 13% (CEA 2014) of the total installed capacity of around 254 GW from all sources. However, the prospect of India's RE sector is bright. The government targets a cumulative RE capacity of more than 175 GW by 2022, as shown in Table 5.4.

Four years later, it is reported that India's plans to attain a total capacity of 225 GW by 2022 and has also surpassed its target pledged at the Paris Agreement as at 175 GW. India becomes the fourth most attractive RE market at the global level (CEA 2018). India can be the first country to have a completely renewable power grid by 2050, given its enormous renewable resources. According to a report by the University of Technology (LUT) in Finland, it is possible if the world uses the latest technologies.

5.6 CONCLUSION

India's RE journey is evidence of ambition, innovation, and resilience. As a major global energy consumer and a significant contributor to greenhouse gas emissions, it is important for India to adopt sustainable energy systems to fulfill its development agenda and combat global warming. India's RE portfolio includes solar and wind, hydroelectric, and biomass sources. The policy frameworks have accelerated RE growth through initiatives like the National Solar Mission and the National Wind-Solar Hybrid Policy. The country's commitment to international agreements and participation in international RE alliances further reflect its leadership in fostering global collaboration on sustainable energy solutions. Still, challenges such as intermittency in solar and wind energy, solutions to storing energy, grid modernization, lack of funds for finance, regulatory barriers, and acquisition of land continue to constrain India in the pursuit of achieving renewable energy.

Note

1 The 26th summit of nations (also referred to as the Conference of Parties or COP) that signed the United Nations Framework Convention on Climate Change (UNFCCC) treaty in 1994.

REFERENCES

Ahmed, M, Shuai, C & Ahmed, M 2023, 'Analysis of energy consumption and greenhouse gas emissions trend in China, India, the USA, and Russia', *Int. J. Environ. Sci. Technol.*, 20, 2683–2698.

Akadiri, SS & Adebayo, TS 2022, 'Asymmetric nexus among financial globalization, non-renewable energy, renewable energy use, economic growth, and carbon emissions: impact on environmental sustainability targets in India', *Environ. Sci. Pollut. Res.*, 29(11), 16311.

Amita, U & Soni, MS 2011, 'Concentrating solar power – Technology, potential and policy in India', *Renew. Sustain. Energy Rev.*, 15, 5161–5175.

Atteridge, A, Shrivastava, MK, Pahuja, N & Upadhyay, H 2012, 'Climate policy in India: What shapes international, national and state policy?' *Ambio*, 41(1), 68–77.

Batra, N, Gulati, M, Singh, A & Srivastava, MB 2013, 'It's Different: Insights into home energy consumption in India', In *Proceedings of the 5th ACM Workshop on Embedded Systems For Energy-Efficient Buildings*, 11, 1–8.

Birol, F & Kant, A 2022, 'India's clean energy transition is rapidly underway, benefiting the entire world', Commentary, IEA, IEA.

CEA 2014, 'Executive Summary, Power Sector', September 2014, Central Electricity Authority, Govt. of India, New Delhi. Available at: https://cea.nic.in/executive/executive-summary-of-power-sector-executive-summary-january-2014/?lang=en. Accessed on 18 Dec 2024.

CEA 2018, 'Government of India', Available at: www.cea.nic.in. Accessed on 18 Dec 2024.

CEA 2020, Central Electricity Authority. 'All India Installed capacity of power stations' Available at: https://web.archive.org/web/20200512013135/http://www.cea.nic.in/reports/monthly/installedcapacity/2020/installed_capacity-03.pdf. Accessed on 13 Dec 2024.

CEA 2023, 'Power Sector at a Glance ALL INDIA', Central Electricity Authority, Ministry of Power, Government of India.

Climate Trasnparency 2022, 'Climate Transparency Report: Comparing G20 Climate Action', Available at: https://www.climate-transparency.org/media/climate-transparency-report-2022-2. Accessed on 14 Dec 2024.

Das, A, Hardik KJ, Nagababu, G & Kachhwaha, SS 2020, 'A comprehensive review of wind–solar hybrid energy policies in India: Barriers and recommendations', *Renew. Energy Focus*, 35, 108–121.

Deshpande, T, Mukherji, R & Sastry, M 2023, 'Policy styles and India's national action plan on climate change (NAPCC)', *Policy Studies*, 46(1), 46–63.

Energy Institute 2024, 'Statistical Review of World Energy 73rd Edition', EI, Energy Institute.

Financial Express 2020, 'Bigger than Singapore, Bahrain! World's biggest, hybrid renewable energy park in Gujarat to attract huge investment', Available at: https://www.financialexpress.com/business/infrastructure-bigger-than-singapore-bahrain-worlds-biggest-hybrid-renewable-energy-park-in-gujarat-to-attract-huge-investment-2150904/. Accessed on 11 Dec 2024.

Gautam, AK, Verma, NMP & Linda, N 2024, 'An econometric analysis of electricity consumption from renewable energy in India', *Soc. Sci. J. Adv. Res.*, 4(1), 47–51.

Ghosh, U, Das, D, Banerjee, D, Karmarkar, S, & Das, J 2021, 'Biomass energy potential in India: A review', *Int. J. Eng. Res. Technol. Nceter*, 9(11), 42–45.

GRID-INDIA 2023, 'Report on Electricity Demand Pattern Analysis', Grid Controller of India Limited, Available at: https://posoco.in/wp-content/uploads/2024/01/All-India-Regional-Vol-I_v1.pdf. Accessed on 14 Dec 2024.

Haldar, S & Sharma, G 2022, 'Impact of urbanization on per capita energy use and emissions in India', *Int. J. Energy Sector Manag.*, 16(1), 191–207.

Hassan, R, Das, BK & Hasan, M 2022, 'Integrated off-grid hybrid renewable energy system optimization based on economic, environmental, and social indicators for sustainable development', *Energy*, 250, 123823.

IBEF 2024, 'Power', India Brand Equity Foundation, Available at: https://www.ibef.org/download/1728036699_Power_August_2024.pdf. Accessed on 17 Dec 2024.

ICED 2024, 'India Climate & Energy Dashboard', ICED manual, India Energy, NITI Aayog, Available at: https://iced.niti.gov.in/energy, ICED.

IEA 2011, International Energy Agency. 2011. Available at: https://web.archive.org/web/20120113032718/http://www.iea.org/Textbase/npsum/solar2011SUM.pdf. Accessed on 10 Dec 2024.

IEA 2021, 'Net Zero by 2050: A roadmap for the global energy system', International Energy agency, UN, Available at: https://iea.blob.core.windows.net/assets/deebef5d-0c34-4539-9d0c-10b13d840027/NetZeroby2050-ARoadmapfortheGlobalEnergySector_CORR.pdf. Accessed on 14 Dec 2024.

Indian Exim Bank 2022, 'Indian Solar Sector: Fostering Growth and Sustainable Development', Available at: www.eximbankindia.in/. Accessed on 10 Dec 2024.

Kapoor, G 2023, 'Power Distribution Challenges in India: Overcoming the Obstacles', Power Distribution in India, Power Distribution, Available at: https://powerdistribution.in/power-distribution-challenges-in-india/. Accessed on 16 Dec 2024.

Kumar, A, Kumar, K, Naresh, K, Sharma, S & Mishra, S 2010, 'Renewable energy in India: Current status and future potentials', *Renew. Sustain. Energy Rev.*, 14, 2434–2442.

Kumar, JCR & Majid, MA 2020, 'Renewable energy for sustainable development in India: current status, future prospects, challenges, employment, and investment opportunities', *Energ. Sustain. Soc.* 10, 2.

Kumar, JCR, Kumar, V & Majid, MA 2019, 'Wind energy programme in India: Emerging energy alternatives for sustainable growth', *Energy Environ.*, 30, 1135–1189.

LCIPP 2024, 'Local Communities and Indigenous Peoples Platform', Advance unedited version, UNCC, United Nations Climate Change, LCIPP, Available at: https://lcipp.unfccc.int/homepage. Accessed on 23 Dec 2024.

Memon, SA & Patel, RN 2021, 'An overview of optimization techniques used for sizing of hybrid renewable energy systems', *Renew. Energy Focus*, 39, 1–26.

Ministry of Power 2008, Government of India. 'Hydro Power Policy 2008', Available at: https://aerc.gov.in/subpage/1672144572.pdf. Accessed on 20 Dec 2024.

MNRE 2011, 'Strategic plan for new and renewable energy sector for the period', Ministry of New and Renewable Energy, Govt. of India, Available at: www.india.gov.in/strategic-plan-new-and-renewable-energy-sector-2011-2017. Accessed on 15 Dec 2024.

MNRE 2019, Ministry of New and Renewable Energy, PIB Delhi: Government of India, Available at: https://pib.gov.in/Pressreleaseshare.aspx?PRID=1564039. Accessed on 9 Dec 2024.

MNRE 2023, Ministry of New and Renewable Energy. PIB Delhi, Available at: https://pib.gov.in/PressReleasePage.aspx?PRID=1944759. Accessed on 15 Dec 2024.

MNRE 2024a, Ministry of New and Renewable Energy, Available at: https://mnre.gov.in/en/bio-energy-overview/. Accessed on 23 Dec 2024.

MNRE 2024b, Ministry of New and Renewable Energy, PIB Delhi, Available at: https://pib.gov.in/PressReleaseIframePage.aspx?PRID=2073038#:~:text=Solar%20power%20leads%20the%20way,wind%20corridors%20across%20the%20country. Accessed on 10 Dec 2024.

Mohit & Pawar, R 2024, 'Driving sustainable energy in India: The role of demand-side management in power optimization and environmental conservation', *Energy Sources A Recovery Util. Environ. Eff.*, 46(1), 14089–14117.

Mubarak, U 2024, 'Unleashing India's solar potential: A review of the National Solar Mission and the Path to Sustainable Energy in India', *J. Technol. Innov. Renew. Energy*, 3, 49–63.

NITI Aayog 2024, 'India Climate & Energy Dashboard', Available at: https://iced.niti.gov.in/energy/electricity/generation/pipeline-capacity/hydro. Accessed on 23 Dec 2024.

Pandit, M, Manish, K, Singh, G & Chowdhury, A 2023, 'Hydropower: A low-hanging sour-sweet energy option for India', *Heliyon*, 9(6), e17151.

Planning Commission 2012, 'Eleventh Five-year Plan, 2007-12: Agriculture, Rural Development, Industry, Services and Physical Infrastructure, Vol. III', Planning Commission, Government of India, Available at: https://www.niti.gov.in/sites/default/files/2023-08/11th_vol3.pdf. Accessed on 20 Dec 2024.

Purohit, I, Purohit, P & Shekhar, S 2013, 'Evaluating the potential of concentrating solar power generation in Northwestern India', *Energy Policy*, 62, 157–175.

Reddy, BS 2016, 'India's energy system transitiond-Survival of the greenest', *Renew. Energy*, 92, 293–302.

Sahu, S 2008, 'Trends and patterns of energy consumption in India', *Munich Personal RePEc Archive*, MPRA, Munich.

Schmid G 2012, 'The development of renewable energy power in India: Which policies have been effective?', *Energy Policy*, 45, 317–326.

Shukl, AK 2017, 'Renewable energy resources in South Asian countries: Challenges, policy and recommendations', *Resour. Effic. Technol.*, 3, 342–346.

Singh, J, Panesar, B & Sharma, S 2008, 'Energy potential through agricultural biomass using geographical information system – A case study of Punjab', *Biomass Bioenergy*, 32(4), 301–307.

Singh, N, Kumar, A & Rai, S 2014, 'Potential production of bioenergy from biomass in an Indian perspective', *Renew. Sustain. Energy Rev.*, 39, 65–78.

Thapar, S & Sarangi, GK 2024, 'Role of Indian power markets in energy transition', *Energy Sustain. Dev.*, 82, 101538.

UNFCCC 2024, 'Concept note and Agenda – COP 29 Outcomes Webinar Series Part 2: Inclusive climate action', Daily Programmes, Regional Collaboration Centre Bangkok, United Nations Climate Change, United Nations Climate Change, Bangkok.

Chapter 6

Dielectric energy storage and energy harvesting using lead-free piezoelectric materials for sustainable energy solutions

Vijayeta Pal and Vimal Kumar

6.1 INTRODUCTION

The growing demand for environmentally friendly energy solutions has driven the progress of innovative materials for energy harvesting and storage applications. In recent times, technology focused on electric energy storage has become important in the modern electric industry, mainly in medical amenities, mobile communication, and autonomous transportation applications. Thus, it is important to progress research in electric energy storage technologies. The fuel cells, supercapacitors, batteries, and capacitors (dielectric) have turned into an extended research objective among the different energy storage structures, as displayed in Figure 6.1. Dielectric and piezoelectric materials have emerged as viable alternatives for the applications due to their ample energy storage and conversion efficiency. Lead-free piezoelectric materials, in particular, have received widespread interest as environmentally benign alternatives to traditional lead-based ceramics, harmonizing with global initiatives to eliminate environmental dangers. Dielectric energy storage uses materials with high dielectric constants and breakdown strengths to obtain high energy densities. These features are essential for capacitors and other high-performance energy storage devices. Dielectric capacitors have the highest power density (P_D > 10^4 W/kg) due to their quick charging and discharging (in microseconds), making them ideal for pulse power supplies and electric vehicles. They also have exceptional temperature stability, long cycle life, and safety. Thus, they are essential for high-power discharge storage, both theoretically and practically. Over the years, dielectric capacitors have been made from a variety of materials, both inorganic and organic (Li et al., 2015; Chen et al., 2018). Dielectric capacitors are often made of ceramics or polymers due to their superior properties (Kim et al., 2020b). They are an essential part for communication devices and industrial gadgets, meeting demands for thermal stability, high capacity, radio frequency, and ultra-miniaturization, driven by the growth of IoT and smart technologies. Biaxially orientated polypropylene (BOPP) is commonly used in commercial pulse power systems, such

 DOI: 10.1201/9781003581536-6

as EV inverters, due to its processability, impact resistance, and insulating qualities. However, its low dielectric constant ($\varepsilon_r = 2.2$), restricted energy density (W_{rec} = 2–5 J/cm^3), and narrow functioning temperature range (T < 105°C) require big capacitors. This emphasizes the importance of dielectric materials with wider temperature ranges and increased energy storage properties for pulse capacitor applications. Piezoelectric materials, on the other hand, allow for energy harvesting by turning mechanical vibrations into electrical energy, making them perfect for powering sensors, wearable devices, and Internet of Things applications. Research on diverse dielectric capacitors, as shown in the inset of Figure 6.1, has been continuously growing in recent years (Wang et al., 2021; Yang et al., 2019). Lead-free dielectric materials, such as bismuth-based and alkali niobates, provide a sustainable solution for these applications. Their non-toxic nature and equivalent performance metrics to lead-based competitors make them ideal for large-scale incorporation into environmentally friendly energy systems. Advances in bulk ceramics, thick films, and thin films attempt to increase energy density while decreasing energy loss. Bulk ceramics are inexpensive and suitable for tuning their properties through various mechanisms in research, whereas thin films excel in structural and electrical analysis but are costly. Thick-film ceramics find a balance by combining scalability and great industrial potential (Pan et al., 2020).

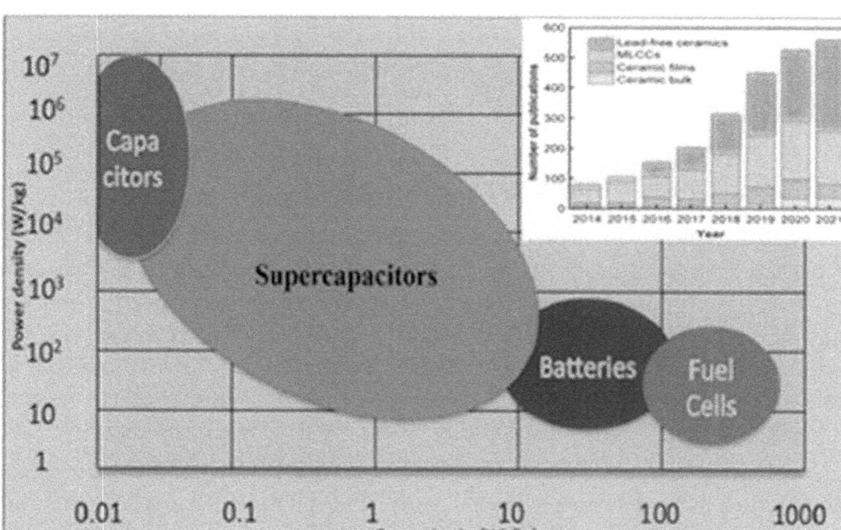

Figure 6.1 Ragone plot comparing energy storage technologies (capacitors, supercapacitors, batteries, fuel cells). Inset displayed bar graph of dielectric capacitor publications (2014–2021, Web of Science). MLCCs, multilayer ceramic capacitors.

Advances in synthesis techniques of lead-free dielectric materials, including sol-gel, semi-wet (Pal et al., 2014a,b), and solid-state processes (Pal et al., 2023), have allowed for exact control of their microstructure and characteristics. Despite the potential, there are still obstacles to improving the energy storage characteristics (density and efficiency) of these dielectric materials. The future of dielectric capacitors lies in their ability to operate under extreme conditions (temperature: –50°C to 250°C, frequency: 1–10^3 Hz) while achieving high performance, including energy-related characteristics (recoverable energy; Wrec) above 10 J/cm^3 and storage efficiency (η) exceeding 90% (Zhang et al., 2022). Figure 6.1 shows the Ragone plot comparing energy storage technologies (capacitors, supercapacitors, batteries, and fuel cells).

To obtain improved performance, researchers are working on refining their composition, incorporating dopants, and building composites and advanced materials. The expanding global need for renewable energy solutions has underlined the importance of effective energy storage and harvesting systems.

6.1.1 Fundamentals of dielectric energy storage

Dielectric energy storage is a critical technology for modern energy systems, allowing for effective storage and rapid release of electrical energy. It uses dielectric materials that polarize under an electric field to store energy as electrostatic charge. This mechanism underlies high-performance capacitors, which are widely utilized in a variety of applications ranging from electronics to renewable energy systems. They provide rapid functionality and great power density.

6.1.2 Principles of dielectric energy storage

An insulating dielectric layer separates the two metal electrodes that make up a dielectric capacitor. As seen in Figure 6.2, the dielectric polarizes and stores energy as electrostatic charge between the electrodes when an external electric field is applied (Palneedi et al., 2018; Yao et al., 2017). As demonstrated in Figure 6.2 polarization-electric field (P-E) loop, charge is stored in the electric field and released back into the circuit when the field is eliminated. Charging and discharging occur swiftly, with little energy conversion or material movement, resulting in a high-power density. The capacitance (CV-1) of an ideal capacitor is proportional to the applied voltage (V) and can be written as follows: $Q = C \cdot V$, where C is the proportionality constant. In honor of the physicist Michael Faraday (1791–1867), who made important contributions to electricity, the capacitance of 1 C V-1 is called 1F (Palneedi et al., 2018). Numerous factors, including interelectrode

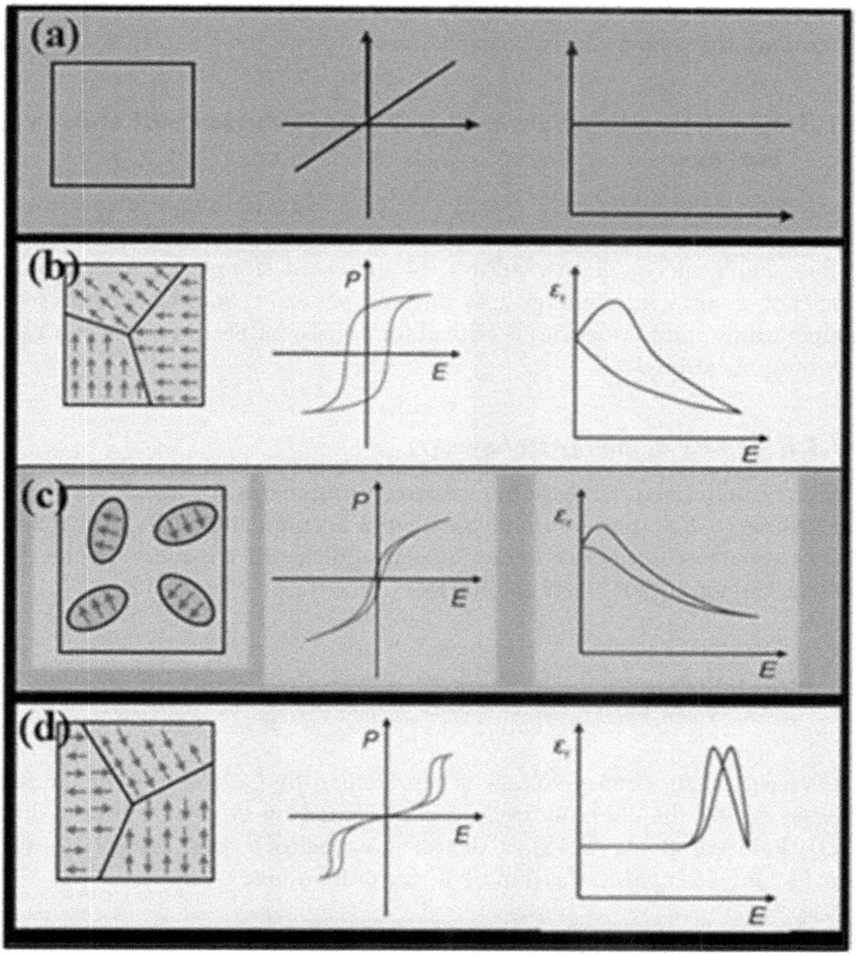

Figure 6.2 Differences between (a) linear, (b) FE, (c) relaxation FE, and (d) AFE dielectrics in terms of domain structure, polarization characteristics, and dielectric constant response to electric field.

distance, dielectric type, and electrode size, affect capacitance. In real-world applications, capacitance is inversely proportional to the dielectric constant, the distance between electrodes, and the electrode plate area, as stated in Equation (1). (6.1):

$$C = \frac{A}{\varepsilon^{*} d} \tag{6.1}$$

where the electrode plate area, dielectric constant, and interelectrode distance are denoted by A, ε, and d, respectively. It goes without saying that ε determines the properties of the capacitor.

6.1.3 Key criteria for assessing the performance of energy storage

The performance of dielectric energy storage is assessed using energy storage density, energy efficiency, dielectric breakdown strength, stability, and reliability. High energy density allows for increased storage capacity, while efficiency assures little energy loss during operation. Stability at different temperatures and frequencies is critical for consistent performance in a variety of applications.

6.1.3.1 Density of energy storage (W)

Dielectric material quality can be evaluated using energy storage density (W). According to Eq. (6.2), the voltage-charge accumulation (Q) connection can be used to describe the energy stored in dielectric capacitors while the voltage stays constant (Li et al., 2021).

$$w = \int_0^{Q_{max}} V.dq \tag{6.2}$$

The charge at maximum voltage is represented by Q_{max}, while the charge change during the charging cycle is represented by dq. According to Eq. (6.3) (Palneedi et al., 2018), a dielectric capacitor's total energy storage density (W_{st}) is equal to the sum of W per unit volume ($A.d$).

$$w_{st} = \frac{w}{A.d} = \int_0^{Q_{max}} \frac{V.dq}{A.d} \tag{6.3}$$

$$\int_0^{D_{max}} E.dD \tag{6.4}$$

where D_{max} is the electrical displacement vector and E is the electric field. The electric displacement (D) is influenced by the dielectric constant and can be expressed as follows:

$$D = \varepsilon_0 E + P \tag{6.5}$$

where P is the dielectric's polarization. Nevertheless, D is nearly supplied by P for dielectrics with large dielectric constants, such ceramics, that is:

$$D = \varepsilon_0 E = \varepsilon_0 \varepsilon_r E \tag{6.6}$$

In this case, ε_r is the relative permittivity and ε_0 is the vacuum permittivity. The following is a new way to express the connection between electrical displacement and energy storage density:

$$W_{st} = \int_0^{P_{max}} E.dP = \int_0^{E_{max}} \varepsilon_0 \varepsilon_r E.dE \tag{6.7}$$

In this case, E_{max} stands for the maximum applied electric field, and P_{max} for the polarization associated with that field. The energy storage density W_{st}, which is dependent on the electric field (E) and polarization (P) values, is frequently calculated using Eq. (6.7). According to several studies, the area above the polarization-electric field (P-E) loop, bounded by the polarization axis, as shown in Figure 6.2, represents the recovered energy density of a dielectric capacitor (Damjanovic, 1998).

Figure 6.2 illustrates how the micro-electrical structure of dielectrics affects their polarization behavior (Palneedi et al., 2018). Significant ferroelectric energy loss is introduced by the polarization-electric field (P-E) loop hysteresis for a typical ferroelectric (FE) ceramic, as seen in Figure 6.3b.

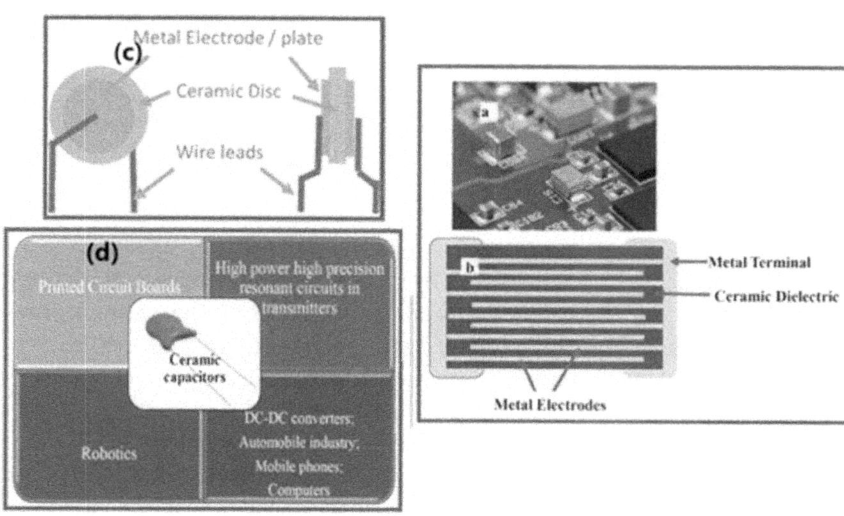

Figure 6.3 (a) Photographical image of MLCC, (b) multilayer ceramic capacitors, (c) disc capacitor, and (d) diverse applications of ceramic capacitors.

Through chemical alteration, this hysteresis can be reduced, resulting in the formation of unique relaxor-FE (RFE) features, as illustrated in Figure 6.3c. RFE ceramics are ideal for energy storage applications because of their low polarization hysteresis and high dielectric constant ε_r (Zhao et al., 2020). Saturated polarization occurs when antiferroelectric (AFE) ceramics undergo a shift from the AFE phase to a stable FE phase at high electric fields. However, residual polarization (P_r) is negligible at lower electric fields.

The polarization behaviour of dielectrics varies according to their micro-electrical structure, as demonstrated in Figure 6.2 (Palneedi et al., 2018). For a typical ferroelectric (FE) ceramic, as depicted in Figure 6.3b, the polarization-electric field (*P-E*) loop exhibits hysteresis, which introduces significant ferroelectric energy loss. This hysteresis can be lessened through chemical modification, leading to the appearance of distinct relaxor-FE (RFE) characteristics, as shown in Figure 6.3c. RFE ceramics are ideal for energy storage applications because of their low polarization hysteresis and high dielectric constant ε_r (Zhao et al., 2020). Saturated polarization occurs when antiferroelectric (AFE) ceramics undergo a shift from the AFE phase to a stable FE phase at high electric fields. However, residual polarization (P_r) is negligible at lower electric fields. A field-induced phase shift is primarily responsible for the energy loss in AFE ceramics, as shown in Figure 6.3d. Equations (6.8) and (6.9), respectively, can be used to determine the stored energy density (W) and recovered energy density (W_{rec}) of FE, RFE, and AFE ceramics due to their different energy loss characteristics.

$$W = \int_{0}^{P_{amax}} E.dP \tag{6.8}$$

$$W_{rec} = \int_{P_r}^{P_{max}} E.dP \tag{6.9}$$

In conclusion, the dielectric constant, breakdown field strength, polarization hysteresis, and energy loss all have a significant impact on the energy storage density. Researchers must take into account and thoroughly address each of these issues in order to create high-energy-density dielectric materials.

6.1.3.2 Energy storage efficiency (η)

Energy storage efficiency in dielectric capacitors is characterized as the ratio of recoverable energy to total energy input, which emphasizes the system's ability to minimize energy loss, as written in the Eqs. (6.10) and (6.11).

$$\eta = \frac{W_{rec}}{W} \tag{6.10}$$

$$W = W_{rec} + W_{loss} \tag{6.11}$$

where W_{loss} is the energy loss density and W_{rec} is the recoverable energy density. Linear ceramics are the most advantageous for improving η because of their small P_r, delayed polarization rise, modest polarization loss, and other phenomena that help to optimize η (Zhao et al., 2020; Pal et al., 2023).

6.1.3.3 Stability of dielectric materials

Stability at varying temperatures and frequencies is essential for ensuring reliable performance of dielectric materials in diverse applications. Materials must maintain their dielectric properties and energy storage capabilities under extreme conditions, such as high temperatures or wide frequency ranges. This is particularly important in advanced technologies like electric vehicles, renewable energy systems, and high-frequency electronics, where fluctuations in operating conditions are common. When developing high-performance dielectric capacitors, consistent stability is crucial since it improves the system's durability and effectiveness.

6.2 RELIABILITY OF THE MATERIALS

Dielectric capacitors operate in complex systems, influenced by temperature, frequency, and voltage variations during cyclic charging and discharging. High reliability and fatigue resistance are critical for preserving energy storage performance in demanding environments. This is commonly examined using polarization hysteresis loops. BOPP, a prevalent polymer in EV power inverters, requires supplementary cooling systems because to its low thermal stability (<105°C), which increases weight, volume, and energy expenses. Developing materials with greater dependability and thermal stability is critical for overcoming these limits.

6.2.1 Dielectric breakdown strength

When exposed to high voltages, dielectric breakdown-a physical phenomenon involving dielectric materials and their corresponding capacitors—causes material and electrical flux ionization. The term "dielectric breakdown strength" (E_b) refers to an electric field threshold that dielectric materials cannot withstand and lose insulation, thereby exhibiting resistance properties similar to semiconductors or conductors, negatively affecting the dielectric capacitors' dependability and performance. Since dielectric breakdown strength is affected by both internal (band gap) and external (microdefect, microstructure, etc.) factors, it is an essential parameter for predicting energy storage performance (McPherson et al., 2003; Zhou et al., 2018). The failure phenomena of dielectric capacitors have been explained by a number of causes, but the main ones are partial discharge, electric puncture, thermal runaway, and electromechanical breakdown. According to Wang et al. (2021), the breakdown process of dielectric capacitors is significantly influenced by either a single mechanism or the interaction effect of several concerns. For instance, thermal breakdown and aging of dielectric insulators are likely to occur when electron hopping and internal heat buildup brought on by a high-voltage electric field combine. To verify intrinsic breakdown strength, testing data is analysed using the Weibull distribution function with two parameters, as shown below Eq. (6.12):

$$P = 1 - \exp\left[-\left(\frac{E}{E_b}\right)^{\beta}\right] \tag{6.12}$$

In this equation, E represents the measured breakdown strength, β is the dispersion of the experimental data, and P is the cumulative probability of electric failure. By adjusting both intrinsic and extrinsic properties, dielectric materials can be made to resist dielectric breakdown, which happens at high electric fields. Dielectrics' insulating ability is directly correlated with their band gap (E_g), with larger E_g values resulting in higher breakdown strength. In general, chemical doping is a proven method for improving breakdown strength by adjusting E_g. External factors like porosity, grain size, flaws, and secondary phases can impact dielectric breakdown strength. Optimizing breakdown qualities presents both opportunities and challenges. Avoiding porosity improves compactness. Secondary phases can improve breakdown strength by utilizing effective tactics like core-shell construction, even though they are often detrimental to other breakdown properties. Elements related to internal and external should work together to improve breakdown strength, rather than being equally incompatible.

6.2.2 Comparison of energy storage characteristics of distinct lead-free ceramics

Dielectric capacitors (ceramic), known for their low energy losses, hold significant potential for applications in systems that are microelectronic and tiny power devices (Guo et al., 2021). Researchers are actively focusing on enhance the energy storage related characteristics of lead-free ceramics, leading to frequent breakthroughs in this field. Extensive work has been conducted on systems such as $BaTiO_3$ (BT) (Lin et al., 2019; Badapanda et al., 2017), $Ba_{1-x}Ca_x\ Zr_yTi_{1-y}O_3$ (BCTZ) (Pal et al., 2023, 2017, 2018; Zhan et al., 2018; Mezzourh et al., 2021), $Na_{0.5}Bi_{0.5}TiO_3$ (NBT) (Chen et., 2021; Pal et al., 2014a), $BiFeO_3$ (BFO) (Lu et al., 2020; Wang et al., 2020; Qi et al., 2019), $NaNbO_3$ (NN) (Chen et al., 2021), and $K_{0.5}Na_{0.5}NbO_3$ (KNN) (Zhang et al., 2020a), among others. Table 6.1 provides a summary of important developments in the creation of lead-free ceramics.

Table 6.1 Comparison of energy storage properties (W_{rec} and η) in lead-free piezoelectrics

Materials	*W_{rec} (J/cm³)*	*η (%)*	*References*
$BaTiO_3$	0.450	28.23	Lin et al. (2019)
$BaZr_{0.05}Ti_{0.95}O_3$	0.218	72	Badapanda et al. (2017)
$Ba_{0.7}Ca_{0.3}TiO_3$	—	58	Fu et al. (2008)
$Ba_{0.85}Ca_{0.15}Zr_{0.10}Ti_{0.90}O_3$	0.0104	59.9	Mezzourh et al. (2021)
$Ba_{0.95}Ca_{0.05}Zr_{0.20}Ti_{0.80}O_3$	0.4100	72	Zhan et al. (2018)
$Ba_{0.955}Ca_{0.045}Zr_{0.17}Ti_{0.83}O_3$	0.495	72.8	Puli et al. (2013)
$Ba_{0.85}Ca_{0.15}Zr_{0.10}Ti_{0.90}O_3$	0.014	80	Hanani et al. (2019)
$0.92BaTiO_{3\text{-}0.08}K_{0.73}Bi_{0.09}NbO_3$	2.51	86.9	Lin et al. (2019)
$0.9BaTiO_{3\text{-}0.1}Bi(Mg_{2/3}Nb_{1/3})O_3$	1.13	95	Wang et al. (2015)
$0.6BaTiO_{3\text{-}0.4}Bi(Mg_{0.5}Ti_{0.5})O_3$	4.49	93	Hu et al. (2020)
$0.85BaTiO_{3\text{-}0.15}Bi(Mg_{0.5}Zr_{0.5})O_3$	4 2.9	86.8	Yuan et al. (2018)
$AgNb_{0.45}Ta_{0.55}O_3$	6.3	90	Luo et al. (2020)
$BaTiO_{3\text{-}0.06}Bi_{2/3}(Mg_{1/3}Nb_{2/3})O_3$ 520 4.55 91	4.55	91	Yang et al. (2020)
$0.78NaNbO_{3\text{-}0.22}Bi(Mg_{2/3}Ta_{1/3})O_3$	5.01	86.8	Shi et al. (2020)
$0.75[0.9NaNbO_{3\text{-}0.1}Bi(Mg_{0.5}Ta_{0.5})O_3]\text{-}0.25(Bi_{0.5}Na_{0.5})_{0.7}Sr_{0.3}TiO_3$	8	90.4	Chen et al. (2021)
$0.8NaNbO_{3\text{-}0.2}SrTiO_3$	3.02	80.7	Zhou et al. (2018)
$0.9NaNbO_{3\text{-}0.1}Bi(Mg_{2/3}Nb_{1/3})O_3$	2.8	82	Ye et al. (2019)
$0.85K_{0.5}Na_{0.5}NbO_{3\text{-}0.15}Bi(Zn_{0.5}Zr_{0.5})O_3$	3.5	86.8	Zhang et al. (2020)
$0.57BiFeO_{3\text{-}0.30}BaTiO_{3\text{-}0.13}Bi(Li_{0.5}Nb_{0.5})O_3$	3.64	74	Wang et al. (2020)
$0.57BiFeO_{3\text{-}0.33}BaTiO_{3\text{-}0.10}NaNbO_3$	8.12	90	Qi et al. (2019)
$0.99(0.4Na_{0.5}Bi_{0.5}TiO_{3\text{-}0.6}SrTiO_3)\text{-}0.01SnO_2$	2.35	80	Wang et al. (2020)
$0.55Na_{0.5}Bi_{0.5}TiO_{3\text{-}0.45}Sr_{0.7}La_{0.2}TiO_3$	4.14	92.2	Qiao et al. (2020)
$0.90(Na_{0.5}Bi_{0.5})_{0.7}Sr_{0.3}TiO_{3\text{-}0.10}Bi(Ni_{0.5}Sn_{0.5})O_3$	4.18	83.64	Li et al. (2020)
$0.75Bi_{0.58}Na_{0.42}TiO_{3\text{-}0.25}SrTiO_3$	5.63	94	Yan et al. (2020)
$0.78\ Na_{0.5}Bi_{0.5}TiO_{3\text{-}0.22}NaNbO_3$	7.02	85	Qi et al. (2019)
$0.62Na_{0.5}Bi_{0.5}TiO_{3\text{-}0.30}Sr_{0.7}Bi_{0.2}TiO_{3\text{-}0.08}Bi(Mg_{2/3}Nb1/3)O_3$	7.5	92	Ji et al. (2021)
$0.6BaTiO_{3\text{-}0.4}Bi(Mg_{0.5}Ti_{0.5})O_3$	4.49	93	Hu et al. (2020)
$BaTiO_{3\text{-}0.06}Bi_{2/3}(Mg_{1/3}Nb_{2/3})O_3$	4.55	91	Yang et al. (2020)
$0.5BiFeO_{3\text{-}0.4}SrTiO_{3\text{-}0.1}BiMg_{2/3}Nb_{1/3}O_{3\text{-}0.03}Nb$	8.2	74.1	Lu et al. (2020)
$0.65(0.84Na_{0.5}Bi_{0.5}TiO_{3\text{-}0.16}K_{0.5}Bi_{0.5}TiO_3)\text{-}0.35(Bi_{0.2}Sr_{0.7}TiO_3)$	4.06	87.3	Hu et al. (2020)
$Ba_{0.95}Ca_{0.05}Ti_{0.89}Sn_{0.11}O_3$	0.1072 0.1249	51.36 74.63	Hadouch et al. (2022)
(x)BMT-(1-x){0.50BZT-0.50BCT}; $x = 0.10$	0.153	82	Pal et al. (2023)

6.3 ENERGY HARVESTING USING PIEZOELECTRIC MATERIALS

Using transduction processes to convert ambient energy sources such as mechanical, solar, thermal, wind, or fluid flow—into electrical energy is known as energy harvesting, or EH. Typical techniques include electromechanical transducers for mechanical vibrations, thermoelectrics for temperature gradients, and photovoltaics for solar energy (Naqvi et al., 2022). Because of their intrinsic polarization-based transduction, piezoelectric materials are especially useful for mechanical EH (Sezer and Koç, 2021). Unlike electrostatic, electromagnetic, or triboelectric approaches, they don't require an external voltage, magnetic field, or material contact. Here, some articles have been referred to for an in-depth study on energy harvesting (Li and Lee, 2022; Naqvi et al., 2022; Sezer and Koç, 2021).

6.3.1 Energy harvesting mechanisms

Piezoelectric energy harvesting (PEH) relies on the direct class of piezoelectric effect, where mechanical stress or strain induces an electric charge in piezoelectric materials. The key mechanisms used in PEHare are displayed in Table 6.2.

6.3.2 Advantages of lead-free piezoelectrics in energy harvesting

Lead-free piezoelectrics offer several advantages over lead-based ceramic systems for energy harvesting applications. Firstly, they are environmentally

Table 6.2 Key mechanisms of piezoelectric energy harvesting

Mechanism	*Description*	*Applications*
Bending mode	A cantilever beam with a piezoelectric layer bends under force, inducing strain and generating charge.	MEMS, IoT sensors, low-power devices
Axial mode	Axial compression or elongation generates charge along the material's axis.	Force sensors, structural health monitoring
Vibrational mode	Captures ambient vibrations (machinery, vehicles, or human motion) and converts them into electricity.	Wireless sensor nodes, embedded electronics
Impact mode	Impact forces compress the material, generating high-frequency electric charges.	Footwear sensors, smart pavements, impact harvesters
Shear mode	Shearing forces applied parallel to the material surface generate electric charges.	Specialized sensing devices, industrial use
Stacked mode	Multiple piezoelectric layers stacked to amplify charge output under compression.	High-force environments, higher power output
Hybrid mode	Combines piezoelectric harvesting with other methods (triboelectric, electromagnetic) for efficiency.	Wearables, hybrid systems, high-efficiency devices

friendly, avoiding the toxic and hazardous effects associated with lead, and comply with global regulations such as RoHS and WEEE. This makes lead-free systems a sustainable choice, aligning with the increasing demand for eco-friendly technologies. In terms of material performance, lead-free piezoelectrics, such as $BaTiO_3$, $Na_{0.5}Bi_{0.5}TiO_3$, and $K_{0.5}Na_{0.5}NbO_3$, exhibit comparable or even superior piezoelectric coefficients in specific compositions, ensuring efficient energy conversion.

6.4 LEAD-FREE PIEZOELECTRIC MATERIALS

Lead-free piezoceramics have emerged as a vital alternative to traditional lead-based materials due to their environmentally friendly nature and potential for high-performance applications. Key types include perovskites, such as potassium sodium niobate (KNN) and barium titanate (BT), bismuth-based materials like sodium bismuth titanate (NBT), and alkali niobates. These materials exhibit promising piezoelectric and ferroelectric properties, making them suitable for various industrial applications (Yang et al., 2019).

The piezoelectric coefficient can be raised by altering the phase boundary composition and shifting the phase transition temperatures by adding dopant elements to the A-site and/or B-site in the KNN perovskite structure. Metal oxides can serve as sintering aids, increasing material densification and improving piezoelectric qualities, according to Zhang et al. (2020a).

A morphotropic phase barrier forms in barium calcium zirconium titanate (BCTZ), a BT-based material with calcium replaced at the A-site and zirconium at the B-site. It exhibits a high piezoelectric coefficient (d_{33} = 500–650 pC/N) and outstanding piezoelectric characteristics. It is a very promising material with a low coercive field and a big remnant polarization. The Curie temperature of BCTZ, however, is comparatively low at 83°C–100°C. Further enhancement of BCTZ's piezoelectric characteristics can be achieved by changing its transition temperature from tetragonal to orthorhombic close to the operational (ambient) temperature (Pal et al., 2018.

6.4.1 Lead-free versus lead-based materials: environmental and performance

The excellent piezoelectric characteristics of lead-based materials, such as lead zirconate titanate (PZT), have led to their widespread use. However, lead-free alternatives have been developed as a result of environmental concerns related to lead toxicity. Lead-free piezoceramics, while still improving, offer comparable performance without the environmental risks. The shift towards these materials helps reduce lead contamination and supports sustainable practices.

6.4.2 Perovskites, bismuth-based, and alkali niobates based key families

6.4.2.1 Perovskites

This family includes materials like KNN and BT, known for their excellent dielectric properties and potential for high-temperature applications. KNN, with its morphotropic phase boundary, shows enhanced piezoelectric performance. BT, on the other hand, is renowned for its high dielectric constant and application in capacitors.

6.4.2.2 Bismuth-based systems

NBT represents this category, notable for its high Curie temperature and robust ferroelectric properties. Despite challenges like high conductivity and dielectric losses, NBT remains a strong candidate for replacing lead-based materials, especially in high-temperature electromechanical systems.

6.4.2.3 Alkali Niobates-based ceramic systems

This group includes various compositions that exhibit good piezoelectric and ferroelectric properties. These materials are still under extensive research to optimize their performance and expand their applicability.

The development of lead-free piezoceramics marks a significant step towards environmentally sustainable and high-performance materials in the field of piezoelectric applications. Continuous research and a lot of innovation are being carried out to realize their potential and replace lead-based counterparts fully.

6.5 GENERAL METHODOLOGY AND CHARACTERIZATION OF LEAD-FREE PIEZOCERAMICS

Lead-free piezoceramics are synthesized using methods like solid-state reaction, sol-gel process, and hydrothermal techniques. Characterization of the ceramics involves various techniques to evaluate their electrical and mechanical properties, ensuring they meet the desired standards. Important methods for characterisation include scanning electron microscopy (SEM), X-ray diffraction (XRD), and dielectric and piezoelectric measurements. These combined methodologies and characterization techniques ensure the development of high-performance, environmentally friendly piezoceramics for diverse applications.

6.5.1 Solid-state reaction method

A popular technique used to prepare ferroelectric oxides with exceptional dielectric and piezoelectric characteristics is the solid-state reaction synthesis method. This irreversible technique employs the heat energy from milling to fuse powder compacts into dense polycrystalline ceramics. The decrease in interfacial free energy is the primary driving force. Since solid-state synthesis increases material density and hence improves piezoelectric and dielectric performance, it has been widely used to prepare KNN, BT, BNT, and BT-BF ceramics(Pal et al., 2023; Guo el al., 2024). In general, this process involves mixing and grinding raw powdered materials. After that, the mixture is heated to high temperatures in order to cause chemical reactions, followed by calcination and sintering to form dense ceramics. It is cost-effective and suitable for large-scale production, yielding ceramics with good mechanical strength and stability. The step-by-step process of the solid-state reaction method is illustrated in the flow chart for making lead-free ceramics.

6.5.1 Sol-gel process: In this process, metal alkoxides are hydrolysed to form a colloidal solution (sol). The sol undergoes gelation, forming a gel that is dried and calcined to obtain ceramic powder, which is then sintered. This offers precise control over the chemical composition and homogeneity of ceramics, producing fine, uniform powders at lower temperatures compared to solid-state reactions (Wang et al., 2021). The flow chart summarizes the steps involved in the sol-gel process for synthesizing lead-free piezoceramics.

6.5.2 Hydrothermal process

In this method, materials crystallize from aqueous solutions at elevated temperatures and pressures in a sealed vessel (autoclave), allowing for the development of individual crystals and nanostructured ceramics. In other words, the synthesis of complex and nanostructured lead-free piezoceramics is made possible by this processing method, and it offers exquisite control over crystallization (Fang et al., 2013). It is perfect for cutting-edge applications since it produces materials with improved qualities and is eco-friendly. The complete steps are summarized in the flow chart for synthesizing lead-free piezoceramics.

6.6 CHARACTERIZATION OF LEAD-FREE PIEZOCERAMICS

Lead-free piezoceramics are characterized using a variety of tools to assess their mechanical, electrical, and structural characteristics. An outline of important characterisation techniques is provided in Table 6.3.

These characterisation methods aid in the optimization of lead-free piezoceramics for particular uses, including energy harvesters, sensors, and actuators.

6.7 APPLICATIONS OF LEAD-FREE PIEZOELECTRICS IN ENERGY SOLUTIONS

Because of their superior functional qualities and environmentally friendly nature, lead-free piezoelectrics are essential to the advancement of energy

Table 6.3 Details characteristics, experimental tools with description of the properties for lead-free piezoceramics

Sl. No.	*Characteristics*	*Experimental Techniques*	*Details description of the properties*
1.	Crystal structure	X-ray diffraction (XRD)	Provides the lattice parameters, phase purity, and crystal structure.
2.	Microstructure	SEM, TEM	Examines grain size, grain boundaries, porosity, and overall microstructural features.
3.	Dielectric properties	Impedance spectroscopy	Measures dielectric constant, loss tangent, and frequency-dependent behaviour.
4.	Piezoelectric properties	d_{33} Meter and impedance analyser	Determines piezoelectric coefficients (d_{33}, d_{31}) and electromechanical coupling factors (k).
5.	Ferroelectric behavior	P-E Hysteresis loop analysis	Evaluates ferroelectric properties, remanent polarization (P_r), coercive field (E_c), and energy density.
6.	Thermal stability	DSC, TGA	Assesses thermal stability, phase transitions, and decomposition temperature.
7.	Mechanical properties	Nanoindentation, Vickers hardness test	Evaluates hardness, toughness, and elastic modulus.

solutions. They provide dependable performance and improved energy storage capacities in the area of energy storage devices, and they are used in capacitors and high-energy-density applications. Because of their high piezoelectric response, which makes it possible to convert mechanical-based energy into electrical energy efficiently, these materials are also excellent for usage in energy harvesting devices, including sensors, wearable electronics, and Internet of Things (IoT) systems.

6.7.1 Energy storage devices for ceramic capacitors and high-energy-density usages

For gadgets that store energy, especially ceramic capacitors used in high-energy-density applications, lead-free piezoelectric materials are gaining increasing interest. In general, ceramic materials act as dielectrics in the ceramic capacitors, whereas electrodes are made of conducting metals. Because of their easily accessible raw materials, straightforward design, affordability, and extensive range of electrical capabilities, these capacitors are frequently found in electronic equipment. Based on the dielectric materials utilized, they are mainly divided into Class I and Class II varieties. Class I ceramic capacitors, sometimes referred to as high-frequency ceramic capacitors, are appropriate for temperature compensation, filters, and resonant circuits due to their low dielectric loss, strong insulation resistance, and linear dielectric constant variation with temperature. Conversely, ferroelectric materials are used in Class II ceramic capacitors, sometimes referred to as low-frequency or ferroelectric ceramic capacitors. Higher specific capacitance and non-linear capacitance variations with temperature are features of ceramic capacitors, which use ceramics as dielectrics. They are perfect for bypass or coupling applications in electronic devices because of these features, as well as higher losses. They are divided into ceramic disk capacitors and multilayer ceramic capacitors (MLCCs) according to their structure and form. Recent studies have focused on enhancing MLCC performance. These materials are perfect for quick charge-discharge cycles and dependable performance because of their high dielectric permittivity, minimal energy loss, and superior thermal stability. They are commonly used in renewable energy systems, power electronics, and electric cars, where effective energy distribution and storage are essential. Lead-free alternatives with minimal hysteresis loss and strong breakdown strength, like potassium sodium niobate and barium titanate (BT), promise environmental safety. They are essential to the development of high-efficiency and sustainable energy storage technologies because of their exceptional performance and environmental friendliness.

6.7.2 Ceramic disc capacitors

Since they are inexpensive and straightforward to solder, they are widely utilized in conventional electrical circuits. The area of the ceramic disc or dielectric and the distance between the silver electrodes determine the capacitance of these capacitors. While multilayer ceramic discs are necessary for high-capacitance ceramic disc capacitors, a single ceramic disc coated with silver contacts is adequate for disc capacitors with lower capacitance. Ceramic discs are prepared using conventional ceramic processing techniques. The dielectric powder is first disaggregated by ball milling. The powder is then made ready for formation by adding an organic binder. The powder is placed into an automatic granulator to form it into a raw disk after it has been spray-dried. The final product is produced by sintering the ceramic after two firing cycles. The first of these cycles burns out the binder. The electrodes coated on both sides of the ceramic dielectric with an outer protective layer are visible in the cross-sectional image of a ceramic disc capacitor when seen under a scanning electron microscope. These capacitors use a ceramic disc, on which a silver coating is deposited on both sides, which act as electrode plates. These capacitors are made from 10 pF to 100 μF and have voltage ratings from 16 V to 20 kV or above. A single disc of 3–6 mm is generally used for low capacitance values. Several layers can be used for higher capacitance values. Ceramic discs were initially made with holes in the center, a design now obsolete. Disc capacitors have voltage ratings from 16 V to several kV, as shown in Figure 6.3. The capacitors are encapsulated and hermetically sealed in epoxy resins (Pan and Randall, 2010). Figure 6.3 shows the (a) Photographical image of MLCC, (b) Multilayer Ceramic Capacitors, (c) Disc capacitor, and (d) diverse applications of ceramic capacitors.

6.8 MULTILAYER CERAMIC CAPACITORS (MLCC)

MLCC capacitors are made by creating several layers of ceramic material, interspersed with alternating metal layers, thus forming a stack of several capacitors in parallel. The layers are extremely thin, at the micron level. Capacitors formed this way are very small, and end connections are directly soldered to printed circuits. MLCC may use over 500 layers, employing thin film and multilayer technologies, with a thickness as low as 0.5 μm. With the improvement in technology, thinner layers and smaller sizes have become possible. These capacitors can replace electrolytic capacitors in a number of places, where the nonpolar nature of MLCC is an advantage. Developments over decades have brought down costs to pennies level, and sizes to unimaginably low levels. Ceramic capacitors have an extensive range of applications. Capacitors of the size of a grain of sand are available. A few of the applications of ceramic capacitors are displayed below. However, the list is never-ending. From coupling/decoupling, filtering, resonance, and so many other functions are served by these capacitors in an enormous array of applications all over the world (Park et al., 2023; Lv et al., 2024).

6.9 ENERGY HARVESTING DEVICES: WEARABLE ELECTRONICS, SENSORS, AND IOT

In modern technology, energy harvesting devices are essential, especially for wearable electronics, sensors, and the Internet of Things (IoT). Small electronic systems can be powered by the electrical energy produced by these devices, which gather and transform ambient energy from sources like light, heat, and motion. Energy harvesting in sensors improves dependability and lowers maintenance by allowing continuous operation without the need for regular battery replacements. Energy harvesting offers wearable electronics like smartwatches and fitness trackers a sustainable power source, enhancing user convenience and gadget longevity. Energy harvesting ensures reliable data gathering and communication in Internet of Things applications by enabling the placement of several sensors and devices in hard-to-reach or remote locations without the need for conventional power sources. Traditional batteries restrict the capabilities of wearable electronics, which are revolutionizing robotics, informatics, and health monitoring. Energy harvesting is a sustainable solution that converts ambient energy from the body or the environment into electrical power. Wearable energy harvesting systems, utilizing body-based energy sources such as light, metabolic chemicals, biomechanics, and heat, are essential for powering portable electronics, particularly for sensors and health monitoring.

Electronics and sensors for fitness tracking, health monitoring, and portable gadgets are powered by wearable energy harvesters. Power management units (PMUs) maximize energy conversion and distribution for wearable applications by storing the energy harvester's output in capacitors or batteries (Xu et al., 2021; Dai et al., 2023).

TENGs (triboelectric nanogenerators) and PEGs (piezoelectric generators) convert mechanical deformation into high-impedance AC power. TENGs exhibit delayed output due to surface charge buildup, while PEGs generate power instantly upon deformation. When mechanically deformed, TENGs and PEGs both generate high-impedance AC power; however, TENGs exhibit delayed output because of surface charge accumulation. PV cells and BFCs generate low-impedance DC power from light and biofluid due to their variable input; this power is categorized as intermittent. TEGs continually produce low-impedance DC electricity using body heat, which is influenced by environmental and physiological factors. The type, materials, and structure of harvesters affect how long they work. In addition to powering electronics, wearable energy harvesters can also be utilized as self-powered sensors. Wang and associates (2020). Since these sensors get their energy from the user's body or the surroundings, they don't need an external power source.

6.9.1 Wearable Triboelectric Energy Harvesting

Triboelectric nanogenerators (TENGs) are effective wearable energy harvesters that use the triboelectric effect to transform biomechanical energy from motions like bending, tapping, or sliding into electrical power. They have straightforward designs and a high output power. TENGs maximize charge transfer through material selection and design by utilizing multiple operating modes, including contact-separation, sliding, single-electrode, and freestanding, as shown in the Figure 6.4. For instance, a wearable sweat sensor with a freestanding-mode TENG employed arm movement to generate power for a PSoC module and enable Bluetooth data transmission (Song et al., 2021). This power was controlled by a PMIC and discharged at a voltage threshold. The capacitor's charge/discharge cycle lasted between 2.1 and 3.7 minutes while the treadmill ran at 9 km/h, demonstrating the viability of triboelectric energy harvesting for continuous operation of wearable sensors.

6.9.2 Wearable piezoelectric energy harvesting

In addition to triboelectric nanogenerators (TENGs), piezoelectric nanogenerators (PENGs) are potential biomechanical energy harvesters (Zou et al., 2020; Mendez-Lira et al., 2021; Choi el al., 2017). PENGs create an electric potential difference through lattice deformation, converting mechanical stress into electric charges. This process is based on the piezoelectric phenomenon seen in materials such as zinc oxide, quartz, ceramics,

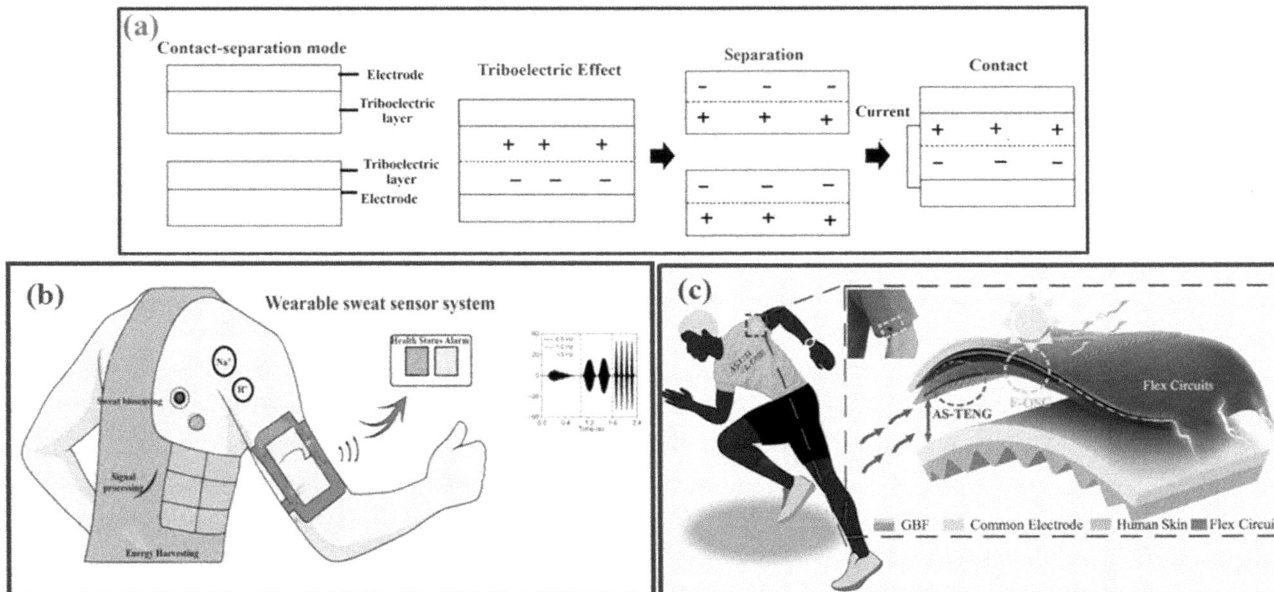

Figure 6.4 Wearable biomechanical energy harvesting: (a) contact-separation-mode structure and operation TENGS. (b) Freestanding mode wearable sweat sensor powered by TENG (Song et al., 2021). (c) Triboelectric all-textile sensor array for physiological monitoring (Fan et al., 2020). (d) Flexible hybrid energy harvester combining TENG and organic PV(**d**) Schematic representation of the organic photovoltaic cell, TENG, and flexible electronic circuitry that make up the flexible hybrid energy harvester. (Source: Reproduced with permission from Ren et al., 2020.)

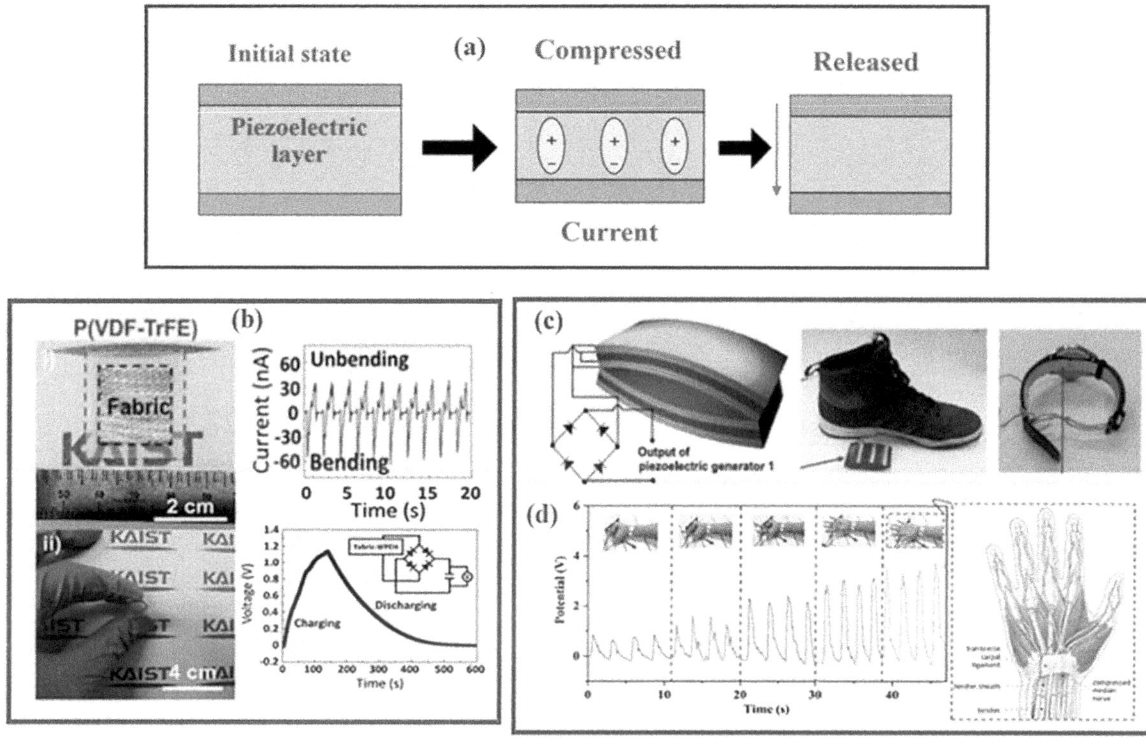

Figure 6.5 (a) PENGs' schematic layout and operational principles. (b) A fabric-based wearable PENG's charging curve for a 1 μF capacitor, output current, and photograph (Kim et al., 2020). Its dimensions and bendability are shown in (i) and (ii), respectively. (c) Curved multilayer PENG's schematic construction and applications in watch straps and shoe insoles. (Source: Reproduced with permission from Jung et al., 2015.) (d) Nanowire PENG-based self-powered sensor to track finger movements. (Source: Reproduced with permission from Waseem et al., 2021.)

and polymers (Figure 6.5a). Beca-se of their high crystallinity, dipole configuration, and flexibility, flexible piezoelectric polymers like polyvinylidene fluoride (PVDF) and P(VDF-TrFE) are frequently utilized in wearable applications (Jung et al., 2015; Lee et al., 2014). Kim and colleagues (2020a) developed a fabric-based wearable PENG that produced a piezoelectric d_{33} coefficient of −32 pC/N using conductive fabrics and a sandwich construction of P(VDF-TrFE) (75/25). The PENG produced a maximum voltage of 4.5 V and a current of 69 nA when finger pressure and elbow bending were applied, and it remained robust for 15,000 cycles. It effectively charged a 1-µF capacitor at about 3 Hz with a peak output of 16.83 nW·cm^2 at 60 MΩ load resistance, demonstrating its potential for wearable electronics.

By incorporating PVDF multilayers into a curved framework, Jung et al. (2015) created a curve-based piezoelectric generator for wearable-type applications that achieved a maximum output of 120 V and 0.7 mA. In 10 seconds, the device charged a 47 µF capacitor to 5 V with an instantaneous power density of 3.9 mW·cm^2. Applications showed adaptability by including PENG installation in watch bands, chest mounts, and insoles, as well as battery charging with power management circuits. Waseem et al. (2021) studied a GaN:Mg/ZnO nanowire-based PENG (as shown in Figure 6.5d) with enhanced performance by ZnO shell deposition and nanowire diameter reduction, attaining outputs of 40 µA and 66 V. The PENG has shown potential for biomechanical applications by successfully sensing finger movement.

A hybrid piezoelectric–pyroelectric type energy harvester employing flexible PDMS-CNT electrodes and micropatterned P(VDF-TrFE) was presented by Lee et al. (2014). Using a coupling effect between pyroelectric and piezoelectric phenomena, this skin-conformable device produced separate and combined outputs from heating-cooling and stretching-releasing cycles for enhanced energy harvesting.

6.10 CHALLENGES AND FUTURE PROSPECTS

Although lead-free piezoelectric materials hold promise, their widespread application is hampered by a variety of issues. Because their dielectric and piezoelectric properties usually fall short of those achieved by their lead-based counterparts, their efficacy in high-performance applications is limited. High energy density, thermal stability, and long-term dependability remain challenges. Additionally, producing and processing lead-free materials can be challenging and costly, requiring precise control over composition and microstructure. Scalability for industrial applications and compatibility with existing device designs remain significant obstacles. These limitations need to be overcome by material engineering and advanced production methods to fully realize their potential as sustainable energy sources.

Scaling lead-free piezoelectric for commercial energy systems requires resolving problems with material production, performance optimization, and cost-effectiveness. Adoption of these materials is contingent upon establishing suitable production procedures, guaranteeing consistent quality, and integrating them into current designs. Reliability, environmental compliance, and application-specific performance enhancements are equally crucial. By utilizing sustainable manufacturing processes and collaborating with industry, politicians, and researchers, companies can expedite commercialization. These advancements position lead-free piezoelectric as a key element of efficient and ecologically friendly energy solutions for global sustainable development.

Enhancing energy conversion efficiency using cutting-edge materials, creative designs, and hybrid systems that combine several energy sources is necessary to maximize output power in wearable energy harvesters (Zou et al., 2020; Xu et al., 2021; Chen et al., 2021; Jung et al., 2015). Maintaining mechanical integrity and reliable performance in small-scale devices is one of the problems of miniaturization, which calls for stable energy output strategies and strong, flexible materials (Song et al., 2021; Ren et al., 2020; Zhao et al., 2020).

Flexible, long-lasting materials that can withstand repeated usage without losing their functionality are necessary to ensure wearability and durability in wearable energy harvesters (Choi et al., 2017; Yuan and Zhu, 2020). To increase longevity under a variety of circumstances, standardized manufacturing procedures and cutting-edge designs are crucial. To ensure optimal operation and extend device lifespan, advanced PMUs and algorithms that balance irregular energy intake and optimize storage are essential for effective power management (Yin et al., 2022).

In order to provide consistent data without requiring external power, real-time monitoring mainly depends on reliable self-powered sensors, which require complex energy management and stability (Wang et al., 2020). Furthermore, the multifunctional integration of energy harvesters, sensors, batteries, and circuits requires innovative designs that maintain functionality while ensuring compactness and lightweight systems (Yin et al., 2022). Future studies should concentrate on improving lead-free piezoelectric performance through the creation of composites, improved doping, and nano-structuring.

The development of hybrid energy systems that incorporate scalable, cost-effective synthesis methods along with storage and harvesting capabilities is crucial. The development of sustainable energy solutions will prioritize application-specific enhancements, environmental compliance, and robustness.

6.11 CONCLUSION

The transition to sustainable energy systems is driving the development of new materials capable of efficiently storing and harvesting energy. Dielectric and piezoelectric materials have become critical to satisfy the increased need for high-performance energy devices. Due to their high energy density and dependability, dielectric materials excel at storing energy, while piezoelectric materials enable mechanical-to-electrical energy conversion, making them indispensable for energy harvesting applications. A critical step toward ecologically friendly solutions is the creation of lead-free piezoelectric materials. These materials, which are devoid of the hazardous effects of lead, offer a sustainable option without sacrificing performance. Bismuth-based and alkali niobate piezoceramics are among the most promising possibilities, with outstanding qualities for use in capacitors, sensors, and hybrid power systems. By providing practical, environmentally friendly solutions for energy harvesting and storage, lead-free piezoelectric and dielectric materials are transforming sustainable energy systems. By lowering environmental impact and facilitating the switch to sustainable and renewable energy systems, their integration supports worldwide energy goals.

ACKNOWLEDGEMENTS

The author (V. Pal) acknowledges the I-DAPT Hub Foundation for funding the DAPT Chanakya Postdoctoral Fellowship at IIT (BHU), Varanasi.

REFERENCES

Badapanda, T., Chaterjee, S., Mishra, A., Ranjan, R., & Anwar, S. (2017). Electric field induced strain, switching and energy storage behaviour of lead free Barium Zirconium Titanate ceramic. *Phys. B Condens. Mater.* 521, 264–269.

Chen, H. Y., Shi, J. P., Chen, X. L., Sun, C. C., Pang, F. H., Dong, X. Y., Zhang, H. L., & Zhou, H. F. (2021). Excellent energy storage properties and stability of $NaNbO_3$-$Bi(Mg_{0.5}Ta_{0.5})O_3$ ceramics by introducing $(Bi_{0.5}Na_{0.5})_{0.7}Sr_{0.3}TiO_3$. *J. Mater. Chem. A.* 9(8), 4789.

Chen, J., Wang, Y. F., Yuan, Q. B., Xu, X. W., Niu, Y. J., & Wang, Q. (2018). Multilayered ferroelectric polymer films incorporating low-dielectric-constant components for concurrent enhancement of energy density and charge–discharge efficiency. *Nano Energy* 54, 288–297.

Choi, Y.-M., Lee, M., & Jeon, Y. (2017). Wearable biomechanical energy harvesting technologies. *Energies* 10, 1483.

Dai, K., Huo, Z.-Y., Miao, X., Xiong, P., Zhang, H., Wang, X., & Kim, S.-W. (2023). Self-powered triboelectric functional devices and microsystems in healthcare applications: An energy perspective. *EnergyChem* 5, 100109.

Damjanovic, D. (1998). Ferroelectric, dielectric and piezoelectric properties of ferroelectric thin films and ceramics. *Rep. Progr. Phy.* 61(9), 1267–1324.

Fang, X.-Q., Liu, J.-X., & Gupta, V. (2013). Fundamental formulations and recent achievements in piezoelectric nano-structures: A review. *Nanoscale* 5, 1716.

Fu, D., Itoh, M., & Koshihara, S. (2008). Crystal growth and piezoelectricity of $BaTiO_3$-$CaTiO_3$ solid solution. *Appl. Phys. Lett.* 93, 012904.

Guo, B., Yan, Y., Tang, M. Y., Wang, Z. Y., Li, Y., Zhang, L. Y., Zhang, H. B., Jin, L., & Liu, G. (2021). Energy storage performance of $Na_{0.5}Bi_{0.5}TiO_3$ based lead-free ferroelectric ceramics prepared via non-uniform phase structure modification and rolling process. *Chem. Eng. J.* 420, 130475.

Hadouch, Y., Moumen, S. B., Mezzourh, H., Mezzane, D., Amjoud, M., Asbani, B., Razumnaya, A. G., Gagou, Y., Rozic, B., Kutnjak, Z., & Marssi, M. E. (2022). Electrocaloric effect and high energy storage efficiency in lead-free $Ba_{0.95}Ca_{0.05}Ti_{0.89}Sn_{0.11}O_3$ ceramic elaborated by sol-gel method. *J. Mater. Sci. Mater. Electron.* 33, 2067–2079.

Han, Z., Mezzane, D., Amjoud, M., Razumnaya, A. G., Fourcade, S., & Marssi, M. E. (2019). Phase transitions, energy storage performances, and electrocaloric effect of the lead-free $Ba_{0.85}Ca_{0.15}Zr_{0.10}Ti_{0.90}O_3$ ceramic relaxor. *J. Mater. Sci. Mater. Electron.* 30, 6430–6444. https://doi.org/10.1007/s10854-019-01136-8

Hanani, Z., Mezzane, D., Amjoud, M., Razumnaya, A. G., Fourcade, S., Gagou, Y., Hoummada, K., Marssi, M. E., & Goune, M. (2019). Phase transitions, energy storage performances and electrocaloric effect of the lead-free $Ba_{0.85}Ca_{0.15}Zr_{0.10}Ti_{0.90}O_3$ ceramic relaxor. *J. Mater. Sci. Mater. Electron.* 30, 6430–6438.

Hu, D., Pan, Z. B., Zhang, X., Ye, H. R., He, Z. Y., Wang, M. K., Xing, S., Zhai, J. W., Fu, Q., & Liu, J. J. (2020). Greatly enhanced discharge energy density and efficiency of novel relaxation ferroelectric BNT–BKT-based ceramics. *J. Mater. Chem. C.* 8(2), 591.

Hu, Q. Y., Tian, Y., Zhu, Q. S., Bian, J. H., Jin, L., & Wei, X. Y. (2020). Achieve ultrahigh energy storage performance in $BaTiO_3$-$Bi(Mg_{1/2}Ti_{1/2})O_3$ relaxor ferroelectric ceramics via nano-scale polarization mismatch and reconstruction. *Nano Energy* 67, 104264.

Ji, H. F., Wang, D. W., Bao, W. C., Lu, Z. L., Wang, G., Yang, H. J., Mostaed, A., Li, L. H., Feteira, A., Sun, S. K., Xu, F. F., Li, D. J., Ma, C. J., Liu, S. Y., & Reaney, I. M. (2021). Ultrahigh energy density in short-range tilted NBT-based lead-free multilayer ceramic capacitors by nanodomain percolation. *Energy Storage Mater.* 38, 113.

Jung, W.-S., Lee, M.-J., Kang, M.-G., Moon, H. G., Yoon, S.-J., Baek, S.-H., & Kang, C.-Y. (2015). Powerful curved piezoelectric generator for wearable applications. *Nano Energy* 13, 174–181.

Kim, J., Byun, S., Lee, S., Ryu, J., Cho, S., Oh, C., Kim, H., No, K., Ryu, S., Lee, Y. M., et al. (2020a). Cost-effective and strongly integrated fabric-based wearable piezoelectric energy harvester. *Nano Energy* 75, 104992.

Kim, J., Saremi, S., Acharya, M., Velarde, G., Parsonnet, E., & Martin, L. W. (2020b). Ultrahigh capacitive energy density in ion-bombarded relaxor ferroelectric films. *Science* 369(6499), 81–85.

Lee, J.-H., Lee, K. Y., Gupta, M. K., Kim, T. Y., Lee, D.-Y., Oh, J., Ryu, C., Yoo, W. J., Kang, C.-Y., Yoon, S.-J., et al. (2014). Highly stretchable piezoelectric-pyroelectric hybrid nanogenerator. *Adv. Mater.* 26, 765–769.

Li, D., Lin, Y., Yuan, Q. B., Zhang, M., Ma, L., & Yang, H. B. (2020). A novel lead-free $Na_{0.5}Bi_{0.5}TiO_3$-based ceramic with superior comprehensive energy storage and discharge properties for dielectric capacitor applications. *J. Materiomics.* 6(4), 743.

Li, L. S., Fan, P. Y., Wang, M. Q., Takesue, N., Salamon, D., Vtyurin, A. N., Zhang, Y. J., Tan, H., Nan, B., Lu, Y., Liu, L. J., & Zhang, H. B. (2021). Review of lead-free Bi-based dielectric ceramics for energy-storage applications. *J. Phys. D Appl. Phys.* 54(29), 293001.

Li, Q., Chen, L., Gadinski, M. R., Zhang, S., Li, H., & Wang, Q. (2015). Flexible high-temperature dielectric materials from polymer nanocomposites. *Nature 523*(7562), 576–580.

Li, T., & Lee, P. S. (2022). Piezoelectric energy harvesting technology: From materials, structures, to applications. *Small Struct.* 3, 2100128.

Lin, Y., Li, D., Yuan, Q., Zhang, M., Ma, L., & Yang, H. (2019). Excellent energy-storage properties achieved in $BaTiO_3$-based lead-free relaxor ferroelectric ceramics via domain engineering on the nanoscale. *ACS Appl. Mater. Interfaces* 11(40), 36824–36833.

Lu, Z. L., Wang, G., Bao, W. C., Li, J. L., Li, L. H., Mostaed, A., Yang, H. J., Ji, H. F., Li, D. J., Feteira, A., Xu, F. F., Sinclair, D. C., Wang, D. W., Liu, S. Y., & Reaney, I. M. (2020). Superior energy density through tailored dopant strategies in multilayer ceramic capacitors. *Energy Environ. Sci.* 13(9), 2938.

Luo, N. N., Han, K., Cabral, M. J., Liao, X. Z., Zhang, S. J., Liao, C. Z., Zhang, G. Z., Chen, X. Y., Feng, Q., Li, J. F., & Wei, Y. Z. (2020). Constructing phase boundary in $AgNbO_3$ antiferroelectrics: Pathway simultaneously achieving high energy density and efficiency. *Nat. Commun.* 11(1), 1.

McPherson, J. W., Kim, J. Y., Shanware, A., Mogul, H., & Rodriguez, J. (2003). Thermochemical description of dielectric breakdown in high dielectric constant materials. *Appl. Phys. Lett.* 82(13), 1771–1778.

Mendez-Lira, O., Spinelli, E. M., & Gonzalez-Landaeta, R. (2021). Battery-less power management circuit powered by a wearable piezoelectric energy harvester. *Energy Technol.* 9, 2100520.

Mezzourh, H., Belkhadir, S., Mezzane, D., Amjoud, M., Choukri, E., Lahmar, A., Gagou, Y., Kutnjak, Z., & Marssi, M. E. (2021). Enhancing the dielectric, electrocaloric and energy storage properties of lead-free $Ba_{0.85}Ca_{0.15}Zr_{0.1}Ti_{0.9}O_3$ ceramics prepared via sol-gel process. *Phys. B Condens. Matter.* 603, 412760.

Naqvi, A., Ali, A., Altabey, W. A., & Kouritem, S. A. (2022). Energy harvesting from fluid flow using piezoelectric materials: A review. *Energies* 15, 7424.

Pal, V., Dwivedi, R. K., & Thakur, O. P. (2014a). Effect of neodymium substitution on structural and ferroelectric properties of BNT ceramics. *Mater. Res. Bull.* 51, 189–196. https://doi.org/10.1016/j.materresbull.2013.11.015

Pal, V., Dwivedi, R. K., & Thakur, O. P. (2014b). Synthesis and ferroelectric behavior of Gd doped BNT ceramics. *Curr. Appl. Phys.* 14, 99–107. https://doi.org/10.1016/j.cap.2013.11.001

Pal, V., Dwivedi, R. K., & Thakur, O. P. (2018). Structural investigation of Ca/Zr co-substituted $BaTiO_3$ through XRD and Raman spectroscopy. *J. Alloys Compd.* 741, 707–714.

Pal, V., Kumar, A., Thakur, O. P., Dwivedi, R. K., & Prasad, N. E. (2017). Preparation, microstructure and relaxor ferroelectric characteristics of BLNT–BCT lead-free piezoceramics. *J. Alloys Compd.* 714, 725–735.

Pal, V., Kumar, D., & Singh, A. K. (2023). Crystal structure, microstructure, and ultra-high energy storage properties of lead-free $(x)Bi(Mg_{0.5}Ti_{0.5})O_{3-(1-x)}[0.50Ba(Zr_{0.2}Ti_{0.8})O_{3-0.50}(Ba_{0.7}Ca_{0.3})TiO_3]$ ternary ceramics. *Mater. Sci. Eng. B* 288, 116194. https://doi.org/10.1016/j.mseb.2022.116194

Palneedi, H., Peddigari, M., Hwang, G. T., Jeong, D. Y., & Ryu, J. (2018). High-performance dielectric ceramic films for energy storage capacitors: Progress and outlook. *Adv. Funct. Mater.* 28(42), 1803665.

Pan, M.-J., & Randall, C. A. (2010). A brief introduction to ceramic capacitors. *IEEE Electr. Insul. Mag.* 26, 44–50.

Pan, Z. B., Wang, P., Hou, X., Yao, L. M., Zhang, G. Z., & Wang, J. (2020). Fatigue-free aurivillius phase ferroelectric thin films with ultrahigh energy storage performance. *Adv. Energy Mater.* 10(31), 2001536.

Park, Y., Park, J. J., Park, K. S., Hong, Y. M., Lee, E. J., Kim, S. O., & Lee, J. H. (2023). Enhancement of mechanical properties of multilayer ceramic capacitors through a $BaTiO_3$/polydopamine cover layer. *Polymers* 15, 4014.

Puli, V. S., Pradhan, D. K., Chrisey, D. B., Tomozawa, M., Sharma, G. L., Scott, J. F., & Katiyar, R. S. (2013). Structure, dielectric, ferroelectric, and energy density properties of (1-x)BZT-xBCT ceramic capacitors for energy storage applications. *J. Mater. Sci.* 48, 2151–2157.

Qi, H., Xie, A. W., Tian, A., & Zuo, R. Z. (2019). Superior energy-storage capacitors with simultaneously giant energy density and efficiency using nanodomain engineered $BiFeO_3$-$BaTiO_3$-$NaNbO_3$ lead-free bulk ferroelectrics. *Adv. Energy Mater.* 10(6), 1903338.

Qiao, X. S., Zhang, F. D., Wu, D., Chen, B., Zhao, X. M., Peng, Z. H., Ren, X. D., Liang, P. F., Chao, X. L., & Yang, Z. P. (2020). Superior comprehensive energy storage properties in $Bi_{0.5}Na_{0.5}TiO_3$-based relaxor ferroelectric ceramics. *Chem. Eng. J.* 388, 124158.

Ren, Z., Zheng, Q., Wang, H., Guo, H., Miao, L., Wan, J., Xu, C., Cheng, S., & Zhang, H. (2020). Wearable and self-cleaning hybrid energy harvesting system based on micro/nanostructured haze film. *Nano Energy* 67, 104243.

Sezer, N., & Koç, M. (2021). A comprehensive review on the state-of-the-art of piezoelectric energy harvesting. *Nano Energy* 80, 105567.

Shi, J. P., Chen, X. L., Li, X., Sun, J., Sun, C. C., Pang, F. H., & Zhou, H. F. (2020). Realizing ultrahigh recoverable energy density and superior charge-discharge performance in $NaNbO_3$-based lead-free ceramics via a local random field strategy. *J. Mater. Chem. C.* 8(11), 3784.

Song, Y., Mukasa, D., Zhang, H., & Gao, W. (2021). Self-powered wearable biosensors. *Acc. Mater. Res.* 2, 184–197.

Wang, G., Lu, Z. L., Li, Y., Ji, H. F., Feteira, A., & Reaney, I. M. (2021). Electroceramics for high-energy density capacitors: Current status and future perspectives. *Chem. Rev.* 121(10), 6124–6141.

Wang, G., Lu, Z. L., Yang, H. J., Ji, H. F., Mostaed, A., Li, L. H., Wei, Y. Q., Feteira, A., Sun, S. K., Sinclair, D. C., Wang, D. W., & Reaney, I. M. (2020). Fatigue resistant lead-free multilayer ceramic capacitors with ultrahigh energy density. *J. Mater. Chem. A.* 8(22), 11414.

Wang, T., Jin, L., Li, C. C., Hu, Q. Y., & Wei, X. Y. (2015). Relaxor ferroelectric $BaTiO_3$-$Bi(Mg_{2/3}Nb_{1/3})O_3$ ceramics for energy storage application. *J. Am. Ceram. Soc.* 98(2), 559.

Wang, Y., Wu, H., Xu, L., Zhang, H., Yang, Y., & Wang, Z. L. (2020). Hierarchically patterned self-powered sensors for multifunctional tactile sensing. *Sci. Adv.* 6, eabb9083.

Waseem, A., Johar, M. A., Abdullah, A., Bagal, I. V., Ha, J.-S., Lee, J. K., & Ryu, S.-W. (2021). Enhanced performance of a flexible and wearable piezoelectric nanogenerator using semi-insulating GaN:Mg/ZnO coaxial nanowires. *Nano Energy* 90, 106552.

Xu, C., Song, Y., Han, M., & Zhang, H. (2021). Portable and wearable self-powered systems based on emerging energy harvesting technology. *Microsyst. Nanoeng.* 7, 25.

Yan, F., Huang, K. W., Jiang, T., Zhou, X. F., Shi, Y. J., Ge, G. L., Shen, B., & Zhai, J. W. (2020). Significantly enhanced energy storage density and efficiency of BNT-based perovskite ceramics via A-site defect engineering. *Energy Storage Mater.* 30, 392.

Yang, H. J., Lu, Z. L., Li, L. H., Bao, W. C., Ji, H. F., Li, J. L., Feteira, A., Xu, F. F., Zhang, Y., Sun, H. J., Huang, Z. C., Lou, W. C., Song, K. X., Sun, S. K., Wang, G., Wang, D. W., & Reaney, I. M. (2020). Novel $BaTiO_3$-based, Ag/nPd-compatible lead-free relaxors with superior energy storage performance. *ACS Appl. Mater. Interfaces* 12(39), 43942.

Yang, L. T., Kong, X., Li, F., Hao, H., Cheng, Z. X., Liu, H. X., Li, J. F., & Zhang, S. J. (2019). Perovskite lead-free dielectrics for energy storage applications. *Progr. Mater. Sci.* 102, 72.

Yao, Z. H., Song, Z., Hao, H., Yu, Z. Y., Cao, M. H., Zhang, S. J., Lanagan, M. T., & Liu, H. X. (2017). Homogeneous/inhomogeneous-structured dielectrics and their energy-storage performances. *Adv. Mater.* 29(20), 1601727.

Ye, J. M., Wang, G. S., Zhou, M. X., Liu, N. T., Chen, X. F., Li, S., Cao, F., & Dong, X. L. (2019). Excellent comprehensive energy storage properties of novel lead-free $NaNbO_3$-based ceramics for dielectric capacitor applications. *J. Mater. Chem. C.* 7(19), 5639.

Yin, L., Kim, K. N., Trifonov, A., Podhajny, T., & Wang, J. (2022). Designing wearable microgrids: Towards autonomous sustainable on-body energy management. *Energy Environ. Sci.* 15, 82–101.

Yuan, Q. B., Li, G., Yao, F. Z., Cheng, S. D., Wang, Y. F., Ma, R., Mi, S. B., Gu, M., Wang, K., Li, J. F., & Wang, H. (2018). Simultaneously achieved temperature-insensitive high energy density and efficiency in domain engineered $BaTiO_3$-$Bi(Mg_{0.5}Zr_{0.5})O_3$ lead-free relaxor ferroelectrics. *Nano Energy.* 52, 203.

Yuan, J., & Zhu, R. (2020). A fully self-powered wearable monitoring system with systematically optimized flexible thermoelectric generator. *Appl. Energy* 271, 115250.

Zhan, D., Xu, Q., Huang, D. P., Liu, H. X., Chen, W., & Zhang, F. (2018). Contributions of intrinsic and extrinsic polarization species to energy storage properties of $Ba_{0.95}Ca_{0.05}Zr_{0.2}Ti_{0.8}O_3$ ceramics. *J. Phys. Chem. Solids* 114, 220–227.

Zhang, G., Zhang, S., & Wang, Q. (2022). Dielectric materials for electrical energy storage. *J. Materiomics* 8(6), 1287–1289.

Zhang, M., Yang, H. B., Li, D., Ma, L., & Lin, Y. (2020a). Giant energy storage efficiency and high recoverable energy storage density achieved in $K_{0.5}Na_{0.5}NbO_3$-$Bi(Zn_{0.5}Zr_{0.5})O_3$ ceramics. *J. Mater. Chem. C.* 8(26), 8777.

Zhao, P. Y., Cai, Z. M., Chen, L. L., Wu, L. W., Guo, L. M., & Wang, H. (2020). Ultra-high energy storage performance in lead-free multilayer ceramic capacitors via a multiscale optimization strategy. *Energy Environ. Sci.* 13(12), 4882–4894.

Zhou, M. X., Liang, R. H., Zhou, Z. Y., Yan, S. G., & Dong, X. L. (2018). Novel sodium niobate-based lead-free ceramics as new environment-friendly energy storage materials with high energy density, high power density, and excellent stability. *ACS Sustain. Chem. Eng.* 6(10), 12755–12765.

Zou, Y., Raveendran, V., & Chen, J. (2020). Wearable triboelectric nanogenerators for biomechanical energy harvesting. *Nano Energy* 77, 105303.

Chapter 7

Advanced substrate treatment methods and additives for enhancing efficiency in biogas plants

Daria Sławczyk, Beata Bień, and Anna Grobelak

7.1 INTRODUCTION

The global use of non-renewable energy sources such as oil, coal, and natural gas is dependent on the availability of a specific resource. The availability of fossil fuels, in turn, is declining due to overexploitation (Dhull et al., 2024). In addition, the use of non-renewable energy sources leads to the emergence of environmental problems, contributing to climate change and environmental degradation (Obileke et al., 2024).

Anaerobic digestion, the basis for producing electricity, heat, and organic fertilizers from agricultural substrates in biogas plants, is a practical, environmentally safe, and economically profitable process. However, the efficiency of biogas production depends on the feedstock composition, applied pre-treatment or co-fermentation technology, and the process conditions, which should be properly controlled and modified in order to optimize the fermentation process. Moreover, there is no comprehensive summary of the effectiveness of using above-mentioned techniques in relation to a specific type of feedstock (Kulichkova et al., 2020).

Agricultural biogas production currently plays a significant role in the ongoing energy transformation, which encourages the broader adoption of renewable energy sources (RES). It should be noted that one of the beneficial RES is biomass and biogas (Venturin et al., 2019). Moreover, obtaining good-quality digestate as a result of microbiological anaerobic processing of agri-food waste allows it to be used as a fertilizer, which contributes to soil conditioning and, at the same time, is in line with the idea of a closed-loop economy (CE) (Qayyum et al., 2024). According to this idea, the consumption of primary raw materials with limited resources should be reduced, and the level of waste processing, especially that with energy potential, should be increased (Kulichkova et al., 2020).

Traditional biogas production processes are associated with many problems that arise during substrate processing. These include, among other things, long hydrolysis resulting from limited accessibility of microorganisms to the surface of unpretreated substrate and the appearance of reaction

 DOI: 10.1201/9781003581536-7

inhibitors such as ammonia, hydrogen sulphide, ethylene, halogen aliphatic hydrocarbons, and aliphatic nitro compounds (Czatzkowska et al., 2020; Gu et al., 2024). Limited conversion of lignocellulosic biomass (LCB) necessitates the development of economically advantageous and efficient methods, along with optimized process conditions (Venturin et al., 2019). Biogas upgrading technologies should also be improved in order to inject the produced biomethane into energy grids (Zhang et al., 2020).

7.2 MODERN METHODS OF PROCESSING FEEDSTOCK IN BIOGAS PLANTS

Every year, agriculture generates billions of tons of organic waste. Their proper management is important due to the threats that may result from the lack of an implemented waste disposal system (Gu et al., 2024). In this aspect, it is worth considering the possibilities of using agricultural waste as a feedstock in biogas plants for the production of agricultural biogas, as shown in Figure 7.1. Although plant residues are not strictly intended for energy purposes, they provide an effective source of biomass for processing. It should also be mentioned that waste storage is associated with the release

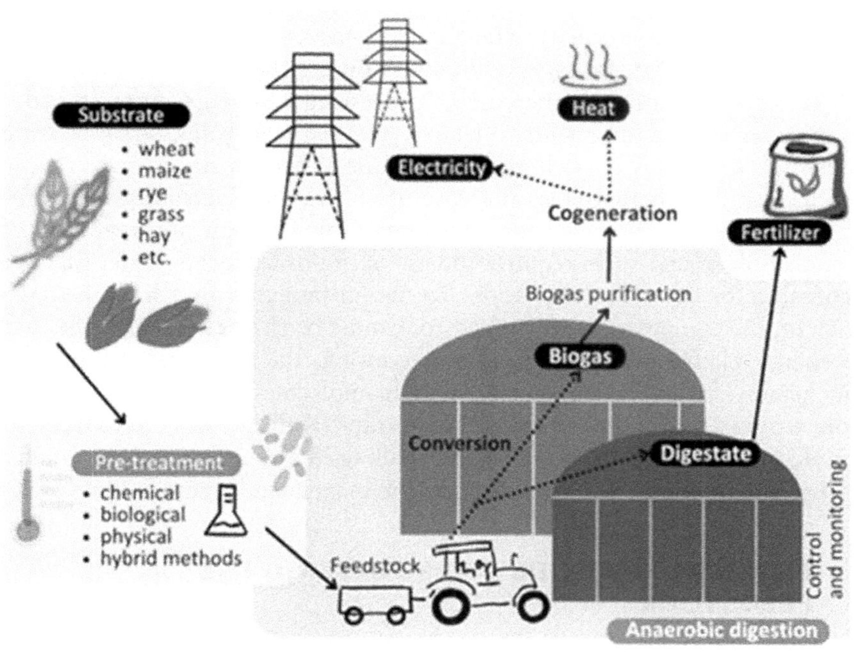

Figure 7.1 Simplified scheme of feedstock processing and agricultural biogas plant operation.

of methane into the atmosphere, so in order to limit environmental impact, waste should be properly managed (Kulichkova et al., 2020).

In order to increase the efficiency of waste processing, a common procedure is to use various substrate pre-treatment techniques (Figure 7.1). This allows for enhancing the degree of decomposition of lignocellulose contained in agricultural waste and to obtain larger amounts of biogas in a shorter period of time. Pre-treatment technologies based on physical, chemical, thermal, and biological processes are of particular importance (Obileke et al., 2024).

Current work on improving pre-treatment methods should focus on combining known methods used in feedstock preparation for anaerobic digestion.

Satisfactory results can also be achieved by using co-fermentation. For example, studies conducted by Qayyum et al. (2024) confirm the importance of co-fermentation of substrates such as agricultural and food waste and manure. By adding cattle manure to the waste, higher biogas production could be achieved due to the increased content of available nutrients for fermentation bacteria. The desired effect of co-fermentation is possible because the C/N ratio is regulated. This is crucial, especially in the processing of lignin, whose structure is rich in carbon (Xie et al., 2022).

The scheme of substrate processing and its conversion into biogas, as presented in Figure 7.1 indicates the possibilities of using biogas and digestate after anaerobic digestion. The use of cogeneration systems allows for the production of heat and electricity, which can also meet the needs of biogas plants, reducing their energy dependence on other sources (Jarosz et al., 2023).

The key issue today is the correct processing of energy crops in order to improve conversion results. Effective pre-treatment is necessary because substrates are highly cost-effective to use due to their features, such as low production cost, high yields, and ease of cultivation. Increasing the availability of feedstock surface for microorganisms through pre-treatment of complex plant structures requires the use of improved technologies that are profitable for use on a large scale. To obtain biogas with a high methane content, the composition of the feedstock must be characterized. In the case of energy plants and residues after harvesting, the origin (plant species), biochemical structure (carbon content in molecules), method of cultivation, storage, and harvest time are important (Kulichkova et al., 2020). It should also be noted that type of feedstock used also influences the quality of digestate obtained and its subsequent management (Figure 7.2).

7.3 PHYSICAL METHODS OF PROCESSING THE FEEDSTOCK

Physical methods of feedstock preparation are aimed at reducing the size of particles and increasing the surface of biomass accessible to microorganisms involved in digestion (Obileke et al., 2024). For this purpose, methods based

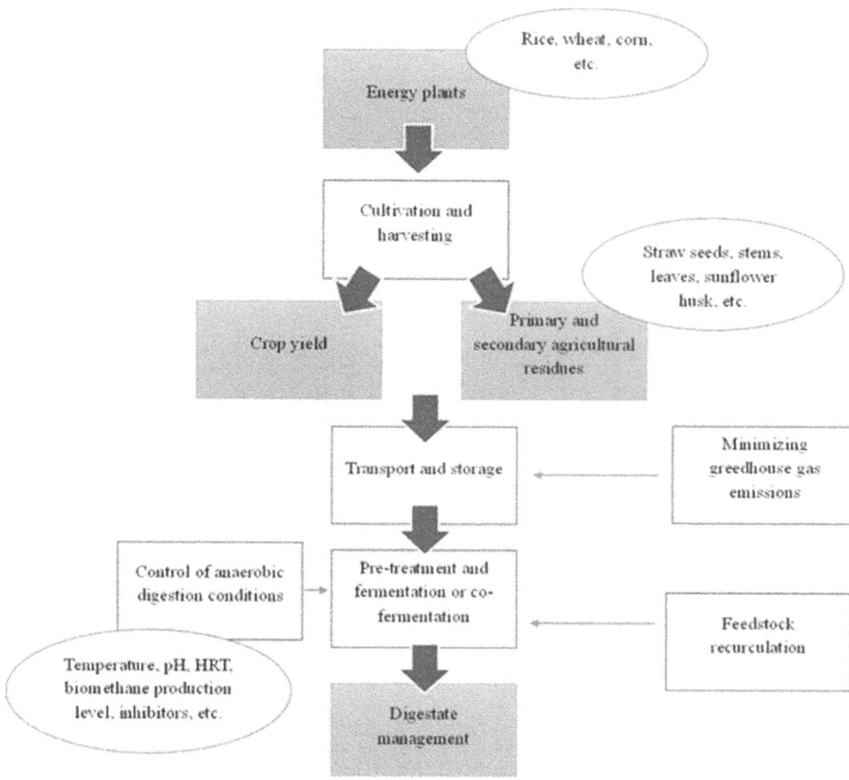

Figure 7.2 Scheme of utilization of energy crops and agricultural waste in the production of digestate. (Based on: Kulichkova et al., 2020.)

on the ultrasonic cavitation process, grinding by milling, extrusion, and sieving are used (Table 7.1). Dell'Omo and Spena (2020) reported that the methane yield was more than twice as high after physical processing of giant reed. An increase in the rate of biochemical transformation of substrate was also observed in the pre-treated substrate. Some of the other advantages of physical methods are the low cost of the process and the lack of toxic substances produced as a result. For this reason, they are commonly used for pre-treatment, also in combination with other methods, which allows for achieving higher substrate degradation (Venturin et al., 2019).

7.4 CHEMICAL METHODS OF PROCESSING

The use of chemical treatment increases the decomposition of lignocellulose in plant feedstock. With regard to cellulose, such processing contributes

Table 7.1 Comparison of physical processing methods

Type of physical method	*Principle of operation*	*Advantages and possibilities of use*	*Disadvantages and challenges*	*References*
Milling	The use of different types of mills for pre-treatment of feedstock in order to reduce its size. Among the milling techniques, the following can be distinguished: ball milling, wet milling, hammer milling, colloidal milling, rod milling and two-roll milling.	• Biomass particle size reduction • An effective method for plant and algae feedstocks • Improving the conversion of biomass (wheat and rice straw) to biogas • Reduction of cellulose crystallinity and degree of polymerization • Relatively quick and simple process • Shortening the fermentation start-up time	• Rarely used alone, without additional processing methods due to insufficient efficiency • Higher energy consumption on a larger scale	Dell'Omo and Spena (2020), Gallego-García et al. (2023), Obileke et al. (2024), Poddar et al. (2022), Sitotaw et al. (2023), and Venturin et al. (2019)
Extrusion	Destruction of the substrate structure by friction. Extrusion takes place in extruders that contain screw elements. The passage of biomass through the space between the screw and the wall of device causes shear forces	• An effective method of pre-treatment of municipal waste and lignocellulosic biomass • Application in maize and triticale processing (silage) • Reduction of feedstock volume • More methane can be achieved (vines, rice straw and corn straw pre-treatment) • The process does not generate toxic by-products • Mixing stimulation	• Process economics – high capital outlay for operation (energy demand) • Energy consumption depends on the engine power, biomass flow, screw speed and torque	Dhull et al. (2024), Gallego-García et al. (2023), Obileke et al. (2024), and Poddar et al. (2022)

Sieving	Mechanical separation of fractions based on differences in substrate granulometry	• Shorter retention time • Simplicity of the method • Can be successfully used in the initial stages of biomass processing (e.g. *Miscanthus giganteus*)	• Limited separation capacity • Suitable for dry materials, wet substrates require appropriate preparation/ another method applied	Chambon et al. (2021) and Venturin et al. (2019)
Ultrasonic cavitation process	The formation of bubbles in a liquid under the influence of vibration induced by sound waves. Imploding bubbles (under higher pressure) leads to the disruption of organic particles	• Improving substrate accessibility for microorganisms • Higher methane content in biogas • Improving the biodegradability of feedstock (breaking of hydrogen bonds between fibrils) • When combined with chemical methods, the initial application of ultrasound facilitates the subsequent penetration of chemical solvents	• Necessity of optimizing the process for each type of substrate • Biogas production is dependent on density • High energy consumption • Noise generation can be a nuisance to pre-treatment workers	Copenhaver et al. (2022), Dhull et al. (2024), and Venturin et al. (2019)
Hydrodynamic cavitation (HC)	Formation of cavitation bubbles due to the reduction of the liquid flow pressure through the restriction/nozzle. The bubbles implode under higher pressure conditions, which causes the disruption of substrate structure	• Application to the decomposition of lignocellulose structure • Reduction of hydrolysis time • Better enhancement of enzyme action more than as a result of ultrasonic cavitation (in the case of sugar cane) • Technology can be successfully implemented on a large scale • High degree of lignocellulose decomposition with the combination of HC and alkalization technology	• Requires further research on the applicability to different types of feedstocks • High energy demand	Poddar et al. (2022) and Sun et al. (2022)

to the reduction of cellulose crystallinity and polymerization (Poddar et al., 2022). Sulfuric acid and sodium hydroxide are the chemical agents most commonly used in this method (Obileke et al., 2024). Chemical pre-treatment methods are presented in Table 7.2. It should not be forgotten that implementing chemical treatment technology involves challenges such as the need for post-treatment neutralization and corrosion prevention (Olatunji et al., 2021).

7.4.1 Thermal and thermochemical methods of processing

Thermal treatment methods are based on the use of heat, which promotes the mixing of particles and the elimination of pathogens (Venturin et al., 2019). The temperature range for heat treatment is 50°C–240°C (Poddar et al., 2022). However, it should be noted that higher temperatures do not always mean an increase in the degree of degradation. Exceeding the optimum value results in inhibition of the digestion due to the formation of compounds resistant to biodegradation (Venturin et al., 2019). Thermal treatment can be applied to pre-treatment of substrates with high lignin and cellulose content because it results in the breakdown of hydrogen bonds. The use of conventional heating is a common technique for heating biomass, which leads to the disintegration of feedstock structures. However, this process causes the formation of inhibitory compounds, such as phenols and derivatives, which inhibit enzymatic activity. Methods classified as conventional heating techniques include autoclaving, over-treatment, microwave heatin,g and water bath (Poddar et al., 2022). Autoclaving and oven treatment are well-known thermal processing technologies. The first method requires, in addition to high temperature, conditions of increased pressure. The principle of microwave radiation technology is to stimulate the process of cell wall degradation and denaturation using non-ionizing radiation (Gallego-García et al., 2023). This simple method allows for achieving higher amounts of methane produced in anaerobic digestion and increasing the level of yield extraction. In addition, it is assumed that microwave radiation effectively reduces hydrolysis time and improves the substrate surface accessibility for enzymes. However, the use of this technology is associated with specific challenges. It should be noted that the effect of pre-treatment depends on the dielectric properties of the feedstock. What is more, the process may require significant energy inputs if large-scale implementation is planned. Moreover, the selectivity of this method in relation to the substrate is also limited (Dell'Omo & Spena, 2020; Obileke et al., 2024).

In addition to standard methods, thermal pre-treatment technologies also include steam explosion, electromagnetic radiation, and water extraction. In the case of steam explosion, the process requires higher temperatures than conventional methods. However, it leads to improved hydrolysis efficiency. Although toxic inhibitory compounds still pose a threat. A similar problem,

Table 7.2 Comparison of chemical processing methods

Type of physical method	*Principle of operation*	*Advantages and possibilities of use*	*Disadvantages and challenges*	*References*
Acidic and alkaline pre-treatment	Chemical compounds lead to the destruction of complex feedstock structures by cell lysis. The use of alkali causes the biomass to swell and increases its susceptibility to enzyme action.	• Improving biodegradability (especially increased rate of cellulose and hemicellulose degradation) • Technology can be successfully used for the pre-treatment of cereals, e.g. wheat • pH regulation during digestion using alkali • Increasing methane yield after alkaline treatment of giant reed • High efficiency of processing sugar cane bagasse and wheat straw and production of good quality digestate using KOH • Improvement of biomass porosity as a result of using alkalizing agents • Low concentration of sodium hydroxide solution (approximately 6%) improve biogas efficiency	• Dependence of process efficiency on the type of chemical compound used and type of feedstock • It often requires longer processing time and involves additional costs • Strong acids can lead to excessive degradation of volatile solids and may lead to inhibition of microbiological activity • Biomass neutralization requirement • High concentrations of alkalizing solutions may cause decrease in the activity of methanogens	Dell'Omo and Spena (2020), Obileke et al. (2024), Poddar et al. (2022), and Venturin et al. (2019)
Oxidation	The use of oxidizing agents such as hydrogen peroxide to decompose organic compound as a result of oxidative fission, substitution, etc.	• High decomposition potential of hemicellulose, lignin and cellulose • The application of hydrogen peroxide can increase the methane content in biogas • H_2O_2 availability • No need to install special processing equipment	• Maintenance of high temperature is obligatory during pre-treatment • High pressure required • Generation of inhibitors such as sodium acetate • Expensive and difficult to recover oxidants	Obileke et al. (2024), Poddar et al. (2022), Venturin et al. (2019), and Zhou et al. (2023)

(*continued*)

Table 7.2 (Cont.)

Type of physical method	*Principle of operation*	*Advantages and possibilities of use*	*Disadvantages and challenges*	*References*
Ozonation	Use of ozone in the degradation of carbohydrates	• Ozonation does not required the use of high temperature or pressure • No harmful substances are produced as a result of pre-treatment process • Potential for use in the processing of microalgae as feedstock • Increasing methane content in biogas • High potential of combining the ozonation method with enzymatic treatment	• If ozonation is used on a larger scale, the economic profitability of the process decreases – high costs of oxidant consumption • By-products that negatively impact the process may be created • Adequate process control must be ensured due to the properties of ozone	Obileke et al. (2024), Osuna-Laveaga et al. (2020), Poddar et al. (2022), and Venturin et al. (2019)
Gaseous treatment	The use of gas penetrating the feedstock structure and causing the breakdown of structural bonds	• SO_2 causes depolymerization of lignin and has a high potential for use in wood processing • Higher lignin oxidation efficiency can be achieved with ClO_2 than with chlorine • Nitrogen oxide catalyzes the removal of hemicellulose (corn, sorghum processing)	• Worse performance with gas medium • High gas consumption • Long duration of pre-treatment • Production of toxic by-products, e.g. halogens (ClO_2 pre-treatment) • The need for further, precise research on the use of gases in pre-treatment • Dependence of efficiency on biomass composition	Carrasco et al. (2015) and Poddar et al. (2022)
Solvent extraction	Decomposition of lignocellulose using organic solvents such as acetone, butanol, methanol, ethanol, etc.	• Possibility of obtaining different hemicellulose fractions • Combining organic solvents can improve treatment efficiency • High separation capacity of wood and corn straw cellulose • Solvents can be reused after pre-treatment • Increased methane content in biogas (methanol treatment)	• Solvent recovery incurs energy costs • The use of extraction on a larger scale requires further analysis	Oliva et al. (2021) and Poddar et al. (2022)

specifically the formation of furan and lignin derivatives that inhibit the fermentation process, concerns the use of water extraction. The dissociation of water at high temperatures enables the catalysis of hydrolysis reactions, although this process requires, in addition to energy input, the use of high pressure (Poddar et al., 2022).

Thermochemical treatment involves using high temperatures with diluted acids, which increases methane production and reduces the risk of toxicity from strong, undiluted acids (Venturin et al., 2019). However, thermochemical processing has some disadvantages. These include, among others, high demand for reagents and costs resulting from the need to supply a significant amount of energy.

7.4.2 Biological methods of processing feedstock and additives used in the fermentation process in biogas plants

The application of anaerobic digestion additives is intended to improve biogas production efficiency and shorten the hydrolysis time. Microorganisms and enzymes play a special role here (Gu et al., 2024). Bioaugmentation has a high potential for implementation. Introducing microorganisms directly into the digester is a simple and efficient method that gives favourable results (Olatunji et al., 2021). The addition of microbes, such as bacteria and fungi, stimulating the degradation of complex plant structures, including those that are components of cell walls (cellulose, hemicellulose, and lignin), increases the degree of feedstock decomposition in reactors. Nevertheless, beyond bioaugmentation, biological processing itself is essential. Biological pre-treatment of substrates containing lignin and cellulose is a key element that research should focus on. This results from the need to reduce the time and negative impact of agricultural waste storage. Moreover, biological techniques enhance the efficiency of compound decomposition, which is crucial due to the high energy potential of substrates. What is more, the use of biological methods significantly reduces the need for chemical additives (Ferdeş et al., 2020).

Additives that increase the activity of methanogenic bacteria are also used. An important role in anaerobic digestion is played by bacteria of the genera *Methanosarcina* and *Methanobrevibacterium,* which were identified in microbiological analyses by Qayyum et al. (2024) during studies on the search for key bacteria involved in methane fermentation. It has also been proven that bacteria such as *Staphylocuccus* spp. and *Escherichia coli*, which are responsible for the decomposition of volatile fatty acids (VFAs), play their role (apart from methanogens). The analysis of microbial consortia participating in anaerobic digestion is important to understand and optimize the composition of inoculant preparations used to degrade organic compounds contained in feedstock.

The undoubted advantage of using biological pre-treatment is its environmental friendliness and the high efficiency of biogas production. Moreover, it is considered that the use of biological methods requires less energy input compared to other methods of feedstock processing. However, the use of living organisms is often associated with a longer time needed to prepare the substrate for digestion (Obileke et al., 2024).

7.4.3 Enzymes and biocatalysts

Enzymatic additives, in particular hydrolases, such as lipases, cellulases, and amylases, are responsible for accelerating the first stage of anaerobic digestion (Gu et al., 2024). Enzymatic decomposition is based on the hydrolysis of the polymer – cellulose to glucose. This simple carbohydrate is used as an energy source by microorganisms (Ferdeş et al., 2020).

Enzymes cause, among other things, the decomposition of corn straw, a substrate often used in biogas plants (Obileke et al., 2024). Satisfactory results can also be obtained in the enzymatic processing of grass, rye, and hay (Dhull et al., 2024). It is worth mentioning that the process is fast, and the obtained biogas is characterized by high efficiency. For instance, a significant increase in methane content was observed as a result of processing corn stover in the presence of a cellulase biocatalyst (Olatunji et al., 2021). However, it cannot be denied that technologies using enzymes and other biocatalysts are expensive due to the process conditions that must be ensured and the need to purchase the appropriate enzyme, as well as prior preparation of the feedstock, e.g. mechanical pre-treatment (Obileke et al., 2024). Nevertheless, the use of enzymatic preparations allows for the acceleration of the biodegradation process and has a high application potential due to the possibility of reusing the enzymes (Dhull et al., 2024). Enzymes of bacterial and fungal origin can be applied not only at the conditioning stage, but also to single- and two-stage digesters (Figure 7.3).

One of the enzymes used in pre-treatment is laccase. It catalyzes the oxidation reaction of phenols. This, in turn, results in the intensification of digestion, as well as stimulates the multiplication of microorganisms. Additionally, the aromatic compounds in the substrates are also cleaved. Laccase can be obtained from plants, bacteria, and white rot fungi. Interestingly, in plants this enzyme is involved in the lignification process, while in fungi it causes lignin degradation (Ferdeş et al., 2020).

Another enzyme used in the processing of lignocellulosic feedstocks is xylanase, which belongs to the class of hydrolase enzymes. Xylanase degrades xylan, causing increased availability of hydroxyl groups and swelling of the fibres (Copenhaver et al., 2022).

Other significant catalysts in enzymatic processing comprise cellulases. Cellulases include endocellulases, which break β-1,4 glycosidic bonds within the cellulose chain, and exocellulases, which cleave bonds at the ends

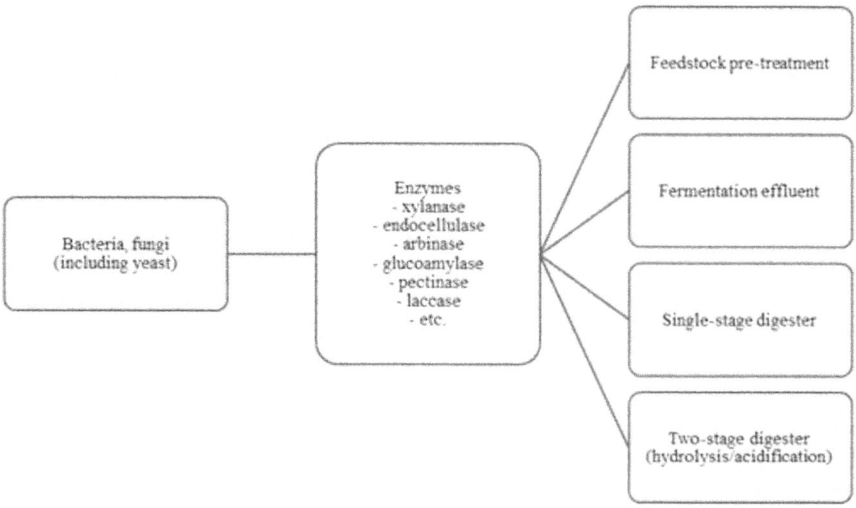

Figure 7.3 Possibilities of enzyme applications to increase lignocellulose degradation and AD efficiency. (Based on: Dhull et al., 2024.)

of the chain, leading to the formation of glucose and cellobiose. Cellulases also encompass β-glucosidases, which are responsible for the conversion of cellobiose into glucose (Yadav et al., 2023). Microorganisms that produce cellulase include fungi, e.g. the filamentous fungus *Aspergillus niger*, and bacteria such as Gram-positive *Bacillus subtilis* or Gram-negative *Acinetobacter junii* (Shrestha et al., 2021).

In addition to cellulases, hemicellulases, and lignin-degrading enzymes, pectinases should be mentioned. The degradation of pectin requires biocatalysts (esterases and depolymerases). The effect of esterase on the pectin substrate is the formation of methanol and pectic acid. At the same time, depolymerases catalyze hydrolysis and trans-elimination reactions leading to the breakdown of glycosidic bonds (Shrestha et al., 2021).

There is significant potential for using enzymatic methods in processing lignocellulosic biomass. More and more reports on the possibility of combining this biological method with other physicochemical pre-treatment techniques appear. For example, Osuna-Laveaga et al. (2020) point to the positive effect of ozonation, which increases the crystallinity of cellulose and the accessibility of enzymes to the feedstock surface.

7.5 MICROBIOLOGICAL ADDITIVES

To improve the efficiency of anaerobic digestion, microbiological additives can be used. These include preparations containing appropriately selected species of bacteria and fungi. Monocultures and mixed cultures are used.

In natural conditions, when oxygen is available, fungi cause the degradation of lignocellulose, which builds plant cell walls. This is possible due to the ability of these microorganisms to produce hydrolytic enzymes, such as cellulases (Obileke et al., 2024). Among the oxidizing enzymes produced extracellularly by WRF, the main ones are manganese peroxidase, laccase, and lignin peroxidase. The delignification process involves the decomposition of phenolic groups forming the substrate, catalysed by enzymes from fungi such as *Streptomyces viridosporus* and *Trametes versicolor* (Ferdeş et al., 2020). In addition to WRF, brown rot fungi (BRF) and soft rot fungi (SRF) can also be applied (Dhull et al., 2024). It should be clarified that brown rot fungi are most effective in the decomposition of cellulose and pectin, while the role of SFR is not clearly defined. However, it is assumed that soft rot fungi mainly participate in delignification and the target of their action is a humid substrate with low lignin content. Generally, it is crucial to focus on selecting the appropriate humidity level and incubation time during fungal pre-treatment (Ferdeş et al., 2020). The operation of BRF is based on the production of enzymes (β-glucosidase, β-xylosidase, endoxylanase, endoglucanase) and on the stimulation of feedstock decomposition using, for instance, quinones and catechols catalyzing the Fenton reaction (Poddar et al., 2022).

Bacteria that stimulate the degradation of lignocellulosic compounds and similar ones can be obtained directly from sources of natural habitation, where the decomposition process takes place, e.g. wood. In addition, their sources include isolated strains available commercially, cultures obtained from previous fermentation cycles taking place in bioreactors, and agricultural by-products. In the case of bacteria, species from the genera *Actinomycetes* and *Streptomycetes* are of fundamental importance (Ferdeş et al., 2020). Bioaugmentation can be used in the processing of corn straw and wheat straw (Obileke et al., 2024). However, the key is the selection of appropriate species of bacteria and, increasingly often, the use of other conditioning methods, for example heat treatment, which, in combination with bioaugmentation, gives enormous results visible in the increased amount of methane produced by even over 200%. *Pseudobutrivibrio xylanivoras*, *Bacteroides cellulosolvens*, and *Enterobacter ludwigii* have been successfully used in bioaugmentation (Olatunji et al., 2021). Palù et al. (2022) used the archaea *Methanoculleus bourgensis* and *Methanothermobacter thermautotrophicus* in bioaugmentation of anaerobic digestion, respectively under mesophilic and thermophilic conditions. Estrada-García et al. (2024) analyzed the effects of probiotic bacteria species *Lactobacillus casei* and *Bacillus subtilis* in the degradation of lignocellulosic wastes. Authors found that lactic acid bacteria (LAB) as inoculants in the biotransformation process increased nitrogen content and stimulated the degradation of cellulose and hemicellulose. All this signifies that probiotic bacteria have the potential to be used in the processing of feedstock intended for energy purposes. Generally, lactic acid bacteria exhibit proteolytic activity and can therefore be applied as biocatalysts in the treatment of agricultural waste (Lee et al., 2023).

In the anaerobic digestion process, microalgae can also be used as a substrate. Typically, for the conversion of biomass to biogas, *Euglena* spp., *Chlorella* spp., and *Oscillatoria* spp. are used as a co-substrate. Microalgae are characterized by high lipid and saccharide content, which makes them desirable for hydrolysis. Microalgae biomass can be obtained from water reservoirs or phytoreactors (Debowski et al., 2020). However, they are resistant to decomposition due to the large amounts of cellulose that they contain. *Bacillus* bacteria, which produce hydrolytic enzymes, are effective in pre-treatment of this feedstock (Ferdeş et al., 2020).

Biological pre-treatment of feedstocks using bacteria, fungi, or enzymes provides numerous advantages. First of all, selecting the optimal conditioning method requires determining key data. It is necessary to take into account the impact of the analysed method on the environment, AD efficiency, biomass characteristics, economic aspects, and its development potential (Figure 7.4).

7.6 BIOCHAR AND MICRONUTRIENTS

Another additive that stimulates the fermentation process is biochar. This adsorbent contributes to the accumulation of bacteria on a porous surface, promotes the formation of biofilm, and stimulates the decomposition of

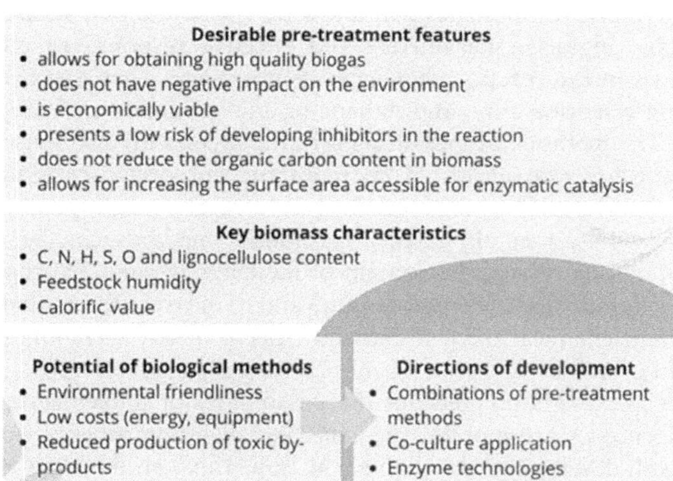

Figure 7.4 The potential of using biological pre-treatment as an efficient method of preparing feedstock for biogas plants.

VFAs (Gu et al., 2024). This sorbent is also able to reduce the concentration of inhibitors, including ammonia and sulphur (Kumar et al., 2021). Moreover, the use of biochar may result in increased methane content in biogas (Zhang et al., 2020). One of the main reasons for using biochar as an additive in anaerobic fermentation is its price and availability. What is more, this additive stabilizes the digestion conditions and enhances the speed of this process. Biochar can be successfully applied, among other applications, to the fermentation of wheat straw and wood waste, resulting in an increased production of methane (Song et al., 2021).

Moreover, adding micronutrients that are components of cofactors and coenzymes is important to maintain the proper growth and functioning of microorganisms in the fermenter. The application of iron, cobalt, and nickel plays a special role, although it should be noted that higher concentrations of added microelements may result in the inhibition of the process (Gu et al., 2024). Microelement supplementation affects the stabilization of the AD process, particularly during the methanogenesis stage. Menon et al. (2017) found that combining micro- and macroelements (such as Ca, Mg, Ni, and Co) with food waste in a fermentation reactor can enhance process conditions and promote methane production.

7.6.1 Optimization of operating parameters in biogas plants

An important aspect of biogas production technology is the control and optimization of process parameters. Among other things, temperature, substrate load, and biogas production must be controlled (Obileke et al., 2024). Therefore, technologies that allow for quick and easy measurement of anaerobic digestion parameters and effective response to unfavourable changes in the reaction sequence are necessary. They can contribute to reducing retention time and enhancing the production of biogas and digestate. The methods of measuring selected factors involve sensors such as a methane flow rate sensor, chromatographic and spectroscopic methods, including Fourier-transformed infrared spectroscopy and photoacoustic spectroscopy (Nsair et al., 2020). Traditional methods can be used to determine, among others, the content of methane, oxygen, hydrogen, and hydrogen sulphide (gas chromatography) and to measure the methanogenic potential and chemical oxygen demand (spectrometric techniques) (Cruz et al., 2021).

Cinar et al. (2021) pointed out the possibilities of implementing artificial intelligence (AI) to monitor the anaerobic digestion process. This technology is still developing, although at the same time, promising in terms of emphasizing possible correlations between data obtained from biogas plants in real time. In addition, AI could be used to create innovative operational models that minimize the risk of failure.

7.6.2 Monitoring fermentation conditions

To optimize the anaerobic digestion process, pH, temperature (depending on the growth requirements and activity of microorganisms), hydraulic retention time (HRT), solid retention time (SRT), C/N ratio, and emerging inhibitors should be controlled and adjusted (Dhull et al., 2024). Before starting the fermentation process, it is necessary to know and characterize the substrate in order to determine the optimal organic loading rates (ORL). Substrates with a higher organic content are usually a richer source for biogas production (Qayyum et al., 2024).

It is assumed that the optimal pH during anaerobic digestion is 6.8–7.2. However, bacteria involved in the first two stages of the process have a greater range of tolerance to pH changes than methanogenic bacteria (Dhull et al., 2024). Maintaining optimal pH is a challenge due to the different ranges of metabolic activity of methane fermentation bacteria as well as the different composition of feedstocks in biogas plants (Ignatowicz et al., 2023).

Due to differences in temperatures resulting from the use of different species of microorganisms, digestion is divided into psychrophilic ($T <$ 10°C), mesophilic (approximately 35°C), and thermophilic ($T >$ 50°C). Stabilized temperature condition affects the efficiency of biogas production. Mesophilic microorganisms are most commonly used in biogas plants because they have a faster metabolism than psychrophilic bacteria and require lower heating costs for fermenters compared to thermophilic fermentation (Dhull et al., 2024). However, thermophilic digestion, which requires high energy consumption, often produces more biogas (Ignatowicz et al., 2023). Increased temperature affects enzymatic activity, gas solubility, and mixing (Kulichkova et al., 2020).

The ratio of carbon to nitrogen (C/N ratio) depends on the type of substrate. The content of these elements directly affects the activity of microorganisms because they use carbon as a source of energy and nitrogen, which is a component of bacterial cells (Dhull et al., 2024). An increase in nitrogen or carbon concentrations results in the inhibition of the methane fermentation process (Figure 7.5). A common solution to optimize the C/N ratio is the use of co-substrates in digestion. Generally, the optimal C/N ratio is considered to be in the range of 20–30:1 (Nsair et al., 2020).

Hydraulic retention time (HRT) is a parameter defining the average retention time of feedstock in the digester. In turn, solid retention time (SRT) refers to the residence time of bacteria in the fermenter. HRT and SRT values for mesophilic fermentation range from 10 to 40 days. However, longer retention times are associated with higher process costs because they require more energy to operate. Additionally, it is worth mentioning that shortening the retention time may result in insufficient substrate processing, a drop in pH, and increased ammonia production, which in turn causes a decrease in digestion efficiency (Dhull et al., 2024).

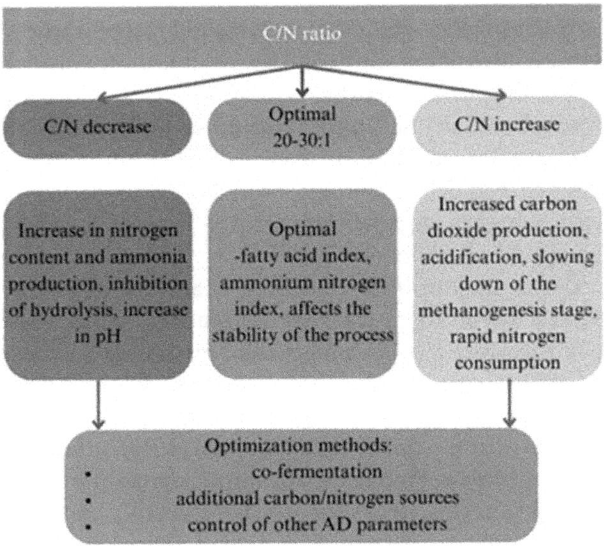

Figure 7.5 C/N ratio – influence on anaerobic digestion and optimization methods. (Based on: Dhull et al., 2024; Nsair et al., 2020; Samadi et al., 2022.)

The choice of the biogas plant operating system itself affects the process parameters. There are single- and two-stage biogas plants. In a single-stage system, the entire fermentation process takes place in a single reactor dedicated for this purpose, while a two-stage system requires the use of separate reactors. The characteristics and advantages of both solutions are shown in Figure 7.6.

The process of biogas production, both on a small scale and on an industrial scale, requires appropriate supervision. Current monitoring systems are advancing towards digitalization and automation. Such development results from the need to ensure stable conditions in the digester, which affects the efficiency of methane production. The Internet of Things (IoT) plays a crucial role in processing large amounts of incoming data, which is vital for identifying changes during the process. The ability to obtain information about the AD process and remotely control conditions increases efficiency and security (Cruz et al., 2021). As of present, monitoring systems are being improved to maximize the effectiveness of biogas plants.

7.6.3 Mixing and recirculation techniques

The mixing process aims to stabilize fermentation conditions, such as pH and temperature, in each part of the digester. As a result, the metabolism

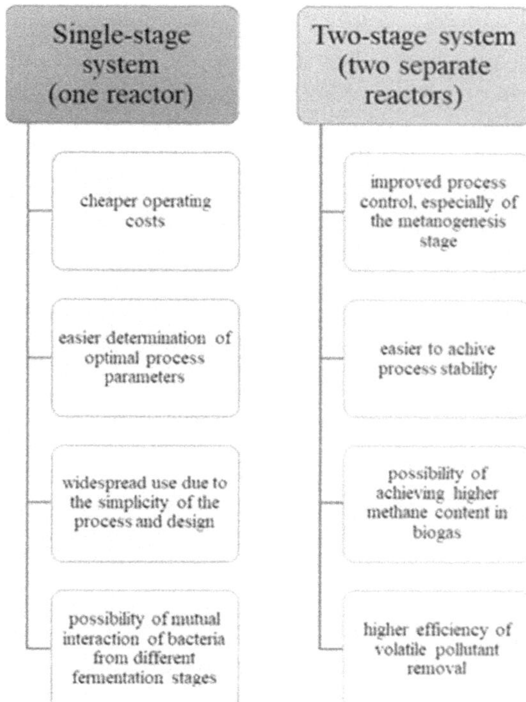

Figure 7.6 Single-stage and two-stage anaerobic systems. (Based on: Nsair et al., 2020.)

of microorganisms decomposing organic compounds is improved. What is more, mixing prevents the formation of sediments and foaming and stimulates the flow of gas to the upper parts of the fermenter (Nsair et al., 2020). According to Singh et al. (2020), the mixing intensity is crucial in the digestion process as it influences the dominant microorganisms at individual fermentation stages. The choice of mixing technology depends on the type of feedstock. This is due to the viscosity and dry mass of substrates. If the substrate contains significant amounts of impurities, then they will be deposited in the reactor as a result of sedimentation (Nsair et al., 2020). In addition, other practical aspects must be taken into account, such as financial outlays, reactor size, and expected processing time (Li et al., 2022). As suggested by Singh et al. (2021), intermittent mechanical mixing at lower mixing intensities gives promising results because it reduces energy costs. Moreover, in the case of mixing time, it should be noted that lowering time by increasing the rotational speed is not necessarily a viable option because it induces shear stresses that have a negative impact on microorganisms.

7.7 CONCLUSIONS

Nowadays, it is indispensable to rise to the environmental and energy challenges and requirements. For this purpose, it is necessary to minimize the dependence on non-renewable energy sources. Instead, the focus should be on effective energy transformation, bringing benefits on both a local and global scale. Anaerobic digestion is key to the circular economy and increasing the share of renewable energy sources (RES) in the energy mix, as it promotes waste management and enables the generation of electricity, heat, and digestate. Improving process parameters, adjusting conditioning and/or co-digestion technology, and using additives that stimulate fermentation are of great importance. Pre-treatment technologies include physical, thermal, chemical, and biological processes, but each method is associated with specific challenges and limitations.

ACKNOWLEDGEMENTS

This research, in part, was funded by the statutory subvention of Czestochowa University of Technology (Faculty of Infrastructure and Environment).

REFERENCES

Carrasco, C., Cuno, D., Carlqvist, K., Galbe, M., & Lidén, G. (2015). 'SO_2-catalysed steam pretreatment of quinoa stalks', *Journal of Chemical Technology & Biotechnology*, *90*(1), pp. 64–71.

Chambon, C. L., Verdía, P., Fennell, P. S., & Hallett, J. P. (2021). 'Process intensification of the ionoSolv pretreatment: Effects of biomass loading, particle size and scale-up from 10 mL to 1 L', *Scientific Reports*, *11*(1), 15383, p. 3.

Cinar, S., Cinar, S. O., Wieczorek, N., Sohoo, I., & Kuchta, K. (2021). 'Integration of artificial intelligence into biogas plant operation', *Processes*, *9*(1), 85, pp. 7, 8, 12.

Copenhaver, K., Li, K., Wang, L., Lamm, M., Zhao, X., Korey, M., Neivandt, D., Dixon, B., Sultana, S., Kelly, P., Gramlich, W. M., Tekinalp, H., Gardner, D. J., MacKay, S., Nawaz, K., & Ozcan, S. (2022). 'Pretreatment of lignocellulosic feedstocks for cellulose nanofibril production', *Cellulose*, *29*(9), pp. 4835–4876.

Cruz, I. A., Andrade, L. R., Bharagava, R. N., Nadda, A. K., Bilal, M., Figueiredo, R. T., & Ferreira, L. F. (2021). 'An overview of process monitoring for anaerobic digestion', *Biosystems Engineering*, *207*, pp. 107–109, 112–115.

Czatzkowska, M., Harnisz, M., Korzeniewska, E., & Koniuszewska, I. (2020). 'Inhibitors of the methane fermentation process with particular emphasis on the microbiological aspect: A review', *Energy Science & Engineering*, *8*(5), pp. 1880–1897.

Debowski, M., Kisielewska, M., Kazimierowicz, J., Rudnicka, A., Dudek, M., Romanowska-Duda, Z., & Zielinski, M. (2020). 'The effects of microalgae biomass co-substrate on biogas production from the common agricultural biogas plants feedstock', *Bioenergy and Biochar*, *79*, p. 80.

Dell'Omo, P. P., & Spena, V. A. (2020). 'Mechanical pretreatment of lignocellulosic biomass to improve biogas production: Comparison of results for giant reed and wheat straw', *Energy*, *203*, 117798, pp. 2, 4. https://doi.org/10.1016/j.energy.2020.117798

Dhull, P., Lohchab, R. K., Kumar, S., Kumarl, M., Shaloo, & Bhankhar, A. K. (2024). 'Anaerobic digestion: Advance techniques for enhanced biomethane/biogas production as a source of renewable energy', *Bioenergy Research*, *17*, pp. 1228–1249. https://doi.org/10.1007/s12155-023-10621-7

Estrada-García, J., Díaz-González, I., Hernández-Aguilar, E., & Méndez-Contreras, J. M. (2024). 'Anaerobic biotransformation process of lignocellulosic waste with Lactobacillus casei and Bacillus subtilis and its effect on the physicochemical parameters case of study: Craft brewers' spent grains', *Biomass Conversion and Biorefinery*, *1–15*, pp. 1, 2, 4, 12. https://doi.org/10.1007/s13399-024-05351-5

Ferdeş, M., Dincă, M. N., Moiceanu, G., Zăbavă, B. S., & Paraschiv, G. (2020). 'Microorganisms and enzymes used in the biological pretreatment of the substrate to enhance biogas production: A review', *Sustainability*, *12*(17), 7205, pp. 2, 6, 9–11, 13–15. https://doi.org/10.3390/su12177205

Gallego-García, M., Moreno, A. D., Manzanares, P., Negro, M. J., & Duque, A. (2023). 'Recent advances on physical technologies for the pretreatment of food waste and lignocellulosic residues', *Bioresource Technology*, *369*, 128397, p. 3, 5, 6, 9, 10. https://doi.org/10.1016/j.biortech.2022.128397

Gu, S., Xing, H., Yu, M., Wang, R., Zhang, L., Shen, S., Zhang, Y., & Li, Y. (2024). 'Additives to enhance the performance and mechanism of biogas production from agricultural wastes: A mini-review', *Water, Air, & Soil Pollution*, *235*(1), 87, pp. 1–7. https://doi.org/10.1007/s11270-024-06897-w

Ignatowicz, K., Filipczak, G., Dybek, B., & Wałowski, G. (2023). 'Biogas production depending on the substrate used: A review and evaluation study – European examples', *Energies*, *16*, 798, pp. 1–4, 10, 12. https://doi.org/10.3390/en16020798

Jarosz, Z., Kapłan, M., Klimek, K., Dybek, B., Herkowiak, M., & Wałowski, G. (2023). 'An assessment of the development of a mobile agricultural biogas plant in the context of a cogeneration system', *Applied Sciences*, *13*(22), 12447, pp. 1, 6, 8. https://doi.org/10.3390/app132212447

Kulichkova, G. I., Ivanova, T. S., Köttner, M., Volodko, O. I., Spivak, S. I., Tsygankov, S. P., & Blume, Y. B. (2020). 'Plant feedstocks and their biogas production potentials', *The Open Agriculture Journal*, *14*(1), pp. 220–225, 227, 228. https://doi.org/10.2174/1874331502014010219

Kumar, A. N., Dissanayake, P. D., Masek, O., Priya, A., Lin, C. S. K., Ok, Y. S., & Kim, S. H. (2021). 'Recent trends in biochar integration with anaerobic fermentation: Win-win strategies in a closed-loop', *Renewable and Sustainable Energy Reviews*, *149*, 111371, pp. 5–7. https://doi.org/10.1016/j.rser.2021.111371

Lee, J. Y., Hong, H., Lee, J. E., Hong, Y. J., Hwang, H. W., Jin, H. S., Shim, H., Hong, Y. D., Park, W. S., Chung, J. O. & Lee, D. W. (2023). 'Valorization of leftover green tea residues through conversion to bioactive peptides using probiotics-aided anaerobic digestion', *Microbial Biotechnology*, *16*(2), pp. 418–431, p. 418. https://doi.org/10.1111/1751-7915.14155

Li, L., Wang, K., Zhao, Q., Gao, Q., Zhou, H., Jiang, J., & Mei, W. (2022). 'A critical review of experimental and CFD techniques to characterize the mixing performance of anaerobic digesters for biogas production', *Reviews in Environmental Science and Bio/technology*, *21*(3), pp. 668–669. https://doi.org/10.1007/s11157-022-09626-z

Menon, A., Wang, J. Y., & Giannis, A. (2017). 'Optimization of micronutrient supplement for enhancing biogas production from food waste in two-phase thermophilic anaerobic digestion', *Waste Management*, *59*, pp. 465–475.

Nsair, A., Onen Cinar, S., Alassali, A., Abu Qdais, H., & Kuchta, K. (2020). 'Operational parameters of biogas plants: A review and evaluation study', *Energies*, *13*(15), 3761, pp. 3–7, 10, 14.

Obileke, K., Makaka, G., Tangwe, S., & Mukumba, P. (2024). 'Improvement of biogas yields in an anaerobic digestion process via optimization technique', *Environment, Development and Sustainability*, *27*, pp. 1–9.

Olatunji, K. O., Ahmed, N. A., & Ogunkunle, O. (2021). 'Optimization of biogas yield from lignocellulosic materials with different pretreatment methods: A review', *Biotechnology for Biofuels*, *14*, pp. 1–34.

Oliva, A., Tan, L. C., Papirio, S., Esposito, G., & Lens, P. N. (2021). 'Effect of methanol-organosolv pretreatment on anaerobic digestion of lignocellulosic materials', *Renewable Energy*, *169*, pp. 1000–1012.

Osuna-Laveaga, D. R., García-Depraect, O., Vallejo-Rodríguez, R., López-López, A., & León-Becerril, E. (2020). 'Integrated ozonation-enzymatic hydrolysis pretreatment of sugarcane bagasse: Enhancement of sugars released to expended ozone ratio', *Processes*, *8*(10), 1274, p. 1, 2, 6, 15, 16.

Palù, M., Peprah, M., Tsapekos, P., Kougias, P., Campanaro, S., Angelidaki, I., & Treu, L. (2022). In-situ biogas upgrading assisted by bioaugmentation with hydrogenotrophic methanogens during mesophilic and thermophilic co-digestion. *Bioresource Technology*, *348*, 126754, pp. 1, 2, 4, 11. https://doi.org/10.1016/j.biortech.2022.126754

Poddar, B. J., Nakhate, S. P., Gupta, R. K., Chavan, A. R., Singh, A. K., Khardenavis, A. A., & Purohit, H. J. (2022). 'A comprehensive review on the pretreatment of lignocellulosic wastes for improved biogas production by anaerobic digestion', *International Journal of Environmental Science and Technology*, *19*, pp. 1–28, 3432–3436, 3438–3441. https://doi.org/10.1007/s13762-021-03248-8

Qayyum, S., Tahir, A., Mian, A. H., Zeb, S., Siddiqui, M. F., & Rehman, B. (2024). 'Optimizing biogas production through anaerobic digestion: Transforming food waste and agricultural residues into renewable energy within a circular economy paradigm', *Biomass Conversion and Biorefinery*, pp. 1–3, 12, 14. https://doi.org/10.1007/s13399-024-05651-w

Samadi, M. T., Leili, M., Rahmani, A., Moradi, S., & Godini, K. (2022). 'Anaerobic co-digestion using poultry slaughterhouse and vegetable wastes to enhance biogas yield: Effect of different C/N ratios', *Biomass Conversion and Biorefinery*, p. 4. https://doi.org/10.1007/s13399-022-03501-1

Shrestha, S., Kognou, A. L. M., Zhang, J. & Qin, W. (2021). 'Different facets of lignocellulosic biomass including pectin and its perspectives', *Waste and Biomass Valorization*, *12*, pp. 4805–4823. https://doi.org/10.1007/s12649-020-01305-w

Singh, B., Singh, N., Čonka, Z., Kolcun, M., Siménfalvi, Z., Péter, Z., & Szamosi, Z. (2021). 'Critical analysis of methods adopted for evaluation of mixing efficiency in an anaerobic digester', *Sustainability*, *13*(12), 6668, pp. 2, 3, 7. https://doi.org/10.3390/su13126668

Singh, B., Szamosi, Z., & Siménfalvi, Z. (2020). 'Impact of mixing intensity and duration on biogas production in an anaerobic digester: A review', *Critical Reviews in Biotechnology*, *40*(4), pp. 508–521. https://doi.org/10.1080/07388551.2020.1731413

Sitotaw, Y. W., Habtu, N. G., Gebreyohannes, A. Y., Nunes, S. P., & Van Gerven, T. (2023). 'Ball milling as an important pretreatment technique in lignocellulose biorefineries: A review', *Biomass Conversion and Biorefinery*, *13*(17), pp. 15593–15616. https://doi.org/10.1007/s13399-021-01800-7

Song, J., Wang, Y., Zhang, S., Song, Y., Xue, S., Liu, L., Lvy, X., Wang, X., & Yang, G. (2021). 'Coupling biochar with anaerobic digestion in a circular economy perspective: A promising way to promote sustainable energy, environment and agriculture development in China', *Renewable and Sustainable Energy Reviews*, *144*, 110973, pp. 3, 5. https://doi.org/10.1016/j.rser.2021.110973

Sun, X., Liu, S., Zhang, X., Tao, Y., Boczkaj, G., Yoon, J. Y., & Xuan, X. (2022). 'Recent advances in hydrodynamic cavitation-based pretreatments of lignocellulosic biomass for valorization', *Bioresource Technology*, *345*, 126251, p. 3, 8. https://doi.org/10.1016/j.biortech.2021.126251

Venturin, B., Bonatto, C., Damaceno, F. M., Mulinari, J., Fongaro, G., & Treichel, H. (2019). 'Physical, chemical, and biological substrate pretreatments to enhance biogas yield', in Helen Treichel and Gislaine Fongaro (Eds.), *Improving Biogas Production. Biofuel and Biorefinery Technologies*, *9*. Springer, Cham, pp. 25–44.

Xie, Y., Zhou, L., Dai, J., Chen, J., Yang, X., Wang, X., Wang, Z., & Feng, L. (2022). 'Effects of the C/N ratio on the microbial community and lignocellulose degradation, during branch waste composting', *Bioprocess and Biosystems Engineering*, *45*(7), pp. 1163–1174.

Yadav, M., Balan, V., Varjani, S., Tyagi, V. K., Chaudhary, G., Pareek, N., & Vivekanand, V. (2023). 'Multidisciplinary pretreatment approaches to improve the bio-methane production from lignocellulosic biomass', *BioEnergy Research*, *16*(1), 228–247, p. 238.

Zhang, L., Kuroki, A., & Tong, Y. W. (2020). 'A mini-review on *in situ* biogas upgrading technologies *via* enhanced hydrogenotrophic methanogenesis to improve the quality of biogas from anaerobic digesters', *Frontiers, Energy Research*, 8, 69, pp. 1–6.

Zhou, Z., Ouyang, D., Liu, D., & Zhao, X. (2023). 'Oxidative pretreatment of lignocellulosic biomass for enzymatic hydrolysis: Progress and challenges', *Bioresource Technology*, *367*, 128208, pp. 2.

Chapter 8

Environmental life cycle analysis of renewable energy

Saurav Kumar Maity, Uplabdhi Tyagi, Krishna Kumar, and Gulshan Kumar

8.1 INTRODUCTION

The world is experiencing a previously unseen shift in energy consumption due to the growing need to address climate change, reduce greenhouse gas emissions, and decrease reliance on depleting fossil fuel resources. Continued economic growth and population expansion ensure that global energy consumption continues to expand, with pressure on the traditional energy systems that have fuelled industrial and social development over time. Figure 8.1 depicts the emissions of CO_2 from different sectors of the earth (Wang et al., 2024). The renewable energy technologies that draw energy from natural processes, which replenish more quickly than they are used, have proven to be a critical component of this approach. Still, although renewables hold significant environmental and economic benefits, they are not devoid of ecological cost. This has established an increased demand for a systematic approach that will enable an evaluation of renewable energy systems based on their entire life cycle, from raw material extraction up to end-of-life disposal. Environmental life-cycle analysis, therefore, serves as one of the critical tools for this purpose, as it aids in gaining a full understanding of environmental impacts associated with renewable energy technologies (Osman et al., 2024).

8.2 OVERVIEW OF RENEWABLE ENERGY TECHNOLOGIES

Renewable sources are energy generators through natural replenishment from sources like sunlight, wind, water, biomass, and geothermal heat. These are clean alternatives to conventional fossil fuels like coal, oil, and natural gas, which have dominated energy supplies for over a century but are finite and highly polluting. Renewables have been promoted due to a decrease in greenhouse gases, the alleviation of air and water pollution, and enhanced energy security through diverse supply sources. The largest renewable energy technologies include solar power, wind energy, hydropower, biomass energy, and geothermal energy, each playing an important role in

 DOI: 10.1201/9781003581536-8

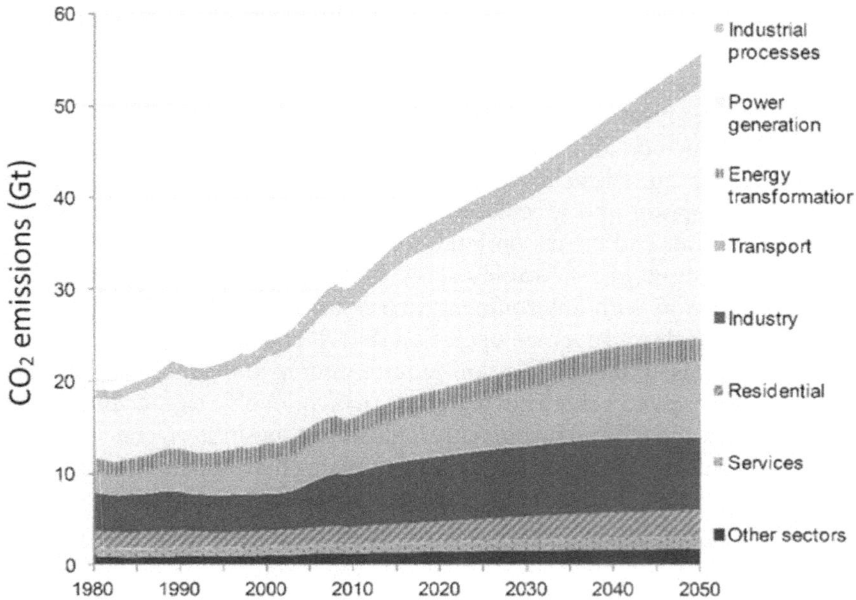

Figure 8.1 Projected global CO_2 emissions from different sectors. (Adapted from an open-access Roy, Mohanty and Misra, 2023.)

the global energy transition (Song et al., 2024). Although they have different energy conversion processes, they share a common benefit: they have minimal or zero direct emissions during operation, utilize abundant natural resources, and can create local employment and economic stimulation. They also have challenges, though. For instance, solar panels and wind turbines require large amounts of materials, including metals and rare earth elements, which must be mined, processed, and transported, leading to environmental effects (Zhao et al., 2024). Biomass is also an energy source that can lead to deforestation and degradation of natural habitats, whereas most large-scale hydropower projects have been known to disrupt ecosystems and displace communities. Even geothermal energy will affect water resources and the subsurface environment, although geothermal energy has a much smaller footprint. These challenges require an assessment that focuses on the actual sustainability of renewable energy technologies (Rotich et al., 2024).

8.3 IMPORTANCE OF *ENVIRONMENTAL LIFE-CYCLE ANALYSIS* (LCA) IN *EVALUATING SUSTAINABILITY*

When evaluating renewable energy technologies, their operational emissions are necessarily not enough. Instead, it is of utmost importance to see the whole life cycle in terms of its environmental performance. That is where environmental life-cycle analysis comes in handy. It is the standardized methodology for assessing the ecological impacts associated with all the stages in a product's life from raw material extraction (cradle) to manufacturing, transportation, operation, and disposal or recycling (grave) (Paul et al., 2024). LCAs include all inputs and outputs in the life cycle of a product, thus providing an all-inclusive picture of environmental impacts from energy consumption and greenhouse gas emissions to resource depletion, waste generation, and effects on land, air, and water quality. Significantly, LCA has been used in renewable energy to compare the benefits of clean-energy production with environmental costs incurred during the life cycle stages that precede and succeed operation (Sikiru et al., 2024). For example, the stages of solar panel production require mining and processing silicon, aluminum, silver, and other raw materials, which involve significant energy input and emissions. Wind turbines are highly intensive in steel, concrete, and fiberglass, meaning these sectors also have their environmental footprints. Hydropower dams, for instance, though completely emission-free in the atmosphere while in operation, can result in massive habitat destruction and loss of biodiversity during construction and operation. For instance, an LCA of a system may conclude that increasing the use of recycled materials in solar panel manufacturing or improving wind turbine blades could significantly reduce the overall environmental impact of these technologies. In addition, LCA (Figure 8.2) is used as the basis for decision support of policies and regulations. Governments and organizations set their sustainability goals, develop environmental standards, and encourage technologies that reduce ecological damage through such guidelines (Morozovska et al., 2024).

Renewable energy systems are also claimed to be less hazardous than fossil fuels. Still, only life-cycle analysis can indicate how much benefit they will provide in mitigating harmful effects on the environment. In many of these cases, the environmental costs of renewable technologies, resource intensity of manufacturing, and large-scale land use changes might challenge the belief that these are intrinsically "green" (Wu and Li, 2024). LCA ensures that world, transitioning to renewable energy, is founded on sound environmental science and the principles of sustainability. Thus, this chapter is devoted to the ecological LCA of renewable energy technologies such as solar, wind, biomass, hydropower, and geothermal energy. A case study is designed to apply LCA to a real-world example, showing that environmental impacts exist from resource extraction through waste disposal (Albizzati et al., 2024). Comparing the various technologies, the main indicators applied will be greenhouse gas emissions and energy return on investment, and could identify trade-offs between or among competing technologies.

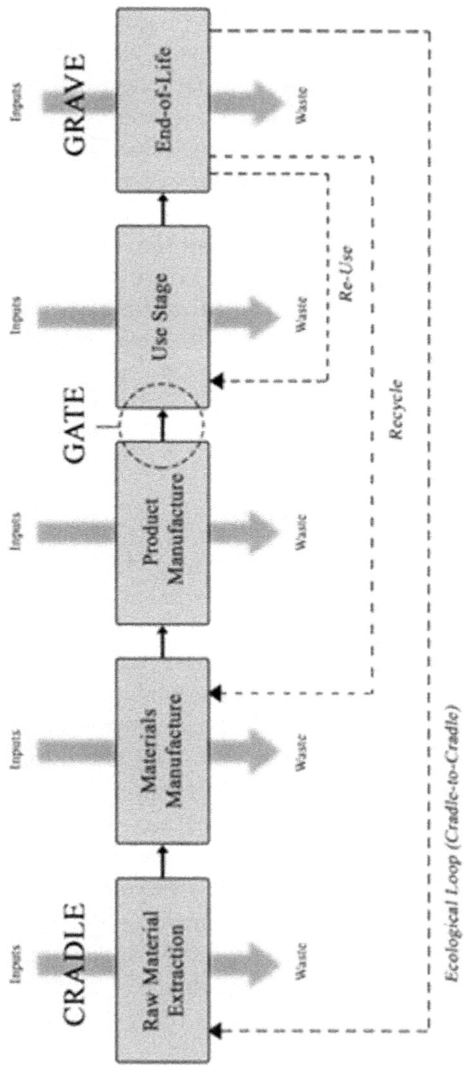

Figure 8.2 Schematic stages of life cycle analysis (LCA).

8.4 FUNDAMENTALS OF LIFE-CYCLE ANALYSIS (LCA)

Life-cycle analysis (LCA) is a systematic method for comparing environmental impacts associated with a product, process, or service through every stage of its life cycle from raw material extraction and processing through manufacturing, use, end-of-life treatment, recycling, and final disposal. LCA seeks to provide a holistic view of all aspects of the environment relevant to a system to ensure that its full environmental burden is clearly understood (Jacquet, le Duigou and Kerbrat, 2024). The approach does not focus solely on the operating phase, as the "evaluation strictly focuses on the operating phase." Instead, it expands the scope to include activities before production, such as material extraction, manufacturing processes, transportation, and end-of-life management by sending it to disposal or recycling sites. This provides a comprehensive view, ensuring that the environmental benefits of renewable energy technologies are neither overstated nor misrepresented. International guidelines, particularly International Organization for Standardization 14040 and 14044, standardize the process of performing an LCA into a structured framework for carrying out these analyses. An LCA may be broadly categorized into four distinct phases: goal and scope definition, inventory analysis, impact assessment, and interpretation (Konradsen et al., 2024).

8.4.1 Goal and scope definition

The first phase of LCA includes determining the purpose of the analysis and well-defined boundaries that clarify what is under investigation, why it's done, and how the results will be used. For renewable energy systems, this phase often comes as specific questions, such as which of two energy sources has the least environmental impact, which of two different technologies is better for a specific application (for example, solar or wind power), or how a given system can be improved. The definition of goals should ensure that results from the LCA are relevant and useful to actual decision-makers (Hannouf et al., 2024). The scope of the study determines the limits of the system and defines what shall be included or excluded in the investigation process. For instance, the life cycle assessment of a PV system entails silicon and metal extraction and processing, manufacturing of photovoltaic modules, transportation to the site, installation and running, and eventually disposal or recycling. At this stage, it will define the spatial and time-related constraints of the assessment, as well as the geographical and temporal scope (Abdelkareem et al., 2024).

8.4.2 Inventory analysis

The second step of LCA is inventory analysis, which involves gathering information about all inputs and outputs related to the life cycle of the product or system under analysis. This stage is very data-intensive and will require a detailed accounting of all material flows, energy inputs, emissions, water use, and waste generation at each stage of the life cycle. This can be stated in renewable energy systems, quantifying the resources consumed during the production of components, for example, the use of steel or silicon in wind turbine blades and solar panels, energy consumption within transportation and installation, and emissions that may occur due to the operation and disposal process (Warrier, Palaniappan and Habert, 2024). The most significant time-consuming part of LCA is inventory analysis, so more information needs to be aggregated from comprehensive sources, which may include material databases, manufacturers, or energy suppliers. In renewable energy technologies, the results of this inventory analysis tend to be very interesting; it can reveal environmental burdens that would otherwise not have been addressed.

8.4.3 Impact assessment

The third stage, impact assessment, transforms the information from the inventory analysis into measurable environmental impacts. This is the estimation of how the inputs and outputs identified in the inventory could contribute to various environmental issues such as climate change, resource depletion, air and water pollution, and ecological disturbances. More common impact categories within LCA are acidification, which involves harmful substances that form acid rain; eutrophication, which is related to water body nutrient pollution; and human toxicity, which assesses the potential harm to human health. Within renewable energy systems, the impact assessment phase can clarify which trade-offs are present in using such technologies (Wahyono et al., 2024). For instance, while sources of wind and photovoltaic power emit very small amounts of greenhouse gases during operation, a life cycle analysis might indicate high rates of resource depletion or energy use during production and transportation phases. Alternatively, renewable systems based on biomass might have a positive carbon footprint but could be associated with land-use change and water consumption. LCA helps analyze the diversified effects and thus provides a better-balanced view of the environmental performance of renewable energy technologies from a different angle (Zagonari, 2024).

8.4.4 Interpretation

The final phase of LCA involves interpreting and comparing the inventory and impact results to draw conclusions that remain within the scope established in the original study. Since this is a critical phase in ensuring the results of an LCA are transparent, meaningful, and valuable, for a given decision, it often involves the identification of significant drivers of environmental impacts and exploration of opportunities for reduction through process improvements, technological advancements, or alternative material choices (Gaffey, Collins and Styles, 2024). Interpretation: For renewable energy systems, interpretation can afford opportunities for some environmental improvements to be carried out. Interpretation also encompasses sensitivity and uncertainty analysis, which validates the robustness of results, giving more confidence in the findings (Acharya, Ranjan Verma and Bolia, 2024).

8.4.5 Importance of LCA in renewable energy systems

LCA, on the other hand, is indispensable in evaluating the actual environmental performance of renewable systems because it brings out the hidden environmental costs that might not be reflected in real life during the operational phase (Kou et al., 2024). It also allows for strategies to mitigate environmental impacts by promoting materials efficiency, using recycled materials, and maximizing the recyclability of parts. In this manner, LCA supports the development of renewable energy systems that not only ensure they are low emitters during their operations but also fully sustainable along their life cycle. Indeed, it will ensure that the global transition to clean energy is fully green and sustainable (N. S. Hassan et al., 2024).

8.4.6 Renewable energy technologies

Since renewable energy technologies use replenished, naturally available resources, they are sustainable alternatives to traditional fossil fuels. Solar power is the most widely used and well-known renewable energy source, with active applications all over the world (Nassar et al., 2024). The country's installed solar PV capacity as of August 2024 was 89,431.98 MW, or more than 58.58% of its total renewable energy capacity (big hydro excluded). India has demonstrated its commitment to expanding its renewable energy base, as evidenced by the addition of over 2.224 GW of solar PV capacity (93.5%) in August alone. India now has more than 152.65 GW of installed renewable energy (not including large hydro), with 89.5% of this being solar and wind power. As demonstrated in Figure 8.3, renewable energy resources such as wind, solar, and biomass energy systems have received positive attention and encouraging investments in the last ten years. Each technology offers unique benefits in decreasing greenhouse gas emissions and ensuring global energy security. However, each of these too has its unique set of environmental impacts, due consideration to which is made at every stage (Q. Hassan et al., 2024).

8.4.7 Solar power

Solar power, as the name suggests, is the use of the sun's power to convert into electricity by photovoltaic (PV) cells or concentrated solar power systems. Although solar energy is an abundant source with no emissions during its

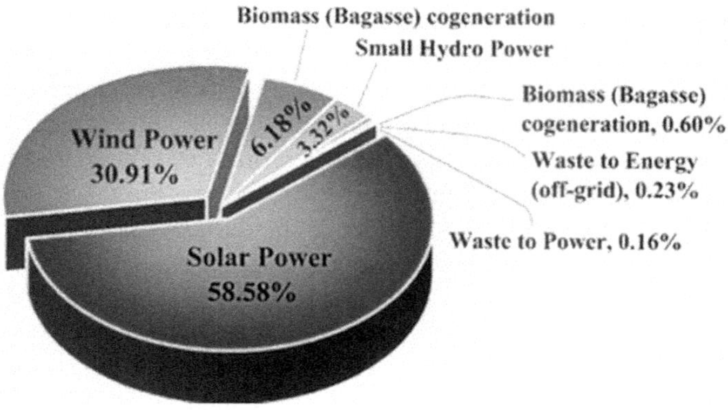

Figure 8.3 Energy capacity of different renewable systems as of August 2024.

operation, the impacts are mainly environmental due to the production, transportation, and disposal of PV panels. The extraction and processing of raw materials like silicon, silver, and aluminum serve as the largest inputs in manufacturing (Ngagoum Ndalloka et al., 2024), consuming significant energy and producing substantial waste and resource depletion. There is also an issue with end-of-life solar panel disposal. There are hazardous materials present in the panels, but recycling programs are beginning to offset this problem. Other factors will include land use, particularly for large solar farms, which could harm ecosystems and agricultural land (Bueno et al., 2024).

8.4.8 Wind energy

Wind energy captures the wind's kinetic energy and converts it into electricity using turbines. It is one of the cleanest energy sources during operation, as it has no direct emissions. However, environmental impacts related to wind energy emanate from materials used in turbine manufacturing, such as steel, concrete, and fiberglass, all of which are energy-intensive to produce (Thakur and Kumar, 2024). Installation and transportation requirements for large turbines also pose significant environmental burdens. For example, onshore wind farms could negatively impact local wildlife, such as birds and bats, by causing collisions with turbine blades. Besides, the visual and noise impacts of the wind turbines might translate into social and environmental concerns, especially in sensitive ecological areas or those that are densely populated (Danovaro et al., 2024).

8.4.9 Hydropower

Hydropower is the generation of electricity using the energy of flowing or falling water. Hydropower is the largest source of renewable electricity globally and is highly valued for its reliability and low operational emissions.

However, hydropower projects, huge dams, would have intense environmental impacts. Dams' constructions often drastically destroy habitat, leading to deforestation and the displacement of inhabitants in a local community (Durvasulu et al., 2024). Large dams also alter river ecosystems, impacting the migratory patterns of fish and water quality. Organic matter decomposition underwater in these big reservoirs can release methane into the environment, an extremely potent greenhouse gas. However, small-scale hydropower projects and improvements in the design of dams are reducing the ecological impact of hydropower systems (Su et al., 2024).

8.4.10 Biomass energy

Biomass energy comes from organic materials such as wood, agricultural residues, and waste. Biomass is burned or converted into biofuels to generate heat and electricity. Biomass is renewable but has questionable sustainability if the feedstock source and management are not proper (Kibria et al., 2024). Poorly managed biomass production may lead to deforestation, land degradation, and loss of biodiversity. Biomass combustion emits carbon dioxide but is considered carbon neutral when the carbon released by combustion is balanced by the amount absorbed in its growth. However, the overall effect of biomass energy depends on other factors, including changes in land use, resource extraction, and transportation (Zitscher and Kaltschmitt, 2024).

8.4.11 Geothermal energy

Geothermal energy exploits the heat of the Earth for power or heat generation. The source of the power is steady and low emission, especially in regions that have a proven geothermal resource. However, most geothermal power plants affect local water resources, emitting trace amounts of greenhouse gases like methane and sulfur compounds during drilling and operation. Ground subsidence, and, in a few cases, even induced earthquakes, are secondary effects of geothermal energy subsurface activities (Korucan et al., 2024). Table 8.1 shows the detailed comparison of renewable energy technologies in terms of efficiency, environmental impact, and material requirements.

8.5 LIFE-CYCLE ASSESSMENT OF SOLAR POWER

The life-cycle assessment (LCA) of solar power, specifically photovoltaic (PV) systems, provides a comprehensive evaluation of the environmental impacts associated with each stage of the system's life, from raw material extraction to manufacturing, use, and disposal. Solar power is widely considered one of the cleanest energy sources, but understanding its

Table 8.1 Detailed comparison of renewable energy technologies in terms of efficiency, environmental impact, and material requirements

Renewable energy source	*Energy conversion technology*	*Efficiency range (%)*	*Capacity factor range (%)*	*Energy density (W/m^2)*	*Geographic limitations*	*Environmental impact*	*Major materials used*
Solar power	Photovoltaic (PV), concentrated solar power (CSP), solar thermal	PV: 15–22	PV: 10–25	5–20 (PV)	Requires high solar irradiance, typically in desert or tropical regions	Land use (panel fields), resource use (silicon, rare metals), waste management (toxic materials)	Silicon, silver, cadmium telluride, gallium arsenide, indium, glass
		CSP: 30–40	CSP: 30–35	10–50 (CSP)			
Wind power	Horizontal-axis wind turbines (onshore and offshore), vertical-axis turbines	35–45	Onshore: 20–35	2–3 W/m^2 (onshore)	Strong, consistent wind resources (offshore/coastal, high-altitude areas)	Land use (turbine spacing), wildlife disruption (bird/bat mortality), noise	Steel, copper, fiberglass, rare earth magnets (neodymium, dysprosium), concrete
			Offshore: 35–45	3–10 W/m^2 (offshore)			
Hydropower	Large-scale dams, run-of-river, pumped storage	80–90	40–60	5–10 W/m^2	Proximity to large water bodies (rivers, lakes), elevation differences (for pumped storage)	Ecosystem disruption (fish migration, sediment build-up), water use, methane emissions from reservoirs	Steel, concrete, copper, aluminum

(*continued*)

Table 8.1 (*Cont.*)

Renewable energy source	*Energy conversion technology*	*Efficiency range (%)*	*Capacity factor range (%)*	*Energy density (W/m^2)*	*Geographic limitations*	*Environmental impact*	*Major materials used*
Biomass	Direct combustion, gasification, anaerobic digestion, biofuels (e.g., ethanol, biodiesel)	25–35	60–85	0.5–1 W/m^2	Availability of agricultural or forestry residue, competition with food production	CO_2 emissions (combustion), land use (deforestation, agricultural impact), water use (crops)	Organic waste, agricultural residues, forest residues, biofuels (ethanol, biodiesel)
Geothermal	Flash steam, dry steam, binary cycle	Oct-20	60–90	50–100 W/m^2	Limited to tectonic plate boundaries or volcanic areas with high geothermal gradients	Land use (drilling sites), water and chemical use (for extraction and cooling), induced seismicity (rare)	Steel, copper, cement, drilling equipment
Tidal and wave power	Tidal barrages, tidal stream generators, wave energy converters	Tidal: 20–40 Wave: 15–25	Tidal: 30–45 Wave: 10–30	2–5 W/m^2 (tidal) 5–15 W/m^2 (wave)	Proximity to coastlines, predictable tides, or strong wave patterns	Marine ecosystem disruption (fish migration, sediment disturbance), visual impact, potential collision risks for shipping	Stainless steel, concrete, polyethylene, fiberglass
Hydrogen (from renewables)	Electrolysis (using renewable electricity)	60–70 (electrolysis)	Dependent on source (renewable electricity)	Depends on storage technology and application	Requires renewable electricity (solar, wind) or water (in some cases)	Water use, CO_2 emissions (if fossil fuels are used in hydrogen production), energy-intensive	Platinum (for fuel cells), electrolyzers, compressed gas storage, metal hydrides

Source: Gayen, Chatterjee and Roy (2024), Kumar et al. (2024), and Z. Wang et al. (2024).

whole life cycle is essential to capture the broader environmental trade-offs involved (Pattanaik et al., 2024).

8.6 LIFE-CYCLE STAGES OF SOLAR POWER

Starting from the extraction of raw materials, particularly silicon, silver, aluminum, and glass, a solar PV system's life cycle can be derived. Silicon stands as the most critical material for PV cells, extracted from quartz. It is mined and purified through the Czochralski method, which involves significant energy consumption. Resource depletion, emissions, and land use changes result from the energy consumption of extracting and refining these raw materials (Qi et al., 2024).

During manufacturing, raw materials are processed, whereby they are transformed into parts such as PV cells, modules, and frames. Another reason is that manufacturing involves chemicals and solvents whose emissions are not controlled at appropriate levels. This is the phase of the highest energy consumption, which leads to greenhouse gas emissions, though the footprint is now slowly coming down due to new and efficient manufacturing techniques (Z. Liu et al., 2024).

No direct emission occurs during the use phase. That is when PV systems achieve the maximum level of environmental benefit since they are producing renewable energy, displacing electricity made from fossil fuels, and emitting less greenhouse gas. A solar panel generally lasts for 25–30 years, during which its environmental impact is minimal. However, panels degrade slightly over time by reducing efficiency (Preet and Smith, 2024).

The last stage is disposal, recycling, or reusing the solar panels. Failure to manage properly at the end-of-life stage in the disposal of solar panels causes harm to the environment, primarily through lead, cadmium, and other metals inside the components of the panel. Recycling programs for PV panels are still in their infancy and under development, with very little infrastructure for large-scale recycling established to date. This can recover valuable materials, like silicon and metals, thus helping reduce the amount of raw material extraction and mitigation of waste-related impacts (Long et al., 2024).

8.7 ENVIRONMENTAL IMPACTS: RESOURCE USE, ENERGY INPUT, EMISSIONS, AND WASTE MANAGEMENT

Some of the major environmental impacts of solar power systems include resource use mainly in material extraction especially silicon and rare metals. Energy input into silicon production and PV module manufacture is high, but of course, it leads to emissions. Energy payback time is usually in the range of one to four years, though it will depend on geography and technology. For this period after the energy payback point, the net output of energy by the system is positive (T. Liu et al., 2024).

The majority of emissions, being greenhouse gases and pollutants, occur in the manufacturing stage. However, over the lifetime of a system, they are dwarfed by avoided emissions from displacing fossil fuel-based electricity generation. Waste management is also another important issue, as at the end of their life, PV panels must be disposed of in ways that do not harm the environment. Recycling solar panels is becoming increasingly acceptable as a solution for waste management and the recovery of valuable materials. This, however, remains the greatest challenge in realizing the recycling goals since separating materials in PV modules is complicated (Goh et al., 2024).

8.8 CASE STUDY: LCA OF PHOTOVOLTAIC SYSTEMS

The case study on LCA of photovoltaic systems in Europe was informative. For instance, it determined that the energy payback time for the PV systems installed in Southern Europe was around 1.5 to two years, while three to four years for those installed in Northern Europe because of less sunlight over there. Over a 30-year lifetime, these systems offset significantly more carbon dioxide than they emitted during production and installation. The study also highlighted the potential of recycling, which can recover up to 80% of the materials in PV panels, thereby minimizing the demand for new raw materials and reducing the burden on waste management (Akhter et al., 2024; Nešović et al., 2024). Table 8.2 shows the energy payback time (EPBT), lifetime, and net energy ratios of different renewable energy systems.

8.9 LIFE-CYCLE ASSESSMENT OF WIND ENERGY

Wind energy is a well-known renewable technology with low greenhouse gas emissions during its operational phase. Life cycle, however, brings to light a much broader range of impacts within the life cycle of wind energy-from extraction of materials and production of turbines through installation, operation, and final end-of-life disposal or recycling. The life-cycle analysis is particularly relevant in the case of both onshore and offshore wind farms, whose different operational conditions and resource demands lead to varying environmental impacts (Maske and Methekar, 2024).

8.9.1 Life-cycle stages of wind energy

The first step of the wind energy life cycle is the extraction of raw materials used in constructing the turbines. The key materials here are steel, aluminum, copper, and rare earth elements for the magnets in direct-drive turbines. Steel and concrete are primarily used in the construction of the towers and

Table 8.2 Energy payback time (EPBT), lifetime, and net energy ratios of different renewable energy systems

Energy technology	*Energy payback time (EPBT) (years)*	*Expected system lifetime (years)*	*Net energy ratio (NER)*	*Primary energy inputs*	*Operational energy inputs*	*Influencing factors*
Solar PV (crystalline silicon)	2–4 (dependent on insolation levels)	25–30	6–15	High energy use during silicon purification, wafer cutting, module assembly energy for mining and transport of materials	Minimal during operation; energy for cleaning panels, inverters, and grid integration	Efficiency improvements, panel degradation rates, cleaning and maintenance
Solar PV (thin film)	1–2.5 (lower for CdTe and CIGS technologies)	20–25	15–30	Lower energy use in thin film deposition, fewer raw materials	Similar to crystalline PV; lower energy for manufacturing	Material toxicity, recyclability, system efficiency improvements
Concentrated solar power (CSP)	3–5 (energy storage dependent)	30–40	8–12	High energy input for mirrors, turbines, and thermal storage (e.g., molten salts)	Energy needed for pumping fluids, heat exchanger operations, cooling	Higher EPBT for systems with thermal storage but better reliability, higher capacity factors
Onshore wind	0.5–1.5	20–25	15–30	Steel production for towers, fiberglass for blades, rare earth metals (neodymium, dysprosium) for magnets	Very low during operation, mostly for maintenance and grid connection	Dependence on wind resources, turbine size, component recyclability
Offshore wind	1–3	20–25	10–20	Higher material needs (steel for underwater foundations), cabling, and installation energy	Higher energy needs for maintenance, access, cable losses due to distance from shore	Higher capacity factors offset by greater maintenance, environmental factors (corrosion, waves)
Hydropower (large-scale)	1–3	50–100	30–100	Initial energy inputs in concrete, steel, and excavation of large dams, turbines	Negligible during operation, but some energy needed for maintenance	EPBT influenced by reservoir size, sediment build-up, methane emissions in tropical climates

(continued)

Table 8.2 (Cont.)

Energy technology	*Energy payback time (EPBT) (years)*	*Expected system lifetime (years)*	*Net energy ratio (NER)*	*Primary energy inputs*	*Operational energy inputs*	*Influencing factors*
Hydropower (run-of-river)	0.5–2	40–80	50–100+	Much lower material needs, smaller turbine sizes, less concrete than large dams	Minimal operational energy, only needed for maintenance and turbine replacements	High variability due to seasonal water flow and regional hydrology
Geothermal	1–2	20–30	10–20	Energy inputs for drilling, steel for piping, and turbine construction	Operational energy mainly for fluid reinjection, pumping, and cooling	Varies with depth of drilling, thermal gradient, and maintenance requirements
Biomass power (combustion)	1–3	10–20	5–10	High energy inputs for crop growth (if dedicated energy crops), transport, and plant construction	Energy for feeding biomass, handling ash, and combustion process	EPBT heavily dependent on biomass type (residues vs. energy crops), processing efficiency
Biogas (anaerobic digestion)	0.5–1.5	10–15	10–20	Inputs for digester construction, feedstock transport, and energy for biogas upgrading	Ongoing energy needed for digester operation, feeding, and biogas processing	Feedstock availability and quality, digester efficiency, methane leakage control
Tidal power	2–4	20–30	8–15	High initial energy for underwater turbines, concrete foundations, and cables	Low operational energy, mainly for turbine maintenance and grid connection	Affected by tidal patterns, marine environment degradation (corrosion, biofouling)
Wave power	2.5–5	20–30	5–12	High energy needs for floating systems, anchoring, turbine equipment	High operational maintenance energy in harsh marine conditions	Highly site-specific based on wave intensity, weather conditions, and maintenance access

Source: Alyssa Ahmad Affandi et al. (2024), Malode, Prakash and Mohanta (2024), and Singh et al. (2024).

foundations, while fiberglass and carbon fiber are used in the blades of the turbines. The extraction, processing, and transportation of these materials are highly energy-intensive, often relying on fossil fuels; such processes consume significant energy and deplete resources (Benitez et al., 2024).

The production of turbines involves creating towers, nacelles (enclosing the generator and gearbox), and blades. In this stage, it is energy-intensive because a large amount of electricity and thermal energy is consumed in material preparation, assembly, and component finishing.

This transportation component of the heavy equipment impacts the environment, particularly with the off-shore installation of wind farm facilities, where specialized ships and cranes must be employed for the erection of structures. Land-based infrastructure, including roads and substations, is required for the site preparation, installation, and operation of wind turbines (Gacu et al., 2024).

In terms of the operating stage, the immediate environmental impact of wind turbines is minimal because most operational units do not emit any pollution. Nonetheless, there are operational issues related to land use, visual impact, noise, and wildlife interactions. Wind farms can always cause problems at local levels especially large onshore projects. Such plants could interfere with local ecosystems and animal habitats, disrupting bird and bat population behaviors. Offshore wind farms would avoid all land-based conflicts but pose dangers to marine ecosystems. Noise pollution is created during the installation process that may disrupt the surrounding marine life (Sander, Jung and Schindler, 2024).

Finally, when at the end of their lifespan, decommissioning and disposal of wind turbines take place. Some parts, like steel, can be easily recycled, while the blades are made of composite materials like fiberglass and carbon fiber that pose complex challenges for disposal based on their durability and high resistance to recycling. Waste disposal methods have long been landfills; however, a myriad of research in blade recycling technologies is underway and gaining momentum (Al-Ghamdi et al., 2024).

8.10 ENVIRONMENTAL CONCERNS: LAND USE, EROI, AND TURBINE DISPOSAL

Land use is one of the major environmental concerns for wind power, particularly in onshore wind farms. Although the footprint of the turbines themselves is extremely small in scale, the distances between turbines and the roads accessing them consume large areas of land, which can lead to habitat fragmentation. On the other hand, wind farms can benefit from dual land use, such as agriculture or grazing, under and around the turbines (Nizamani et al., 2024).

The energy return on investment (EROI) for wind energy is excellent. Wind turbine EROI ranges from 20:1 to 50:1, meaning that the energy it produces far exceeds the amount of energy required in the production, deployment, and maintenance of the turbines over their operational life cycle. Offshore wind farms will have slightly lower EROI compared to onshore installations because the offshore environment requires more energy for installation and maintenance. However, technological advances enhance offshore efficiency (Papagianni et al., 2024).

Another fast-emerging issue is the disposal of turbines because most of the wind turbines have approached or are at the tail end of their operational lifespan, at about 20–25 years of installation. The difficulty in blade disposal is caused by the composite construction, which makes them challenging to recycle. This could be addressed through research into repurposing blades for alternative uses and recycling blade materials (Hasheminezhad et al., 2024).

8.10.1 Case study: LCA of onshore and offshore wind farms

A comparative LCA of an onshore versus an offshore wind farm, therefore, can point out the different environmental trade-offs. An LCA carried out in Northern Europe demonstrates that, although it requires significantly less material and energy for installation and maintenance, the energy payback is higher over the life cycle of an offshore wind farm due to greater and more consistent wind resources. Onshore turbines produce electricity with a carbon footprint averaging 9 gCO_2-eq/kWh. In contrast, offshore turbines have a marginally higher footprint, at 12 gCO_2-eq/kWh, primarily because of the energy-intensive installation and maintenance processes of offshore infrastructure (De Luca Peña et al., 2024).

The case study also showed that any environmental benefit of wind energy, both onshore and offshore, outweighed the impacts at early stages. Energy payback time, the time for a turbine to produce the sum of energy required to manufacture, construct, and maintain that turbine, using that particular turbine, was estimated between six months and two year for onshore systems, and up to two years for offshore systems. It shows the long-term sustainability and environmental advantages of wind energy. As recycling and material use efficiency improve over time, the life-cycle impacts of wind energy continue to reduce (Martín-Betancor et al., 2024).

8.11 LIFE-CYCLE ASSESSMENT OF BIOMASS ENERGY

Biomass is an alternative form of renewable energy derived from organic materials like wood, agricultural residues, and waste. Biomass can be used to generate heat, electricity, or biofuels. The sustainability of this source depends upon how the cycle can be well managed, and it ranges from the cultivation of feedstock, and energy conversion, to energy end-use. Biomass energy life-cycle assessment, therefore, serves to bring forth much-needed critical information and insights into environmental impacts associated with biomass, such as land use, water consumption, and greenhouse gas emissions. Since the LCA reviews each stage, from cultivation to conversion and use, it provides a better understanding of long-term viability and any environmental trade-offs associated with biomass energy (Chopra et al., 2024).

8.11.1 Life-cycle stages of biomass energy

The life cycle of biomass energy begins with cultivation, when crops such as maize, sugarcane, or switchgrass are grown specifically for energy production, or collected agricultural and forestry residues. Dedicated bioenergy crop cultivation involves significant inputs such as fertilizers, pesticides, water, and energy, which contribute to environmental impacts. In particular, expanding land for bio-energy crops might compete with land needed for food production and would likely increase the risk of losing food security and biodiversity. Monoculture tends to deteriorate soil quality over time; sustainable agriculture practices such as crop rotation and intercropping mitigate these adverse effects (Petlickaitė et al., 2024).

The processing stage comprises the harvesting, transportation, and pretreatment of biomass before conversion into energy. Biomass is typically bulkier and more difficult to transport efficiently, which requires increased energy consumption and emissions during this stage. In addition, most pretreatment methods involve drying, pelletizing, or grinding processes that necessarily introduce energy input, thus contributing to the carbon footprint of the biomass supply chain. Alternatively, using locally sourced biomass resources reduces transportation emissions and subsequently will improve the overall environmental balance (Asif et al., 2024).

In the conversion stage, biomass is converted into proper forms of energy either through combustion, gasification, or fermentation. Combustion of biomass produces heat that can be used directly or used to generate electricity. Gasification transforms solid biomass into a gas form and can then be combusted or used in combined heat and power (CHP) systems. Two conversion processes are used in this book for biofuels (Abouemara et al., 2024). Fermentation of crops into ethanol uses crop-based products, like corn or sugarcane, whereas another process to produce biodiesel makes use of oils from plants like soybeans or palm. The difference lies in the environmental burden of each conversion process. Combustion is believed to be a cause of particulate matter and other pollutants, whereas fermentation may result in methane and other greenhouse gases unless managed correctly (Meena et al., 2024).

Most of the energy produced through biomass is utilized in the use stage, either in electricity, heat, or fuel. Biomass energy has long been deemed carbon neutral because it is believed that the CO_2 emitted by burning or converting it is equivalent to the amount of CO_2 consumed when the plant was a growing organism. This balance depends on the sustainable use of the lands used for their growth and efficient management of those lands. Moreover, some emissions like methane and nitrous oxide that have a global warming potential higher than that of CO_2 along the life cycle of biomass complicate the carbon-neutrality narrative (Tiwari et al., 2024).

Key Environmental Impacts: Land Use, Water Consumption, Greenhouse Gas Emissions:

Land use is probably one of the most severe environmental concerns related to biomass energy. For example, significant land areas have to be cultivated for large production of bioenergy crops that can cause deforestation, habitat loss, and changes in land-use systems, which generally lead to increased atmospheric emissions of greenhouse gases, especially when forests or grasslands become agricultural land. Another significant problem in assessing the sustainability of biomass energy is the disruption of food production and ecosystem degradation (Nogueira et al., 2024).

Water is another critical concern. Most bioenergy crops require an excessive amount of water to irrigate, especially in drought-prone regions. The amount of water used depends on the species type, location, and cultivation practices. Efficient use of water and drought-resistant bioenergy crops can be the answer to this (Rolbiecki et al., 2024).

The major contribution of biomass energy's GHG emissions relates to the life cycle of cultivation, processing, and conversion phases. Biomass energy is commonly expected to produce much lower amounts of carbon than fossil fuels. Still, significant contributions to the overall carbon footprint of biomass energy may result from emissions related to the application of fertilizers, soil management, and transport associated with biomass. Secondary indirect land changes, including clearing, can mobilize massive amounts of stored carbon. This may eliminate any benefits linked to GHG from biomass energy (Brás et al., 2024).

8.11.2 Case study: LCA of biofuels and biomass power plants

The case study on LCA of biofuels in the US produced quite complex results regarding the environmental impact of biomass energy. Comparing corn-based ethanol to advanced cellulosic ethanol, it was established that while corn ethanol reduced fossil fuel use, it had higher water and land-use impacts due to intensive cultivation. However, cellulosic ethanol from non-food feedstocks, such as switchgrass, has lower land and water footprints and greater GHG emission avoidance, as this technology utilizes marginal lands and agricultural residues (Matheus et al., 2024).

Similarly, another LCA of European biomass power plants reveals that although substituting fossil fuels with a greater reliance on biomass-fired power plants shifts environmental impacts from the energy conversion process, they are nonetheless greatly dependent on the source of the biomass and the efficiency of the energy conversion process. The report also underscored that using local forestry residues for electricity generation reduces emissions and diminishes transportation-related impacts compared to importing wood pellets from more remote regions. It is also shown that the use of CHP systems, which can produce both electricity and heat simultaneously, brings better overall energy efficiency compared to power plants from biomass that otherwise decrease their environmental impact (Güleç et al., 2024).

8.11.3 Life-cycle assessment of hydropower

Hydropower is one of the most established forms of renewable energy, generating electricity by harnessing the kinetic energy of flowing water. While hydropower is often viewed as a clean energy source, a life-cycle assessment (LCA) reveals significant environmental impacts at various stages, from dam construction to decommissioning. By analyzing these stages, LCA provides a detailed evaluation of hydropower's environmental performance, including ecosystem disruption, greenhouse gas emissions, and sedimentation (Yang et al., 2024).

8.11.4 Life-cycle stages of hydropower

The construction of dams and other infrastructures, such as powerhouses, spillways, and reservoirs, marks the beginning of the life cycle of hydropower. At this stage, substantial environmental and resource impacts are involved since large hydropower schemes utilize huge volumes of materials like concrete and steel. In addition, in the case of reservoirs, quite significant areas of land have to be inundated. The construction phase itself involves land clearing and deforestation that disrupts habitats and the local ecosystem. The energy-intensive production of the infrastructure component also adds to the overall carbon footprint for the project, but these emissions are heavily front-loaded (Wangchuk et al., 2024).

Compared with the construction phase, the operation and maintenance phase of hydropower plants is relatively low in emissions. After the hydropower plant is put into operation, turbines convert water flow into electricity with minimal direct emissions but significant impacts on ecosystems. The adjustment of water flows, temperatures, and oxygen levels adversely affects aquatic habitats and species. Many migratory species, such as salmon, are impeded by dams, which restrict their life cycles and biodiversity. Modern hydroelectric facilities counteract these impacts with fish ladders and bypass systems, but often to no avail (Khaliq et al., 2024).

Even though it is more energy-intensive than the construction, the routine maintenance will avoid horrible damages, for example the floods caused by dam failure.

Decommissioning toward the end of a hydropower plant's life is a process that breaks down infrastructure and, in many cases, restores the flow of a river. Decommissioning may reduce some long-term impacts by allowing the restoration of ecosystems, but can also cause new problems, such as bringing to the surface accumulated sediments or pollutants that were submerged. Sedimentation is an important issue in reservoirs in that the rate of build-up diminishes the storage capacity of water and decreases the efficiency of power generation over time. Planning for material disposal from decommissioned structures further requires ample consideration to prevent further damage to the environment (Rashid, Zarrati and Haun, 2024).

8.11.5 Environmental impacts: ecosystem disruption, greenhouse gas emissions, sedimentation

Among other environmental issues, disruption to ecosystems is perhaps one of the most significant concerns with hydropower. Ecosystem disruption of rivers and resultant loss of biodiversity, interference with fish populations, and changes to terrestrial ecosystems around the reservoirs are all common impacts of building reservoirs. The building of reservoirs forces natural wildlife habitats underwater, and changes in flow regimes lead to erosion and degradation of downstream habitats. Even very small projects have not been found to be less invasive since their impacts can be localized on the ecosystem and biodiversity (Stoffers et al., 2024).

Hydropower is typically regarded as a carbon-free technology, but GHG emissions can still occur, especially at large scales of development. Reservoirs in the tropics are a source of methane and carbon dioxide emissions because of decomposing organic matter submerged in water. The phenomenon is known as "reservoir emissions." It tends to offset some of the usual GHG benefits attributed to hydropower, especially when vast areas of land become submerged (Mohanasundaram et al., 2024).

Another critical environmental effect is sedimentation. Sediments that would have otherwise flowed downstream get trapped in the reservoir, therefore reducing its capacity and efficiency. In the long term, this reduces the power generation potential of the hydropower plant and leads to costly dredging operations. Downstream sediment depletion can then reduce fertile areas of land for agriculture as well as change the riverine ecosystem (Nakai, Masumoto and Asaeda, 2024).

8.12 CONCLUSION

This chapter has presented an extensive analysis of the environmental life cycle explaining the importance of sustainability and the impact of their operations. Indeed, pivotal findings of variability show that life cycle emissions, resource usage, and waste products vary highly among different renewable energy sources. They involve solar, wind, biomass, hydropower, and geothermal systems. Each of the technologies poses a specific, different environmental challenge requiring a tailored approach to mitigation: advances in material efficiency, recycling, and green manufacturing practices. In addition, it was found that although LCA offers an extremely high understanding of the environmental performance of renewable energy systems, issues remain regarding availability, data comparison, standardization, and the necessity of policy integration.

REFERENCES

Abdelkareem, M.A. et al. (2024) 'Hydrogen from waste metals: Recent progress, production techniques, purification, challenges, and applications', *Sustainable Horizons*, 9, p. 100079.

Abouemara, K. et al. (2024) 'The review of power generation from integrated biomass gasification and solid oxide fuel cells: Current status and future directions', *Fuel*, 360, p. 130511.

Acharya, A., Ranjan Verma, A. and Bolia, N.B. (2024) 'Effective collection of end-of-life solar panels through an incentive-based model', *Solar Energy*, 268, p. 112215.

Akhter, M. et al. (2024) 'Sustainable strategies for crystalline solar cell recycling: A review on recycling techniques, companies, and environmental impact analysis', *Sustainability*, 16(13), p. 5785.

Albizzati, P.F. et al. (2024) 'A model to assess the environmental and economic impacts of municipal waste management in Europe', *Waste Management*, 174, pp. 605–617.

Alyssa Ahmad Affandi, N. et al. (2024) 'Analysing temporal factor in dynamic life cycle assessment of solar photovoltaic system', *Solar Energy*, 270, p. 112380.

Asif, M. et al. (2024) 'Techno-economic analysis of biomass pelletization as a sustainable biofuel with net-zero carbon emissions', *Biomass Conversion and Biorefinery* [Preprint].

Benitez, A. et al. (2024) 'Scenario-based LCA for assessing the future environmental impacts of wind offshore energy: An exemplary analysis for a 9.5-MW wind turbine in Germany', *Energy, Sustainability and Society*, 14(1), p. 49.

Brás, I. et al. (2024) 'Valorisation of forest and agriculture residual biomass—The application of life cycle assessment to analyse composting, mulching, and energetic valorisation strategies', *Sustainability*, 16(2), p. 630.

Bueno, P.T. et al. (2024) 'Promoting sustainability in the solar industry: Bibliometric and systematic analysis of alternatives for the end-of-life of photovoltaic modules', *Discover Sustainability*, 5(1), p. 147.

Chopra, J. et al. (2024) 'Life cycle assessment as a key decision tool for emerging pretreatment technologies of biomass-to-biofuel: Unveiling challenges, advances, and future potential', *BioEnergy Research*, 17(2), pp. 857–876.

Danovaro, R. et al. (2024) 'Making eco-sustainable floating offshore wind farms: Siting, mitigations, and compensations', *Renewable and Sustainable Energy Reviews*, 197, p. 114386.

De Luca Pena, L.V. et al. (2024) 'Environmental Life Cycle Assessment of multi-use of marine space: A comparative analysis of offshore wind energy and mussel farming in the Belgian Continental Shelf with terrestrial alternatives', *Journal of Cleaner Production*, 470, p. 143271.

Durvasulu, V. et al. (2024) 'Rationale for adding batteries to hydropower plants and tradeoffs in hybrid system operation: A review', *Renewable and Sustainable Energy Reviews*, 202, p. 114673.

Gacu, J.G. et al. (2024) 'A comprehensive resource assessment for wind power generation on the Rural Island of Sibuyan, Philippines', *Energies*, 17(9), p. 2055.

Gaffey, J., Collins, M.N. and Styles, D. (2024) 'Review of methodological decisions in life cycle assessment (LCA) of biorefinery systems across feedstock categories', *Journal of Environmental Management*, 358, p. 120813.

Gayen, D., Chatterjee, R. and Roy, S. (2024) 'A review on environmental impacts of renewable energy for sustainable development', *International Journal of Environmental Science and Technology*, 21(5), pp. 5285–5310.

Al-Ghamdi, S. et al. (2024) 'Advancements in coffee manufacturing: From dehydration techniques to quality control', *Food Engineering Reviews* [Preprint].

Goh, K.C. et al. (2024) 'Harvesting valuable elements from solar panels as alternative construction materials: A new approach of waste valorization and recycling in circular economy for building climate resilience', *Sustainable Materials and Technologies*, 41, p. e01030.

Güleç, F. et al. (2024) 'Progress in lignocellulosic biomass valorization for biofuels and value-added chemical production in the EU: A focus on thermochemical conversion processes', *Biofuels, Bioproducts and Biorefining*, 18(3), pp. 755–781.

Hannouf, M.B. et al. (2024) 'Social life cycle assessment (S-LCA) of technology systems at different stages of development', *The International Journal of Life Cycle Assessment* [Preprint].

Hasheminezhad, A. et al. (2024) 'A comprehensive review of sustainable solutions for reusing wind turbine blade waste materials', *Journal of Environmental Management*, 366, p. 121735.

Hassan, N.S. et al. (2024) 'Recent review and evaluation of green hydrogen production via water electrolysis for a sustainable and clean energy society', *International Journal of Hydrogen Energy*, 52, pp. 420–441.

Hassan, Q. et al. (2024) 'The renewable energy role in the global energy transition', *Renewable Energy Focus*, 48, p. 100545.

Jacquet, L., le Duigou, A. and Kerbrat, O. (2024) 'A systematic literature review on holistic lifecycle assessments as a basis to create a standard in maritime industry', *The International Journal of Life Cycle Assessment*, 29(4), pp. 683–705.

Khaliq, I. et al. (2024) 'The effect of water temperature changes on biological water quality assessment', *Ecological Indicators*, 159, p. 111652.

Kibria, M.G. et al. (2024) 'Current prospects and challenges for biomass energy conversion in Bangladesh: Attaining sustainable development goals', *Biomass and Bioenergy*, 183, p. 107139.

Konradsen, F. et al. (2024) 'Same product, different score: How methodological differences affect EPD results', *The International Journal of Life Cycle Assessment*, 29(2), pp. 291–307.

Korucan, A. et al. (2024) 'Opportunities and challenges of geothermal energy in Turkiye', *Energy for Sustainable Development*, 79, p. 101417.

Kou, G. et al. (2024) 'Quantum fuzzy decision-making for analyzing the service progress life cycle of renewable energy innovation investments', *Applied Soft Computing*, 163, p. 111884.

Kumar, R. et al. (2024) 'Different energy storage techniques: Recent advancements, applications, limitations, and efficient utilization of sustainable energy', *Journal of Thermal Analysis and Calorimetry*, 149(5), pp. 1895–1933.

Liu, Z. et al. (2024) 'Modeling and assessment of carbon emissions in additive-subtractive integrated hybrid manufacturing based on energy and material analysis', *International Journal of Precision Engineering and Manufacturing-Green Technology*, 11(3), pp. 799–813. https://doi.org/10.1007/s40684-023-00588-3.

Long, H. et al. (2024) 'Advanced processing techniques and impurity management for high-purity quartz in diverse industrial applications', *Minerals*, 14(6), p. 571.

De Luca Peña, L.V. et al. (2024) 'Environmental Life Cycle Assessment of multi-use of marine space: A comparative analysis of offshore wind energy and mussel farming in the Belgian Continental Shelf with terrestrial alternatives', *Journal of Cleaner Production*, 470, p. 143271.

Malode, S., Prakash, R. and Mohanta, J.C. (2024) 'Sustainability assessment of rooftop solar photovoltaic systems: A case study', *Environmental Impact Assessment Review*, 108, p. 107609.

Martín-Betancor, M. et al. (2024) 'Technical-economic limitations of floating offshore wind energy generation in small isolated island power systems without energy storage: Case study in the Canary Islands', *Energy Policy*, 188, p. 114056.

Maske, T. and Methekar, R. (2024) 'Design of pilot-scale delamination plant along with techno-economic and life cycle assessments considering renewable and non-renewable energy sources', *International Journal of Hydrogen Energy*, 55, pp. 1523–1534.

Matheus, T.T. et al. (2024) 'The effect of transportation choices for mitigating climate-related impacts: The case of solid biofuels exported to Europe produced by Latin American countries', *Sustainable Production and Consumption*, 45, pp. 551–566.

Meena, P.K. et al. (2024) 'Thermochemical conversion of agricultural residue for the production of hydrogen, methane, and biofuels: A comprehensive overview', *Industrial Crops and Products*, 221, p. 119340.

Mohanasundaram, S. et al. (2024) 'Green ammonia as peerless entity for realm of clean-energy carrier toward zero carbon emission: Purviews, neoteric tendencies, potentialities and downsides', *Fuel*, 365, p. 131118.

Morozovska, K. et al. (2024) 'Trade-offs of wind power production: A study on the environmental implications of raw materials mining in the life cycle of wind turbines', *Journal of Cleaner Production*, 460, p. 142578.

Nakai, M., Masumoto, T. and Asaeda, T. (2024) 'Strategic siting of hydroelectric power plants to power railway operations with renewable energy', *Sustainability*, 16(16), p. 7086.

Nassar, Y.F. et al. (2024) 'Assessing the viability of solar and wind energy technologies in semi-arid and arid regions: A case study of Libya's Climatic Conditions', *Applied Solar Energy*, 60(1), pp. 149–170.

Nešović, A. et al. (2024) 'Potential of contemporary earth-sheltered buildings to achieve plus energy status in various European climates during the heating season', *Building Simulation*, 17(1), pp. 41–52.

Ngagoum Ndalloka, Z. et al. (2024) 'Solar photovoltaic recycling strategies', *Solar Energy*, 270, p. 112379.

Nizamani, Z. et al. (2024) 'Renewable wind energy resources in offshore low wind speeds regions near the equator: A review', *Ocean Engineering*, 311, p. 118834.

Nogueira, G.P. et al. (2024) 'Supplying the ethanol demand for 2030 in Brazil as a land-based climate change mitigation alternative: Implications on greenhouse gases emissions', *Science of the Total Environment*, 951, p. 175782.

Osman, A.I. et al. (2024) 'Life cycle assessment and techno-economic analysis of sustainable bioenergy production: A review', *Environmental Chemistry Letters*, 22(3), pp. 1115–1154.

Papagianni, S. et al. (2024) 'Review and meta-analysis of energy return on investment and environmental indicators of biofuels', *Renewable and Sustainable Energy Reviews*, 203, p. 114737.

Pattanaik, S.S. et al. (2024) 'Life cycle assessment and forecasting for 30kW solar power plant using machine learning algorithms', *e-Prime – Advances in Electrical Engineering, Electronics and Energy*, 7, p. 100476.

Paul, D. et al. (2024) 'Life cycle assessment of lithium-based batteries: Review of sustainability dimensions', *Renewable and Sustainable Energy Reviews*, 206, p. 114860.

Petlickaitė, R. et al. (2024) 'Evaluation of multi-crop biofuel pellet properties and the life cycle assessment', *Agriculture*, 14(7), p. 1162.

Preet, S. and Smith, S.T. (2024) 'A comprehensive review on the recycling technology of silicon based photovoltaic solar panels: Challenges and future outlook', *Journal of Cleaner Production*, 448, p. 141661.

Qi, X. et al. (2024) 'Applying life cycle assessment to investigate the environmental impacts of a PV–CSP hybrid system', *Renewable Energy*, 227, p. 120575.

Rashid, F., Zarrati, A.R. and Haun, S. (2024) 'Numerical modelling of delta migration and density current motion in a reservoir', *Modeling Earth Systems and Environment*, 10(3), pp. 3517–3528.

Rolbiecki, S. et al. (2024) 'Sustainable water management of drip-irrigated asparagus under conditions of central Poland: Evapotranspiration, water needs and rainfall deficits', *Sustainability*, 16(3), p. 966.

Rotich, I.K. et al. (2024) 'Geothermal energy in Kenya: Evaluating health impacts and environmental challenges', *Energy for Sustainable Development*, 82, p. 101522.

Roy, P., Mohanty, A.K. and Misra, M. (2023) 'Prospects of carbon capture, utilization and storage for mitigating climate change', *Environmental Science: Advances*, 2(3), pp. 409–423.

Sander, L., Jung, C. and Schindler, D. (2024) 'Global review on environmental impacts of onshore wind energy in the field of tension between human societies and natural systems', *Energies*, 17(13), p. 3098.

Sikiru, S. et al. (2024) 'Hydrogen-powered horizons: Transformative technologies in clean energy generation, distribution, and storage for sustainable innovation', *International Journal of Hydrogen Energy*, 56, pp. 1152–1182.

Singh, R.K. et al. (2024) 'Energy matrices, exergo-economic and enviro-economic analysis of an n-sPVT-FPC collector integrated with biogas plant', *Solar Energy*, 273, p. 112537.

Song, C. et al. (2024) 'Unveiling energy transition strategy: A deep dive into China's ambitious renewable energy policy and its impact on carbon emission dynamics', *Journal of Cleaner Production*, 475, p. 143684.

Stoffers, T. et al. (2024) 'Reviving Europe's rivers: Seven challenges in the implementation of the Nature Restoration Law to restore free-flowing rivers', *WIREs Water*, 11(3), p. e1717.

Su, Y. et al. (2024) 'Methane emission from water level fluctuation zone of the Three Gorges Reservoir: Seasonal variation and microbial mechanism', *Science of the Total Environment*, 912, p. 168935.

Thakur, A. and Kumar, A. (2024) 'Technologies based on reusable wind turbine blades', in Inamuddin, Tariq Altalhi, Mohammad Luqman (Eds.), *Wind Energy Storage and Conversion*. Hoboken, USA: Wiley, pp. 133–183.

Tiwari, T. et al. (2024) 'Emerging bio-capture strategies for greenhouse gas reduction: Navigating challenges towards carbon neutrality', *Science of the Total Environment*, 929, p. 172433.

Wahyono, Y. et al. (2024) 'Evaluating the impacts of environmental and human health of the critical minerals mining and processing industries in Indonesia using life cycle assessment', *Case Studies in Chemical and Environmental Engineering*, 10, p. 100944.

Wang, J. et al. (2024) 'Investigating the fast energy-related carbon emissions growth in African countries and its drivers', *Applied Energy*, 357, p. 122494.

Wang, Z. et al. (2024) 'Analysis and evaluation of fuel cell technologies for sustainable ship power: Energy efficiency and environmental impact', *Energy Conversion and Management:* X, 21, p. 100482.

Wangchuk, J. et al. (2024) 'Building information modelling in hydropower infrastructures: Design, engineering and management perspectives', *Infrastructures*, 9(7), p. 98.

Warrier, G.A., Palaniappan, S. and Habert, G. (2024) 'Classification of sources of uncertainty in building LCA', *Energy and Buildings*, 305, p. 113892.

Wu, Q. and Li, S. (2024) 'Decarbonization by digits: How data factors drive nonlinear sustainable dynamics in manufacturing', *Applied Energy*, 374, p. 123967.

Yang, S. et al. (2024) 'Comparison of life cycle assessment between hydrogen production from silicon waste and alkaline water electrolysis', *Science of the Total Environment*, 920, p. 171065.

Zagonari, F. (2024) 'Sustainable business models and conflict indices for sustainable decision-making: An application to decommissioning versus reusing offshore gas platforms', *Business Strategy and the Environment*, 33(2), pp. 180–196.

Zhao, X. et al. (2024) 'Critical raw materials supply: Challenges and potentialities to exploit rare earth elements from siliceous stones and extractive waste', *Resources*, 13(7), p. 97.

Zitscher, T. and Kaltschmitt, M. (2024) 'Sustainable carbon utilization for a climate-neutral economy–framework necessities and assessment criteria', *Energies*, 17(16), p. 4118.

Chapter 9

Smart integrated biorefineries for sustainable development

Sahar Naveed Baig, Kainat, Trisanti Anindyawati, Apridah Cameliawati Djohan, Titin Haryati, Ika Rahmatul Layly, Ahmad Thontowi, and Muhammad Irfan

9.1 INTRODUCTION

Bio-refineries play a very important role in the bio economy and are also helpful in eliminating waste from society. The principle of bio refineries is the same as that of petrochemical refineries, which involves producing a variety of products and a large number of fuels from fossils (Figure 9.1). The most important route for the production of bio economy is industrial bio refinery (Fernando et al., 2006; Octave and Thomas, 2009).

Biorefinery produces these products by a process that involves converting biomass into useful products. As the concept of bio refinery is versatile, it can be applied to a pilot plant in rural areas, which involves the use of agricultural waste. And also, it can be applied to the large scale, which involves utilizing the industrial waste from the surroundings. The process of biorefinery includes utilizing all available and environmentally friendly biomass to produce conventional products, heat, fuel, chemicals, and electricity. The bio refinery is identified by Bozell and Petersen (2010).

In the future large number of biorefinery systems will arise commercially by using flexible technology in order to process available biomass into valuable products, improving the lifestyle of the local population, cost efficiency, and also by using 3D of sustainability (social progress, economic development and environmental protection), reducing environmental harms (Pfaltzgraff and Clark, 2014). Figure 9.2 shows the important factors of sustainability, economic development, social progress, and environmental protection.

9.2 DEFINITIONS OF *BIOREFINERY*

There are different definitions of bio refinery, but the main concept from all definitions is to convert biomass into useful products (chemicals, feed, food, and energy) by using different useful techniques. From all the biorefinery definitions, the most important one is fulfilled by IEA (International Energy Agency) "Bio refineries is environmentally friendly breakdown of biomass

 DOI: 10.1201/9781003581536-9

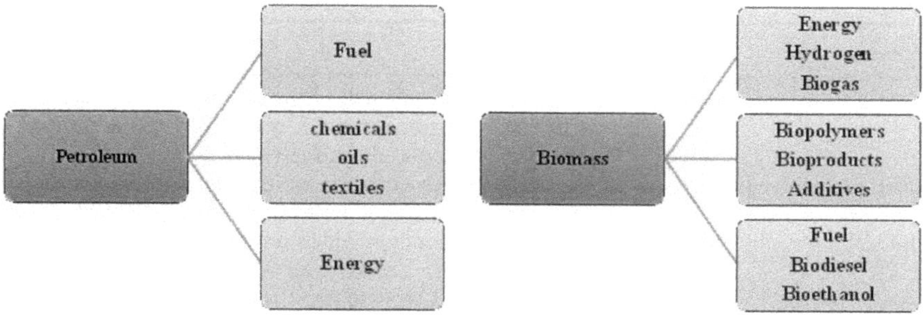

Figure 9.1 Comparison of biorefinery and petroleum refinery.

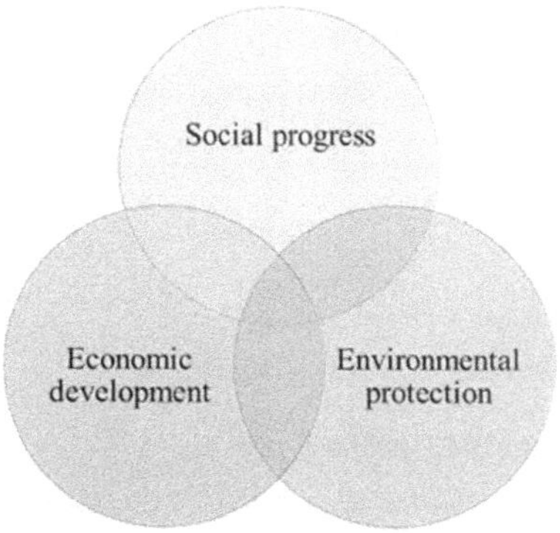

Figure 9.2 Important factors of sustainability; economic development, social progress and environmental protection.

into energy and different useful products" (EA Bioenergy Task 42, 2007). For chemicals and fuel production, biomass is used as a resource (de Jong et al., 2009). According to the "biorefinery Euro View" project, the definition of biorefinery was "Biorefineries are industries, which utilizes biological material. By using different advanced technologies, they make different products as discussed earlier in order to increase the sustainability of economy, society and environment" (Fava et al., 2015). It is an industrial facility that covers a large number of technologies in order to utilize biomass. Most commonly used biomass is crops, solid waste, algae, and lignocellulosic material (Cherubini et al., 2010).

In short, the concept of biorefineries is very easy to understand, that is, an environmentally friendly strategy which focuses on a green economy for a better future (Fava et al., 2015). Table 9.1 shows the different definitions of biorefinery.

Table 9.1 Different definitions of biorefinery

Source	*Definition*
	Like refineries, biorefineries fractionate the raw material into intermediates [proteins, carbohydrates] and then convert it into useful products.
NREL (en.wikipedia.org\wiki\Biorefinery)	It is a facility which converts biomass into chemicals, fuel and power by using different equipment's and processes.
US-DOE	Biorefinery is a concept which process biomass into large number valuable products by using different extraction and conversion techniques.
www.biorenew.iastate.edu\resources\glossary	Biorefineries is a group of industries which produce fuels by using biomass, chemicals, products and materials.

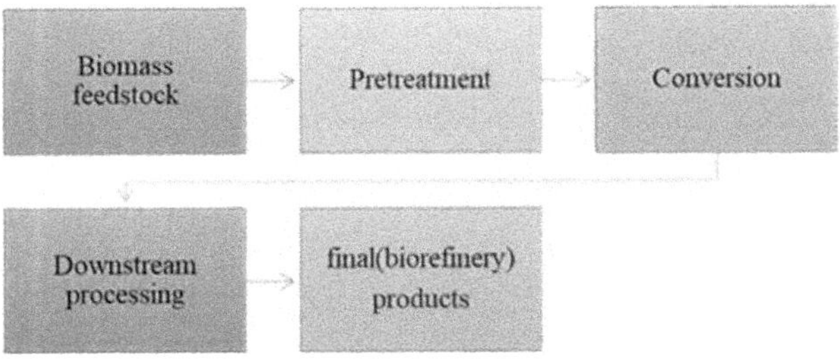

Figure 9.3 General flow sheet of biorefinery process.

9.3 BIOREFINERY CONCEPT

A biorefinery is the same as a petroleum refinery, converting biomass into beneficial products, but it utilizes renewable resources. Multiple chemical components are derived from biomass. Based on four properties, biorefineries are classified into different categories (Cherubini et al., 2009);

This study grouped the biorefinery models on the basis of which conversion technology is used. The public still thinks that the bio refinery is only engaged in biofuel production. However, producing chemicals by the bio-refining process is more sustainable than biofuel production (Mohrand Raman, 2013; Isikgor and Becer, 2015). Chemicals act as an intermediate in order to produce final products like fibers, plastic, and additives. The integrated layout of the biorefinery plant is shown in Figure 9.3 (Cherubini et al., 2009). It involves the following steps as shown in Figure 9.3;

- Pretreatment
- Conversion
- Downstream processing

Different kind of end products are obtained depending upon which process is used and which feedstock is utilized. Among all the stages, pretreatment is very important step whose goal is to decrease the tough and crystallinity structure of biomass into the C5 and C6 sugars to make the extraction and conversion procedure easier and more efficient. The frequently used methodologies of pretreatment include;

- Acid washing
- Ammonia treatment
- Steam explosion
- Grinding/milling

9.4 PROMISING TECHNOLOGIES FOR SUSTAINABILITY

The digital innovation and transformation in companies and industries transform the global system to the new innovative and industrial era, i.e. Industry 4.0 (big data, AI, and IoT) (Dalenogare et al., 2018; Frank et al., 2019; Meindl et al., 2021). These can be supported by recent tools which have broad applications like process design, advanced robotics, and simulation software. Industry 4.0 connects all the industrial sectors by enabling the creation of a cyber-physical system (Benitez et al., 2020).

This industry can also reduce the manufacturing expenses, management, and logistic up to 30% (Nahavandi, 2019). The concept of smart manufacturing is attractive because it connects computational intelligence and different units by wireless networks to increase the sustainability (Wang et al., 2018). Digitalization, such as the use of IoT and AI, brings more innovations for sustainable growth and development (Oztemel and Gursev, 2018; Breque et al., 2021). In order to implement these technologies in industries, it requires infrastructure, economic resources, and also labour. In developing countries, the application of Industry 4 is challenging because of a lack of training and skill, and also because of the cost of sustainable processes (Kumar et al., 2020). Industry 4.0 assists the industry to more efficiently utilize the resources, promoting sustainability and reducing waste (Khan et al., 2021).

9.4.1 Big data analytics

For the new industrial era 4.0, BDA (Big Data Analytics) is an important component distinguished by velocity (involves data that has been generated and then collected), volume (involves data magnitude), and variety (involves data format). When we combine all these ideas with analytics, we can examine and create the knowledge that would be beneficial for us (Klein, 2017). By combining data collected from different industries and fields, BDA helps in improving the decision-making process of bio refinery (Dragone et al., 2020). This increases the capability of farmers so that they can control a large number of crops more efficiently in a short period of time through robotic technology. This also helps to reduce the waste (Ciruela-Lorenzo et al., 2020). It also assists, along with ML technologies, in analyzing and collecting the data of the condition of soils like pH, soil moisture, and nutrients (Hou et al., 2020; Rejeb et al., 2021). Furthermore, another important role of BDA is that it can speed up and improve the product development by virtual lab testing.

9.4.2 Artificial intelligence

Artificial intelligence plays very important role in biorefinery. Artificial intelligence along with its subset that is a ANN (artificial neural network) and ML (machine learning) bring new ideas in bio refinery process but to implement these technologies we need to gather data (parameter, images, yields, etc.) and also we need to create algorithms to unlock the potential application for classification of data, optimization, prediction, and pattern recognition. One of the main purposes of smart integrated biorefinery is to ensure the sustainability in the environment, for example, the removal of detrimental impact from the environment through the efficient production, optimized productivity, and eco-friendly thought design (Coughlan et al., 2022). In order to reduce the detrimental impacts from the environment, we need a life cycle assessment (LCA) methodology, but it is a labor-intensive and time-consuming phenomenon because it requires a large number of variables to process and evaluate. So this problem is alleviated by AI by recognizing and establishing the relationship of patterns and parameters, and also by automation (Culaba et al., 2022). For the development of bio refinery along the process, we also need to consider the background activities like industrial processing or agricultural management as they influence the environmental footprint (Bhuyan et al., 2022). For this purpose, different researchers have proposed a digitalization approach to investigate the environmental impact and optimality of agricultural practices.

When we utilize AI in order to evaluate the different processes, it follows two distinct approaches: one is predicting the product outcomes, and the other is predicting the optimization of process parameters to process the design and to guide technology (Andeobu et al., 2022). Integrated sustainability can be achieved by CAP (computer-aided process), which leverages geometric analysis and logical relationships to optimize the process automatically (Hatim et al., 2020).

9.4.3 Internet of Things (IoT)

IoT technology leverages automation, sensors, and ML to collect, transfer, and process the data to automate the process as well as enabling it to make decisions (Fabris et al., 2020). It acts as a central hub because different devices are connected to it. IoT collects data from all these devices and applies advanced analytics in order to evaluate the data and help in making small decisions and management. In the agriculture field, IoT helps in collecting and analyzing the data to increase the fertility of the soil and increase crop productivity (Patil et al., 2017).

According to a recent study, IoT is also important in microalgae bio refinery. The cultivation of microalgae on a commercial scale at an attractive price is quite challenging. There are also some cons of microalgae bio refinery, including the extraction process, cultivation, and effective monitoring. It has applications in food, pharmacy, cosmetics, and composites (Tang et al., 2023). In the bio refinery process, the use of IoT makes it a more powerful and robust technology as it enables automation and monitoring (KhooKhoo et al., 2020). As it is the advanced technology of using IoT in the biorefinery process, no data has been found in recent literature. It is an interesting gap and also an opportunity to explore and investigate this field.

9.5 EXEMPLARY SMART BIOREFINERIES

Interest in bioenergy has declined until recently due to high-tech advancements that have reduced the cost of fossil fuels. Nonetheless, the unpredictable prices for fossil fuels, deadly air pollution, excessive greenhouse gas emissions, and rapidly increasing demand for transportation fuels worldwide have spurred a great deal of investigation into the development of biomass energy.

Energy produced from any fuel made from biomass is known as bioenergy. Biomass has been investigated as a possible alternative fuel for sustainable energy production in the future because it is a renewable resource. Burning biomass directly to create firewood has long been a way for people to get energy (Naik et al., 2010). A vast variety of feedstocks, such as forestry and agricultural residues, building and municipal solid waste (MSW), and industrial garbage are easily accessible for the manufacture of biofuel in developed nations.

The biofuels generated from these sources fall under the second-generation biofuel category. Second-generation biofuels are produced from biomass residues and lignocellulosic materials such as jatropha, cassava, switchgrass, wood, and straw. Using biofuels as a sustainable energy source lowers emissions of air pollutants (Sims et al., 2010). Until 2050, it is predicted that between 10% and 50% of the total world's energy will come from biomass (Kumar et al., 2015).

According to this prediction, one of the most significant sustainable energy sources will be biomass (Moreira, 2006). The terms "first," "second," "third," and "fourth" refer to the various biofuels generation technologies that are currently in use throughout the world. Surriya et al. (2015) state that the biofuel production with the initial generation of technology has several drawbacks.

9.5.1 Biofuels and bioenergy production

Utilizing biomass has historically been a significant source of energy, impacting the environment and society on local, regional, and worldwide levels (Ebadian et al., 2020; Prasad et al., 2007, 2017). Burning biomass directly for energy has been connected to several conflicts and problems. It has made a substantial contribution to both climate change and GHG emissions. Utilizing contemporary biofuel production technology to recover energy from biomass holds significant promise for maintaining lower trade deficits and ensuring global energy security (Dornburg et al., 2008).

Global estimates from IRENA-2020 indicate that the current total production of bioenergy is approximately 115.7 GW. The European Union produced the most bioenergy (38.5 GW), with Asia coming in second (36.27 GW). China has made the largest contribution to the production of bioenergy in Asia (16.54 GW), with India coming in second (10.23 GW).

The following graph summarizes the variation in bioenergy supply estimates derived from the research utilized in the meta-analysis (Thrän et al., 2012). Figure 9.4 shows the general flow sheet of the biorefinery process.

9.5.1.1 Biomass resources for biofuel production

Biomass that is made up of organic substances like plants, algae, and crops is the source of bioenergy, a renewable energy source. Numerous processes, including photosynthesis, non-photosynthetic activities like food and animal waste, secondary uses of products such as used frying oil, and

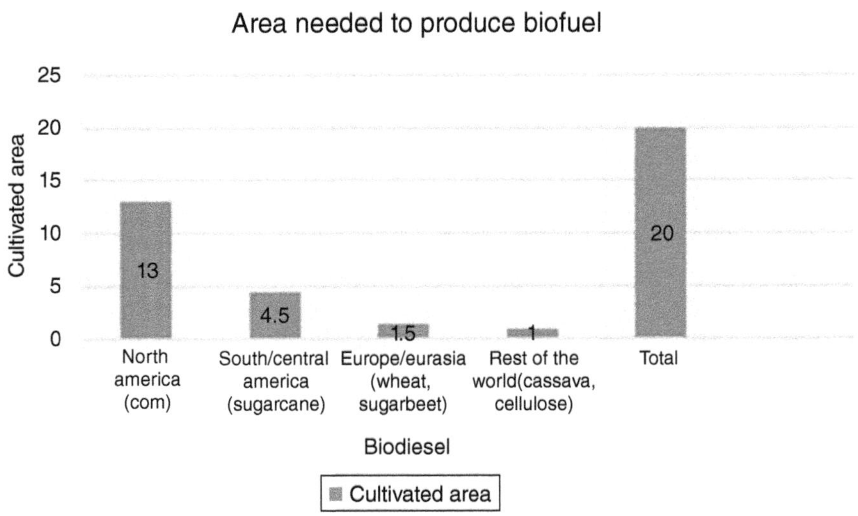

Figure 9.4 Bar chart representing area needed to produce biofuel.

microbiological processes like the production of biogas, can provide these elements (Dahiya, 2020).

A renewable resource, biomass can be easily converted into a different kind of bioenergy, usually applied as feedstocks, or used directly as biofuel. Biomass leftovers and waste products from agriculture, garbage produced by the agro-industry. Biofuels can be produced using sustainable and viable feedstocks, such as municipal solid waste and forest byproducts. Popp et al. (2014) state that municipal sewage wastes, as well as residues from forestry and agriculture, might produce 50–150 Exajoules (EJ) annually. It is anticipated that the global core energy market will grow to 600–1,000 EJ annually by 2050. The possibility of biomass worldwide could be between 200 and 500 EJ annually, primarily from industrial residues (140 EJ), wastes from agriculture (100 EJ), forests (80 EJ), and energy crops (120 EJ).

Members of the Global Biofuel Alliance (GBA), led by India, agreed on September 10, 2023, to promote fossil fuel consumption and biofuel production in industry and transportation to collectively reduce emissions. The GBA's objective is to popularize these environmentally beneficial substitutes, which will significantly lower emissions.

The largest producers of crops worldwide, together with the yields of grain and straw that correspond to them, as well as practical estimations for straw. IRENA (International Renewable Energy Agency) estimates that the global biomass supply will range from 97 to 147 EJ/year (in energy terms) by 2030 (Nakada and Saygin, 2014).

Lipid and fatty acid methyl esters (FAME) can also be found in fried and frying waste oils from lodging establishments. Also, from dining establishments and food processing facilities. The expense of production of biodiesel could be lowered by 60%–90% using the transesterification of used cooking oil as feedstock (Talebian-Kiakalaieh et al., 2013). Furthermore, it is better to use cooking oils to make biodiesel instead of discarding them, which typically has an adverse effect on the environment.

9.5.1.2 Developments in the **generation of biofuel**

Biofuels are crucial to the mix of renewable energy sources since they lower greenhouse gas emissions and reliance on fossil fuels (IRENA, 2024). The main drivers of biofuel research are worries about the volatility of the world's oil prices, the stability of the energy supply, climate change, and the emergence of new farming prospects. Furthermore, the increase in biofuel production has been significantly influenced by energy security, strict government rules, worries about environmentally friendly farming practices, and a drop in CO_2 emissions from the transportation industry.

The increasing use of biofuels as alternatives to fossil fuels is facilitated by the production of biofuels (Assaf et al., 2024). It is important to remember that this increase fluctuates depending on the type of biofuel and the region, and it is influenced by ongoing discussions about sustainability and competition with food production. The use of biofuel is expected to expand fivefold. IRENA's low-carbon road to 2050 estimates that global energy consumption will increase from 130 billion liters in 2016 to almost 650 billion liters by that time. This emphasizes how important it is to continue biofuels research and development in order to fulfill the increasing demand for renewable energy (IRENA, 2024).

The goals of recent advancements in biofuel technology have been to increase the range of feedstocks that may be used to produce biofuels, reduce production costs, and boost the efficiency of the production process (Jiang et al., 2019). Utilizing a variety of microorganisms to produce chemicals and biofuels, for other goods, systems for cocultivation have been created to enhance the production efficiency of biofuels (Akbarian et al., 2022). To meet future industrial demands, several problems related to cocoa-cultivation systems still need to be resolved (Ning et al., 2021). The following image represents the biofuel production worldwide from 2000 to 2022. Taken from Akbarian et al. (2022). Figure 9.5 shows the bar chart representing worldwide production of biofuel from 2000 to 2022.

9.5.1.3 Sustainability standards for biofuels

A lot of work is being done to create sustainability standards and criteria that will guarantee the overall sustainability of biofuels. For

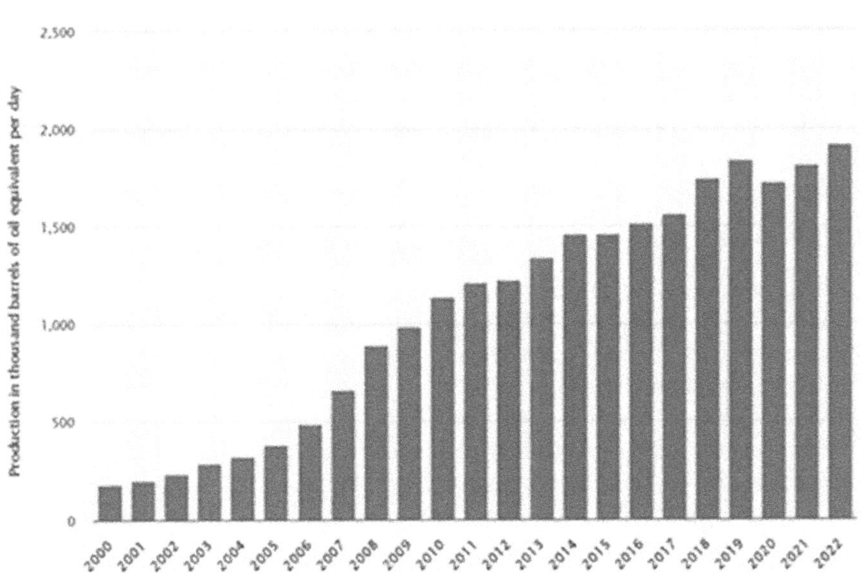

Figure 9.5 Bar chart representing worldwide production of biofuel from 2000 to 2022.

biofuels to contribute to the legally mandated national targets that each Member State must meet by 2020, the EU has established regulations under the Renewable Energy Directive (RED) that specify sustainability requirements. It's possible that the EU will lead the world in sustainability, and other continents might follow. The EU agreed a new biofuels policy in December 2008 as part of the EU Climate and Energy Package's new RED program.

The Standard for Renewable Fuels program is managed by the Environmental Protection Agency (EPA) in the United States (EPA, 2013). By 2022, the yearly volume needs for renewable fuels will have increased to 36 billion gallons. Both US-based local and international manufacturers and distributors of renewable energy are subject to these regulations, in contrast to the fuels made of petroleum that they replace. Advanced biofuels and cellulosic biofuels must demonstrate that they meet minimum GHG reduction goals of 50% and 60%, respectively, based on a life-cycle assessment that accounts for indirect land-use change.

A multidisciplinary approach involving public awareness, corporate sector investment, legislative support, and scientific research is all required to address the issues and optimize the production of biofuel. To advance the sustainable development and application of biofuels, cooperation amongst stakeholders—including scientists, business, policymakers, and environmental organizations—is essential (Shahid et al., 2021).

9.5.2 Waste to bioenergy conversion

One practical way for the production of bioenergy as well as bioproducts in a more enduring way is through the biorefining of garbage. Non-edible biomass and different organic leftovers produced by various economic sectors are considered waste. In a circular bioeconomy paradigm, waste streams present both opportunities and challenges for reuse due to their decentralized and heterogeneous nature (Olguin et al., 2022).

Biomass leftovers and waste are created as byproducts during the planting, processing, and consumption of the intended raw materials, such as biomass farmed specifically for energy (Naik et al., 2010). Specifically, biomass residues fall into one of three categories: primary, secondary, or tertiary. Primary leftovers, such as maize stalks, stems, leaves, and straw, are typically produced when target food crops and forest products are planted in the field (Chen et al., 2015).

Specifically, biomass residues fall into one of the primary, secondary, or tertiary categories. Conversely, tertiary residues become available after human and/or animal consumption of a biomass-derived product. These leftovers might first show up as MSW before to being further processed into wastewater and/or sewage sludge (Li et al., 2007). Among biomass residues and waste, microalgae biomass, Waste cooking oils (tertiary biomass residues), wood, and agricultural leftovers (primary and secondary biomass residues) have all shown encouraging possibilities.

9.5.2.1 Waste cooking oils

While low-cost biodiesel can be made from waste oils, such as residual cooking oils, premium food-grade virgin oils are produced from selected feedstocks. Using leftover cooking oil as feedstock instead of virgin oil is a practical method to minimize the expense of materials used in the production of biodiesel. According to reports, using waste oils might reduce the cost of producing biodiesel by 60%–90%.

Furthermore, reusing used oil reduces the problems associated with discarding large quantities of frying oil that are not fit for human consumption due to their increased concentration of free fatty acids. Phan and Phan (2008) state that engines could run on a mixture of 80% diesel and 20% biodiesel derived from waste oils (B20) without requiring significant changes. Simple Heating and filtration as pre-treatments can be employed to get rid of water and undesirable solid particles before transesterification, and there is no discernible difference in quality between used and unused edible oils (Jacobson et al., 2008).

9.5.2.2 Algal biomass

Macroalgae, also called seaweeds, and microalgae are the two main classes into which algae can be roughly divided. Generally speaking, macroalgae are big, pond-growing multicellular algae that are commonly observed. On the other hand, Microalgae are single-celled, minute algae that commonly grow in water bodies in suspension. Despite having a wider variety of bioactive compounds than microalgae, macroalgae can yield smaller margins for biofuels (Bansemir et al., 2006). Thus, microalgae represent another potential oil source because of their high accumulation of lipids and very quick development rates. Additionally, microalgae do not compete with vast freshwater resources or land that is mostly used for agriculture. Spent microalgae biomass can be converted into biofuels once target goods, such as high-value compounds and oils, have been recovered and processed from it. The following table represents the algal biomass conversion process for biofuel production.

9.5.2.3 Wood and agricultural residues

Waste from wood processing, including wood chips and sawdust, and residual logs from sawmills and lumber processing facilities, can be converted into biofuels. For instance, wood residues and sawdust generated by the saw and paper mill sectors can be utilized as boiler fuels and feedstocks for ethanol production.

Additionally, maize stover—which includes the stalks, cobs, and leaves—has been shown to have the capacity to be converted into fermentable sugars for the production of bio-butanol (Cai et al., 2016). However, sugarcane waste, especially in tropical nations, including sugarcane bagasse and leaves, can be an affordable way to employ raw substrates for the manufacturing of bioethanol and other biofuels like biochar or biogas (Chandel et al., 2012). Palm kernel press cake, a byproduct of palm oil extraction, has shown promise in the fermentation process of producing bioethanol.

9.5.2.4 Agricultural waste into valuable bioproducts

Because modified agricultural waste is more useful and efficient, it cannot be employed without a comprehensive pretreatment (Dey et al., 2021). Agro waste undergoes various treatment procedures before being put to use. Physical, chemical, and biological processes are required for the dissolution of the lignocellulosic components that comprise agro-waste residues (de Souza and Shivakumar, 2021). Generally speaking, the conversion of breaking down complex agro-waste molecule structures into simpler monomers is considered an essential pretreatment step (Saravanan et al., 2022).

The amount of published material strongly predicts future research directions. Different fields of study can be categorized using a list of keywords. The most-cited keywords from 93 publications are shown in network, density, and overlay visualization maps. The production of new value-added goods is the primary focus of current research, according to valuation approaches.

Mechanical, heat, and sonic pretreatments improve the kinetics of wheat straw biodegradation, but at the cost of significant energy consumption (Rahmani et al., 2022). By increasing the biomass's available surface area and decreasing its size, this technique promotes hydrolysis through increased heat and mass transmission (Prasad et al., 2022). Chemical pretreatment (such as wet oxidation, ozonolysis, organosolv, acids, and alkalis) breaks the internal connections of lignin and hemicellulose (Zhou et al., 2022).

When compared to untreated bagasse, pretreatment significantly increased cellulose output and reduced the weight of cassava bagasse. The H_2SO_4 pretreatment produced the highest cellulose yield (Igwo-Ezikpe et al., 2022). Hydrolysis and saccharification are two biological processes (involving bacteria) that are mediated by exoenzymes. It is necessary to ferment whole-cell systems to a solid or liquid substrate or by enzymatic digestion. This process generates no inhibitors, is environmentally benign, and consumes less energy (Cheng and Whang, 2022).

Ionic liquids (ILs), natural deep eutectic solvents, and deep eutectic solvents are all used in green pretreatment. IL pretreatment increased cellulose crystallinity and reduced the concentration of lignin, along with particle size, which improved the generation of glucose from enzymatic hydrolysis (Mudzakir et al., 2022). Hydrogen connections, between and within molecules, are broken, leading to an enhanced degradation of lignin and cellulose (Malolan et al., 2021). These pretreatment techniques increase surface area while lowering operating costs by making enzymes more accessible for hydrolysis.

9.5.3 Biorefinery, sustainability, and circular economy

Wastewater treatment plant operators are primarily concerned with fulfilling their legal obligations to treat wastewater and manage pollution. However, regulatory constraints, increasing service demands, aging infrastructure, growing utility and chemical expenses, and environmental concerns about biosolid disposal methods are just a few of the many issues that WWTPs are currently facing. Wastewater treatment now encompasses materials recovery (Faragò et al., 2021), carbon management (Qandil et al., 2021), emissions reduction, and water reuse (Capodaglio et al., 2020), thanks to the paradigm shift from Water and Resources Recovery Facilities (WRRF) to WWTP.

Biorefining is the sustainable processing of waste sludge biomass, either by itself or in conjunction with biomass of other origins, in order to implement a sludge-based circular economy, also known as the circular bioeconomy (CBE), which is a concept that combines the circular economy and bioeconomy.

9.6 FUTURE PROSPECTS AND OPPORTUNITIES

There are several chances for economic, social, and environmental progress through the creation and marketing of biofuels. The potential of biofuels to reduce carbon emissions and mitigate climate change's effects is one of their most alluring advantages. Achieving global climate targets requires a significant cut in greenhouse gas emissions, and can be achieved by replacing traditional fossil fuels with biofuels (Deora et al., 2022; Rafeeq et al., 2023). The promise of biodiesel made from the oil of rice bran as a sustainable substitute for petrodiesel is reviewed by Hoang et al. (2021). According to the research, the optimal techno-economic composition for the current 80% petrodiesel and 20% biodiesel made from rice bran oil is combined to power diesel engines, which balances cost-effectiveness and environmental advantages.

Along with these advantages for the environment and energy security, biofuels development, and offers financial prospects. Developing biofuels could create new sources of income helps nearby farmers, especially those who cultivate agricultural feedstocks in remote locations (Dineshbabu et al., 2019). The biofuel production, a worldwide commodity, can promote international cooperation on sustainable development objectives and open up new markets for agricultural products. By taking use of these chances, we can hasten the shift to a low-carbon and more sustainable economy (Palomar et al., 2023) and open up new social and economic prospects for people everywhere.

9.7 CONCLUSION

In order to support a sustainable bioeconomy, biorefineries convert biomass into value added products such as biofuels, bioenergy, and bio-based substances. They use industrial and agricultural waste to reduce their effect on the environment, and they operate on both small and large scales. The efficiency and decision-making processes of biorefineries are being improved by technological advancements in order to support a sustainable bioeconomy. Biorefineries convert biomass into value-added products such as biofuels, bioenergy, and bio-based materials.

Better data analysis, more efficient product development, and increased operational efficacy are all made possible by these technologies. The necessity for continued research and innovation to fully realize the potential of biorefineries in fostering a greener future is highlighted by the challenges that remain, particularly in scaling microalgae biorefineries and monitoring cultivation processes.

REFERENCES

Akbarian, A., Andooz, A., Kowsari, E., Ramakrishna, S., Asgari, S. and Cheshmeh Z. A., 2022. Challenges and opportunities of lignocellulosic biomass gasification in the path of circular bioeconomy. *Bioresour Technol.* 362, 127774.

Andeobu, L., Wibowo, S. and Grandhi, S., 2022. Artificial intelligence applications for sustainable solid waste management practices in Australia: Asystematic review. *Sci. Total Environ.* 834, 155389.

Assaf, J. C., Mortada, Z., Rezzoug, S. A., Maache-Rezzoug, Z., Debs, E. and Louka, N., 2024. Comparative review on the production and purification of bioethanol from biomass: A focus on corn. *Processes.* 12(5), 1001.

Bansemir, A., Blume, M., Schröder, S. and Lindequist, U. 2006. Screening of cultivated seaweeds for antibacterial activity against fish pathogenic bacteria. *Aquaculture.* 252, 79–84.

Benitez, G. B., Ayala, N. F. and Frank, A. G., 2020. Industry 4.0 innovation ecosystems: An evolutionary perspective on value cocreation. *Int. J. Prod. Econ.* 228, 1–13.

Bhuyan, B. P., Tomar, R., Singh, T. P. and Cherif, A. R., 2022. Crop type prediction: A statistical and machine learning approach. *Sustain.* 15, 481.

Bozell, J. J. and Petersen, G. R., 2010. Technology development for the production of biobased products from biorefinery carbohydrates-the US department of Energy's "Top 10" revisited. *Green Chem.* 12(4), 539–554.

Breque, M., Nul, L. D. and Petrides, A., 2021. *Industry 5.0 – Towards a Sustainable, Human-centric and Resilient European Industry*, Publications Office of the European Union, Luxembourg.

Cai, D., Li, P., Luo, Z., Qin, P., Chen, C., Wang, Y., et al., 2016. Effect of dilute alkaline pretreatment on the conversion of different parts of corn stalk to fermentable sugars and its application in acetone–butanol–ethanol fermentation. *Bioresour. Technol.* 211, 117–124.

Capodaglio, A. G., 2020. Fit-for-purpose urban wastewater reuse: Analysis of issues and available technologies for sustainable multiple barrier approaches. *Crit. Rev. Environ. Sci. Technol.* 51, 1619–1666.

Chandel, A. K., da Silva, S. S., Carvalho, W. and Singh, O. V., 2012. Sugarcane bagasse and leaves: Foreseeable biomass of biofuel and bio-products. *J. Chem. Technol. Biotechnol.* 87, 11–20.

Chen, D., Yin, L., Wang, H. and He, P., 2015. Reprint of: pyrolysis technologies for municipal solid waste: A review. *Waste Manag.* 37, 116–136.

Cheng, H.-H. and Whang, L.-M., 2022. Resource recovery from lignocellulosic wastes via biological technologies: Advancements and prospects. *Bioresour. Technol.* 343, 126097.

Cherubini, F., 2010. The biorefinery concept: Using biomass instead of oil for producing energy and chemicals. *Energy Convers Managements* 51(7), 1412–1421.

Cherubini, F., Jungmeier, G., Wellisch, M., Willke, T., Skiadas, I., Ree, R. V., et al., 2009. Toward a common classification approach for biorefinerysystems. *Biofuels Bioprod. Biorefin.* 3, 534–546.

Ciruela-Lorenzo, A. M., Del-Aguila-Obra, A. R., Padilla-Meléndez, A. and PlazaAngulo, J. J., 2020. Digitalization of agri-cooperatives in the smart agriculture context. Proposal of a digital diagnosis tool. *Sustainability* 12(4), 1–15.

Coughlan, N. E., Walsh, É., Bolger, P., Burnell, G., O'Leary, N., O'Mahoney, M., Paolacci, S., Wall, D. and Jansen, M. A. K., 2022. Duckweed bioreactors: Challengesand opportunities for large-scale indoor cultivation of lemnaceae. *J. Clean. Prod.* 336, 130285.

Culaba, A. B., Mayol, A. P., Juan, J. L. G. S., Vinoya, C. L., Concepcion II, R. S., Bandala, A. A., Vicerra, R. R. P., Ubando, A. T., Chen, W.-H. and Chang, J.-S., 2022. Smart sustainable biorefineries for lignocellulosic biomass. *Bioresour. Technol.* 334(Part B), 1–13.

Dahiya, A. 2020. *Bioenergy: Biomass to Biofuels and Waste to Energy*, Academic Press, Cambridge.

Dalenogare, L. S., Benitez, G. B., Ayala, N. F., and Frank, A. G., 2018. The expected contribution of Industry 4.0 technologies for industrial performance. *Int. J. Prod. Econ.* 204, 383–394.

de Jong, E., van Ree, R. and Kwant, I. K., 2009. Biorefineries: Adding value to the sustainable utilisation of biomass. *IEA Bioenergy* 1, 1–16.

de Souza, L. and Shivakumar, S., 2021. Conversion of agro-industrial wastes for the manufacture of bio-based plastics. In: Mohammed Kuddus and Roohi (Eds.), *Bioplastics for Sustainable Development*. Springer, Singapore, pp. 177–204.

Deora, P. S., Verma, Y., Muhal, R. A., Goswami, C., and Singh, T., 2022. Biofuels: An alternative to conventional fuel and energy source. *Mater. Today Proc.* 48, 1178–1184.

Dey, T., Bhattacharjee, T., Nag, P., Ghati, A., and Kuila, A., 2021. Valorization of agro-waste into value added products for sustainable development. *Bioresour. Technol. Rep.* 16, 100834.

Dineshbabu, G., Goswami, G., Kumar, R., Sinha, A. and Das, D., 2019. Microalgae–nutritious, sustainable aqua-and animal feed source. *J. Funct. Foods* 62, 103545.

Dornburg, V., Faaij, A., Verweij, P., Banse, M., Van Diepen, K., Van Keulen, H., Langeveld, H., Meeusen, M., Van de Ven, G., Wester, F. and Alkemade, R., 2008. Biomass Assessment. Assessment of global biomass potentials and their links to food, water, biodiversity, energy demand and economy. Inventory and analysis of existing studies.

Dragone, G., Kerssemakers, A. A. J., Driessen, J. L. S. P., Yamakawa, C. K., Brumano, L. P. and Mussatto, S. I., 2020. Innovation and strategic orientations for the development of advanced biorefineries. *Bioresour. Technol.* 302, 1–12.

EA Bioenergy Task 42. 2007. Biorefinery: Processes for converting biomass into energy and chemical products.

Ebadian, M., van Dyk, S., McMillan, J. D. and Saddler, J., 2020. Biofuels policies that have encouraged their production and use: An international perspective. *Energy Pol.* 147, 111906.

EPA, 2013. Renewable Fuel Standard (RFS), Regulations & Standards United States Environmental Protection Agency.

Fabris, M., Abbriano, R. M., Pernice, M., Sutherland, D. L., Commault, A. S., Hall, C. C., et al., 2020. Emerging technologies in algal biotechnology: Toward the establishment of a sustainable, algae-based bioeconomy. *Front. Plant Sci.* 11, 1–22.

Faragò, M., Damgaard, A., Madsen, J. A., Andersen, J. K., Thornberg, D., Andersen, M. H. and Rygaard, M. 2021. From wastewater treatment to water resource recovery: Environmental and economic impacts of full-scale implementation. *Water Res.* 204, 117554.

Fava, F., Totaro, G., Diels, L., Reis, M., Duarte, J., Carioca, O. B., et al., 2015. Biowastebiorefinery in Europe: Opportunities and research & development needs. *New Biotechnol.* 32(1), 100–108.

Fernando, S., Adhikari, S., Chandrapal, C. and Murali, N., 2006. Biorefineries: Current status, challenges, and future direction. *Energy Fuel* 20(4), 1727–1737.

Frank, A. G., Dalenogare, L. S. and Ayala, N. F., 2019. Industry 4.0 technologies: Implementation patterns in manufacturing companies. *Int. J. Prod. Econ.* 210, 15–26.

Hatim, Q. Y., Saldana, C., Shao, G., Kim, D. B., Morris, K. C., Witherell, P., Rachuri, S. and Kumara, S., 2020. A decision support methodology for integrated machining process and operation plans for sustainability and productivity assessment. *Int. J. Adv. Manuf. Technol.* 107, 3207–3230.

Hoang, A. T., Tabatabaei, M., Aghbashlo, M., Carlucci, A. P., Ölçer, A. I., Le, A. T., et al., 2021. Rice bran oil-based biodiesel as a promising renewable fuel alternative to petrodiesel: A review. *Renew. Sustain. Energy Rev.* 135, 110204.

Hou, D., Bolan, N. S., Tsang, D. C. W., Kirkham, M. B. and O'Connor, D., 2020. Sustainable soil use and management: An interdisciplinary and systematic approach. *Sci. Total Environ.* 729, 1–12.

Igwo-Ezikpe, M. N., Ayanshina, A. O. and Babalola, M., 2022. Optimization of chemical pre-treatment of Cassava Bagasse for reducing sugar and bioethanol production. *Univ. Lagos J. Basic Med. Sci.* 3, 14–23.

Isikgor, F. and Becer, C., 2015. Lignocellulosic biomass: A sustainable platform forproduction of bio-based chemicals and polymers. *Polym. Chem.* 6, 4497–4559.

Jacobson, K., Gopinath, R., Meher, L. and Dalai, A. 2008. Solid acid catalyzed biodiesel production from waste cooking oil. *Appl. Catal. B Environ.* 85, 86–91.

Jiang, Y., Wu, R., Zhou, J., He, A., Xu, J., Xin, F. and Dong, W., 2019. Recent advances of biofuels and biochemicals production from sustainable resources using coco-cultivation systems. *Biotechnol. Biofuels* 12(1), 1–12.

Khan, I. S., Ahmad, M. O. and Majava, J., 2021. Industry 4.0 and sustainable development: A systematic mapping of triple bottom line, circular economy and sustainable business models perspectives. *J. Clean. Prod.* 297, 1–16.

Khoo, K. S., Chia, W. Y., Tang, D. Y. Y., Show, P. L., Chew, K. W. and Chen, W.-H., 2020. Nanomaterials utilization in biomass for biofuel and bioenergy production. *Energies* 13(4), 1–19.

Klein, S., 2017. *IoT Solutions in Microsoft's Azure IoT Suite: Data Acquisition and Analysis in the Real World*, Apress, Berkeley, CA.

Kumar, A., Kumar, N., Baredar, P. and Shukla, A., 2015. A review of biomass energy resources, potential, conversion, and policy in India. *Renew. Sustain. Energy Rev.* 45, 530–539.

Kumar, R., Singh, R. K. and Dwivedi, Y. K., 2020. Application of Industry 4.0 technologies in SMEs for ethical and sustainable operations: Analysis of challenges. *J. Clean. Prod.* 275, article no. 124063.

Li, A., Antizar-Ladislao, B. and Khraisheh, M., 2007. Bioconversion of municipal solid waste to glucose for bio-ethanol production. *Bioprocess Biosyst. Eng.* 30, 189–196.

Malolan, R., Gopinath, K. P., Vo, D.-V. N., Jayaraman, R. S., Adithya, S., Ajay, P. S. and Arun, J., 2021. Green ionic liquids and deep eutectic solvents for desulphurization, denitrification, biomass, biodiesel, bioethanol and hydrogen fuels: A review. *Environ. Chem. Lett.* 19, 1001–1023.

Meindl, B., Ayala, N. F., Mendonça, J. and Frank, A. G., 2021. The four smarts of Industry 4.0: Evolution of ten years of research and future perspectives. *Technol. Forecast. Social Change* 168, 1–13.

Mohr, A. and Raman, S., 2013. Lessons from first generation biofuels andimplications for the sustainability appraisal of second generation biofuels. *Energy Policy* 63, 114–122.

Moreira, J. R., 2006. Brazil's experience with bioenergy. In: Hazell, P. and Pachauri, R. K. (Eds.), *Bioenergy and Agriculture: Promises and Challenges*. Focus 14. Int. Food Policy Research Inst, Washington, DC, pp. 1–2.

Mudzakir, A., Rizky, K. M., Munawaroh, H. S. H. and Puspitasari, D., 2022. Oil palm empty fruit bunch waste pretreatment with ben-zotriazolium-based ionic liquids for cellulose conversion to glucose: Experiments with computational bibliometric analysis. *Indones. J. Sci. Technol.* 7, 291–310.

Nahavandi, S., 2019. Industry 5.0 – A human-centric solution. *Sustainability* 11(16), 43–71.

Naik, S. N., Goud, V. V., Rout, P. K. and Dalai, A. K., 2010. Production of first and second generation biofuels: A comprehensive review. *Renew. Sust. Energ. Rev.* 14, 578–597.

Nakada, S., Saygin, D. and Gielen, D., 2014. Global Bioenergy SUPPLY and DEMAND PROJECTIONS A Working Paper for REmap 2030.

Ning, P., Yang, G., Hu, L., Sun, J., Shi, L., Zhou, Y. and Yang, J., 2021. Recent advances in the valorization of plant biomass. *Biotechnol. Biofuels* 14(1), 102.

Octave, S. and Thomas, D., 2009. Biorefinery: Toward an industrial metabolism. *Biochimie* 91(6), 659–664.

Olguín, E. J., Sánchez-Galván, G., Arias-Olguín, I. I., Melo, F. J., González-Portela, R. E., Cruz, L., et al., 2022. Microalgae-based biorefineries: Challenges and future trends to produce carbohydrate enriched biomass, high-added value products and bioactive compounds. *Biology* 11(8), 1146.

Oztemel, E. and Gursev, S., 2018. Literature review of Industry 4.0 and related technologies. *J. Intell. Manuf.* 31(1), 127–182.

Palomar, C. R., García-Alvaro, A., de Almeida Guimarães, V., Hedo, E. B., Muñoz, R. and de Godos Crespo, I., 2023. Biofuels in low carbon economies and societies. In: Bandh, S. A. and Malla, F. A. (Eds.), *Biofuels in Circular Economy*. Springer, Berlin, pp. 31–58.

Patil, G. L., Gawande, P. S. and Bag, R. V., 2017. Smart agriculture system based on IoT and its social impact. *Int. J. Comp. Appl.* 176(1), 1–4.

Pfaltzgraff, L. A. and Clark, J. K., 2014. Green chemistry, biorefineries and second generation strategies for re-use of waste: An overview. In: Waldron, K. (Ed.), *Advances in Biorefineries*. Woodhead Publishing Limited, Cambridge, pp. 3–33.

Phan, A. N. and Phan, T. M., 2008. Biodiesel production from waste cooking oils. *Fuel* 87, 3490–3496.

Popp, J., Lakner, Z., Harangi-Rakos, M. and Fari, M., 2014. The effect of bioenergy expansion: Food, energy, and environment. *Renew. Sustain. Energy Rev.* 32, 559–578.

Prasad, B. R., Padhi, R. K. and Ghosh, G., 2022. A review on key pretreatment approaches for lignocellulosic biomass to produce biofuel and value-added products. *Int. J. Environ. Sci. Technol.* 2022, 1–16.

Prasad, S., Rathore, D. and Singh, A., 2017. Recent advances in biogas production. *Chem. Eng. Process Tech.* 3(2), 1038.

Qandil, M. D., Abbas, A. I., Abdelhadi, A. I., Salem, A. R. and Amano, R. S., 2021. Energy analysis: Ways to save energy and reduce the emissions in wastewater treatment plants. *Int. J. Energy Clean Environ.* 22, 91–112.

Rafeeq, H., Afsheen, N., Rafique, S., Arshad, A., Intisar, M., Hussain, A., et al., 2023. Genetically engineered microorganisms for environmental remediation. *Chemosphere* 310, 136751.

Rahmani, A. M., Gahlot, P., Moustakas, K., Kazmi, A. A., Ojha, C. S. P. and Tyagi, V. K., 2022. Pretreatment methods to enhance solubili-zation and anaerobic biodegradability of lignocellulosic biomass (wheat straw): Progress and challenges. *Fuel* 319, 123726.

Rejeb, A., Rejeb, K. and Zailani, S., 2021. Big data for sustainable agri-food supply chains: Review and future research perspectives. *J. Data Inform. Manage.* 3(3), 167–182.

Saravanan, A., Kumar, P. S., Jeevanantham, S., Karishma, S. and Vo, D.-V. N., 2022. Recent advances and sustainable development of biofuels production from lignocellulosic biomass. *Bioresour. Technol.* 344, 126203

Shahid, M. K., Batool, A., Kashif, A., Nawaz, M. H., Aslam, M., Iqbal, N. and Choi, Y., 2021. Biofuels and biorefineries: Development, application and future perspectives emphasizing the environmental and economic aspects. *J. Environ. Manage.* 297, 113268.

Sims, R. E. H., Mabee, W., Saddler, J. N. and Taylor, M., 2010. An overview of second generation biofuel technologies. *Bioresour. Technol.* 101, 1570–1580.

Surriya, O., Saleem, S. S., Waqar, K., Kazi, A. G. and Oztürk, M., 2015. Bio-fuels: A blessing in disguise. In: Öztürk, M., Ashraf, M., Aksoy, A., and Ahmad, M. S. A. *Phytoremediation for Green Energy*. Springer, Dordrecht, pp. 11–54.

Talebian-Kiakalaieh, A., Amin, N. A. S. and Mazaheri, H., 2013. A review on novel processes of biodiesel production from waste cooking oil. *Appl. Energy* 104, 683–710.

Tang, D., Tan, G. L., Li, G. W., Liang, J. G., Ahmad, S. M., Bahadur, A. and Bououdina, M., 2023. State-of-the-art hydrogen generation techniques and storage methods: A critical review. *J. Energy Storage* 64, 107196.

Thrän, D., Bunzel, K. and Witing, F., 2012. Sustainable Bioenergy Cropping. Presentation, 12th Congress of the European Society for Agronomy. Helsinki, Finland, 20–24 August 20 and author's calculation.

Wang, J., Ma, Y., Zhang, L., Gao, R. X. and Wu, D., 2018. Deep learning for smart manufacturing: Methods and applications. *J. Manuf. Syst.* 48(Part C), 144–156.

Zhou, Z., Ouyang, D., Liu, D. and Zhao, X. 2022. Oxidative pretreatment of lignocellulosic biomass for enzymatic hydrolysis: Progress and challenges. *Bioresour. Technol.* 367, 128208.

Chapter 10

Computational fluid dynamics studies on microfluidic fuel cells

Insights and future directions

Balasubramanian Sivasamy, Ashish Kapoor, Laxmi Deepak Bhatlu, Bonasi Krishna Srihari, Amit Kumar Rathoure, and Dan Bahadur Pal

10.1 INTRODUCTION

Demand for affordable and reliable energy has become important in our society as it supports technological advancements and improves the quality of life. Even though the energy technologies have advanced in the recent past, the global energy production continues to depend widely on fossil fuels. This dependence poses a major challenge to protect our environment and address climate change. To tackle these challenges and neutralize carbon emissions by 2050, transitioning from fossil fuel-based energy systems to renewable sources like solar, wind, and hydrogen has become essential. Due to the periodic nature of renewable energy sources (For instance, wind and solar do not produce power continuously), there is a need to develop ways to produce, store, and transport energy. Fuel cells have emerged as a potential solution to meet these challenges, because it is capable to directly convert chemical energy into electrical energy [1].

Fuel cells offer the following benefits, such as low carbon emission, eco-friendly energy production, reduced fossil fuel dependence, and economic. These systems convert chemical energy into electrical power through electrochemical reactions. Specifically, fuel cells utilize two redox (oxidation and reduction) reactions where fuel undergoes oxidation at the anode and the oxidant reduction takes place at the cathode. An electrolyte between the electrodes plays an essential role in ionic conduction and electrical isolation [2]. The fuel cells are widely used in backup power systems, electric vehicles, and unmanned aerial vehicles. Apart from the above-mentioned benefits, the notable advantages, such as high energy efficiency, superior energy density, and the ability to suppress noise and vibration, make them attractive for modern energy solutions [2].

Researchers have developed various types of fuel cells to further enhance its performance, which include: PEMFCs – proton exchange membrane fuel cells, solid oxide fuel cells – SOFCs, AMFCs – alkaline membrane fuel cells, BFCs – biofuel cells, PFCs – photocatalytic fuel cells and MFCs – microfluidic fuel cells [3]. Of these, MFCs have received greater attention because of

DOI: 10.1201/9781003581536-10

their unique design and advanced capabilities. Compared to traditional fuel cells, MFCs make use of parallel-laminar flows (or low Reynolds number flow) of anode and cathode electrolytes, where the need for proton exchange membranes is eliminated. This configuration overcomes the limitations, such as an increase in ohmic resistance and management of complex hydro-thermal effects, which in turn lowers the efficiency of conventional fuel cells [3]. MFCs are compact, membrane-free, and cost-effective, offering better fuel utilization, lower catalyst needs, and faster response. They operate on diverse fuels under mild conditions, making them ideal for portable and micro-device applications.

Small size, light weight, and portable nature of MFCs design configuration has paved the way for the applications which include: energy supply to portable electronic systems (sensors, portable and wearable electronic gadgets), biomedical tests (lab-on-chip devices), diagnostics (point-of-care), environmental monitoring, and wastewater treatment. Consequently, MFCs have become a prominent subject in emerging technologies, demonstrating significant potential to address future energy challenges [3].

There has been significant work carried out both experimentally and computationally, and it is reported in the literature reviews [1,4–8]. On the other hand, this chapter does not focus on specific emerging sub-classes of microfluidic fuel cells (MFCs), including biofuel cells that utilize microbes or enzymes for catalyzing electrochemical reactions, capillary-driven paper-based systems, or photochemical cells that rely on photocatalytic redox reactions, due to their specialized nature and distinct operational mechanisms, which fall outside the scope of the current discussion.

Although the development of microfluidic fuel cells (MFCs) began with the introduction of the first experimental prototype by Ferrigno et al. [9] and Choban et al. [10], marking the start of experimental research in this area, it was not until 2004 that Bazylak et al. [11] conducted the first computational fluid dynamics (CFD) study on MFCs. They developed a detailed electrochemical model that described the essential electrochemical reactions and established the necessary boundary conditions for the MFC system, which was then analyzed through numerical simulations. Their pioneering work laid the foundation for advancing the research by employing CFD as a tool to better understand the complex performance characteristics of typical MFCs.

10.2 BASIC PRINCIPLES

A microfluidic fuel cell is a compact energy-conversion device that operates using microscale fluidic channels to bring fuel and oxidant streams into close proximity without mixing, typically in a laminar flow configuration. This setup facilitates controlled electrochemical reactions at the anode and cathode, generating electricity directly from the chemical energy of the fuel.

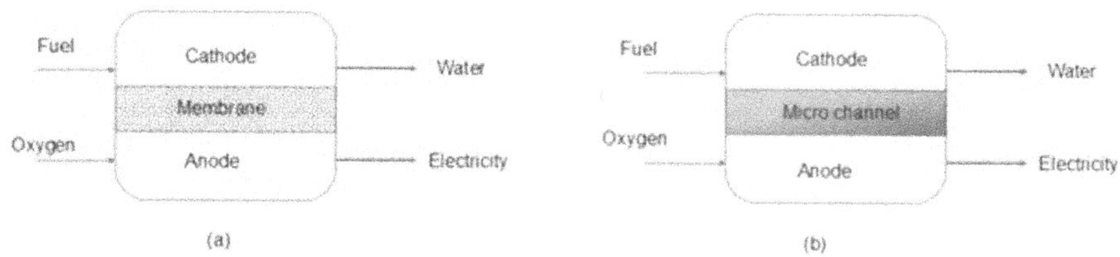

Figure 10.1 Block diagrams illustrating (a) a conventional fuel cell and (b) a microfluidic fuel cell or membraneless fuel cell.

Unlike conventional fuel cells, these systems often do not require a physical membrane for ion separation, making them simpler and more cost-effective for small-scale or portable applications. A typical block diagram that represents the microfluidic device and a conventional fuel cell is shown in Figure 10.1.

The key components in the design of a typical microfluidic cell are the inlet streams, the microchannel, the side walls made up of electrodes (cathode and anode), catalysts, and the outlet streams. A typical microfluidic cell with key components for the energy conversion is presented in Figure 10.2. The two *inlet streams* represent the fuel (e.g., hydrogen, methanol) and oxidant (e.g., oxygen, air) entering the microfluidic channel. These streams are separated by the laminar flow characteristic of microfluidic systems, ensuring no physical mixing but allowing electrochemical interaction. The vertical central region is the *microchannel* (main flow channel) where the fuel and oxidant flow in parallel laminar streams. This channel is designed to maximize the electrochemical reaction area, where fuel oxidation and oxidant reduction occur. The fuel is oxidized to produce Protons and Electrons, i.e. Fuel → Protons (H^+) + Electrons (e^-) [Oxidation Products]. The oxidant is reduced by accepting electrons and reacting with protons to form a product, i.e. Oxidant + Protons (H^+) + Electrons (e^-) → Reduction Products. The side walls of the channel are lined with *electrodes* (anode and cathode). On the anode side, fuel undergoes oxidation, and on the cathode side, the oxidant undergoes reduction. These electrodes are coated with *catalysts* (e.g., platinum or other materials) to facilitate the reactions. The central portion (diffusion layer) allows for ion transfer, specifically protons (H^+), across a separator or membrane (or through the electrolyte in membraneless designs). This drives the electrochemical reaction to generate electricity. The output streams exit the system with reaction products (e.g., water/unreacted fuel and oxidant).

The key advantages of these microfluidic cells are: compact and lightweight, with an extremely small size and a lightweight design, ideal for portable devices. Delivers sustained power with high energy density relative to

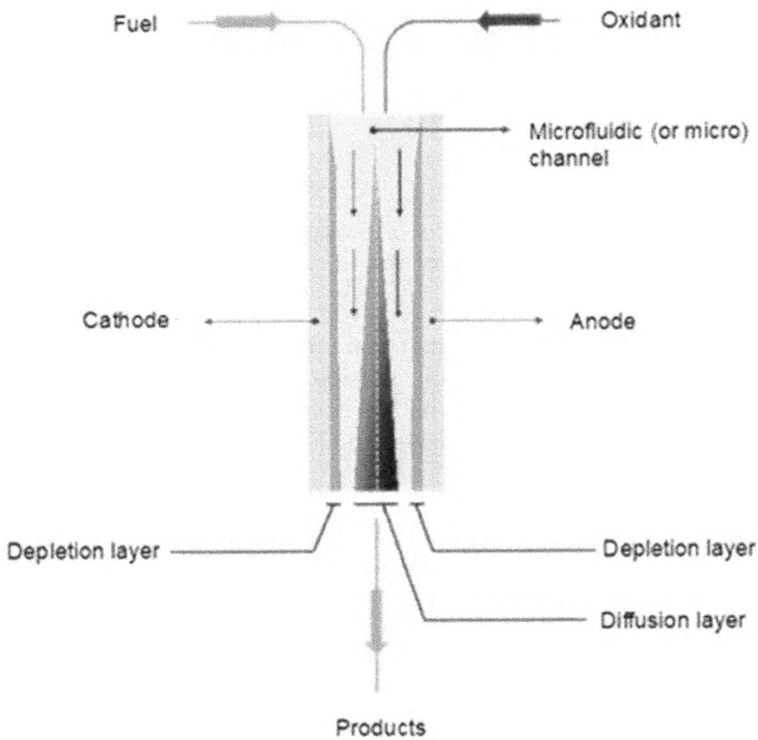

Figure 10.2 A schematic representation of a membraneless microfluidic cell with key components typically designed for the small-scale energy generation, typically used in the portable electronic applications.

its size. This design eliminates the need for a membrane, reducing costs and fouling. Can use a variety of fuels (e.g., hydrogen, methanol) and oxidants (e.g., oxygen). Modular design allows integration of multiple cells for higher power output. Operates efficiently at room temperature or slightly elevated temperatures. Produces by-products (e.g., water, CO_2) and reduces reliance on harmful batteries. As a result, MFCs have a wide range of applications because of their compactness in size, efficiency, and adaptability. The detailed applications of MFCs with a short description are provided in Table 10.1.

10.3 FACTORS AFFECTING MICROFLUIDIC FUEL CELL PERFORMANCE AND THEIR MITIGATION STRATEGIES

The development of MFCs poses many challenges, right from the selection of fuels, oxidants, device design, fabrication, performance, to its application.

Table 10.1 Different applications of MFCs

S. No.	*Applications*	*Description*
1	Biomedical	Act as power sources for implantable devices (e.g., pacemakers, biosensors) and micro-electromechanical systems (MEMS) for medical applications.
2	Educational and Research Tools	Teaching electrochemistry principles and sustainable energy technologies, as well as for experimental research in scientific studies.
3	Environmental Monitoring	Power compact sensors for air and water quality monitoring, as well as remote systems for environmental observation.
4	Lab-on-a-Chip Devices	Provide energy to microfluidic systems used in chemical analysis, DNA sequencing, drug development, and point-of-care diagnostic tools.
5	Military and Aerospace	Power portable communication devices, drones, and remote monitoring equipment in military and aerospace applications.
6	Portable Power Source	Serve as power sources for small electronic devices such as wearable electronics, portable medical devices (e.g., glucose monitors, insulin pumps), and personal gadgets (e.g., smartwatches).
7	Robotics and Automation	Used in small-scale robots and autonomous systems, offering a reliable power source for low-power and remote operations.

Table 10.2 outlines the various factors that influence MFCs' system design and its mitigation strategies for the design of MFCs.

10.3.1 CFD simulations

Computational fluid dynamics (CFD) as a tool plays a vital role in the design and analysis of MFCs. The CFD studies enable the researchers to gain valuable insights into the operational parameters (mass, momentum, heat, and species transport), geometric configuration, overall efficiency, and functionality of the MFCs. In order to conduct the CFD studies for the development of MFCs, the typical steps from 1 to 9, outlined in Figure 10.3, can be followed. These steps allow the researchers to explore the design improvements and the performance metrics of the MFCs.

The steps outlined in Figure 10.3 are grouped into three major stages: (a) pre-processing (steps 1–5), (b) solver (step 6), and (c) post-processing (Steps, 7–9). A comprehensive analysis of all these three stages, comprising various assumptions, governing equations (mass, momentum, species, and charge conservation relations), and the boundary conditions, is reported in Wu et al. [12].

Table 10.2 Key challenges and mitigation strategies in MFC Design

Factors	*Challenges*	*Mitigation strategies*
Fuel and oxidant properties	Low energy density fuels or impurity-related inefficiencies.	Use high-purity fuels and oxidants; explore alternative fuels like formic acid for higher efficiency.
Flow rate	Inefficient flow may limit fuel utilization or cause reactant crossover.	Optimize flow rate using computational fluid dynamics (CFD) simulations to balance delivery and crossover.
Electrode materials	High cost of platinum catalysts and limited durability.	Develop cost-effective alternatives like graphene, metal oxides, or composite materials.
Microchannel design	Suboptimal geometry causing poor reactant distribution or high pressure drops.	Design microchannels with optimized shapes (e.g., serpentine or branched) and ensure fabrication precision.
Diffusion layer efficiency	Fuel and oxidant mixing due to poor diffusion layer design.	Improve material design for the diffusion layer; use porous materials or hydrophobic/hydrophilic coatings.
Operating temperature	Low temperatures reduce reaction rates; high temperatures degrade components.	Maintain optimal temperature with integrated microheaters or cooling systems, depending on requirements.
Electrolyte properties	Poor ionic conductivity and low stability in some electrolytes.	Use ionic liquids, gel electrolytes, or advanced polymers to improve ionic transport and chemical stability.
Fuel crossover	Fuel wastage and reduced efficiency due to unintended mixing.	Optimize channel width, flow velocity, and use specific channel coatings or barriers to minimize crossover.
Durability and stability	Degradation of electrodes and clogging of channels over time.	Develop anti-corrosion coatings, self-cleaning surfaces, and robust fabrication materials for longevity.
Integration and scaling	Complexities in scaling up from prototype to real-world applications.	Adopt scalable fabrication techniques like 3D printing or soft lithography; use modular designs.
Current and power density	Limited by reaction kinetics, mass transport, and electrode surface area.	Enhance catalysts, optimize flow and reactant concentrations, and increase active electrode surface area.
Fuel utilization efficiency	Incomplete fuel oxidation or fuel wastage due to improper flow rates or poor diffusion control.	Use multi-pass flow designs, adjust flow rates, and improve catalyst selectivity to maximize fuel utilization.
Open-circuit voltage (OCV)	Losses due to electrode overpotential, fuel crossover, or poor reaction kinetics.	Use high-performance catalysts, minimize fuel crossover, and improve electrolyte conductivity.

10.3.2 Developments in CFD studies over time

The CFD studies have played a vital role in understanding the optimization of device geometry, complexities in transport phenomena, and improvement of performance metrics, for instance, the current density and fuel utilization.

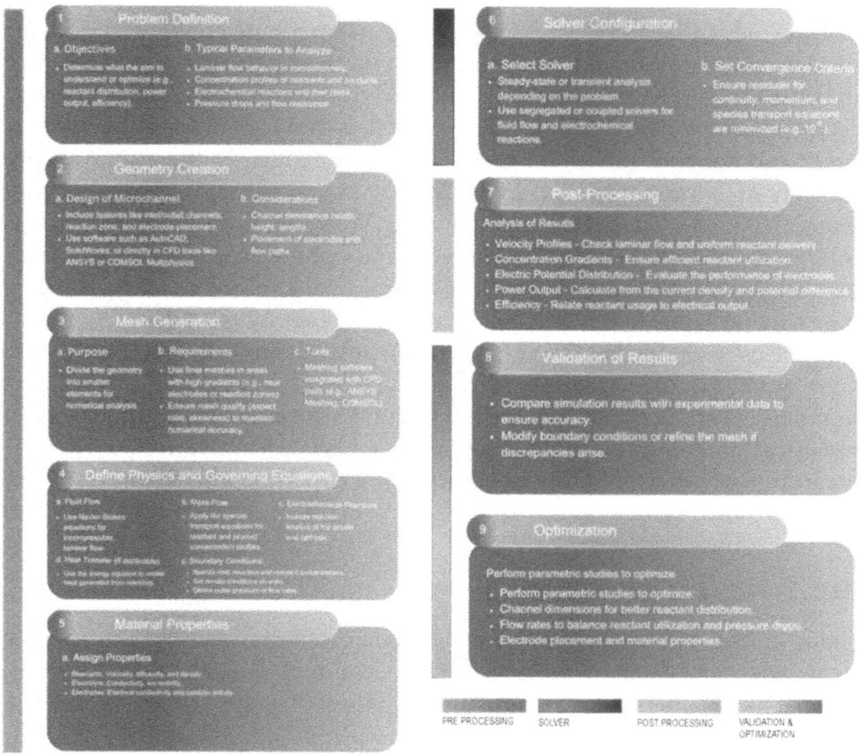

Figure 10.3 Schematic representation of typical steps (1–9) in CFD analysis for the development of MFCs.

10.3.3 Foundational studies

The foundational work on MFCs was started between 2004 and 2007, emphasizing the principles of laminar flow and its role in the design of membraneless designs in the development of MFCs. Bazylak et al. [11,13] was the first to introduce the CFD studies on membraneless liquid-fuel MFCs. In this work, the authors explore the potential of laminar flow for fuel and oxidant separation without the need for a of membrane. In 2006, Chang et al. investigated the effect of laminar flow in Y-shaped microchannels, the analysis was mainly focused on the geometrical design [14]. Lee et al. observed principles of species transport to optimize the performance of MFC [15]. They also highlighted the role of the dynamics of fluid flow in device efficiency. With these foundational concepts established, the focus shifted toward improving geometric designs and performance metrics.

10.3.4 Design improvements

Shaegh et al. and Ahmed et al. investigated air-breathing configurations with porous anodes and the implications of geometric design on the performance of MFCs [16,17]. Krishnamurthy developed computational models of porous electrodes, enhancing the understanding of flow-through designs [18]. In that same year, Sprague and Dutta investigated diffuse charge effects, advancing the theoretical framework [19]. The authors, H. Wang et al., Xuan, Leung, Leung, Ni, et al.; Xuan, Leung, and Leung, Wang, et al., addressed polarization losses through chaotic flow patterns and optimized transport behaviors [20–22]. Xuan et al. analyzed critical operational conditions and the effects of gravity on laminar flow, advancing design reliability [23,24].

The Research further expanded to include energy efficiency analyses, Xuan et al., H. Zhang, Leung, et al., H. Zhang, Xuan, et al. [25–27], and boundary layer control Hassanshahi and Fathipour [28], supported by numerical modeling of channel geometries [29]. The advanced modeling, integration of advanced CFD techniques, and experimental validations were performed between 2014 and 2017.

10.3.5 Advanced modeling

Zhang et al. and Lopez-Montesinos applied three-dimensional simulations to analyze flow-over and bridge-shaped channel designs [30]. Li et al. conducted parametric sensitivity studies in flow through porous electrodes, while Ye et al. modeled alkaline MFCs with cylindrical anodes [31,32]. Xu et al. achieved the highest fuel utilization with greater than 90% in a microfluidic fuel cell using counter flow of formic acid [33]. Li, Nikiforidis, et al. developed an innovative approach, the flow-through porous electrodes with improvements in design flexibility [34]. In 2017, the effect of different channel geometries and electrode configurations on the performance of microfluidic fuel cells (MFCs) was investigated and provided insights into design optimization strategies [35]. Li et al. conducted both experimental and numerical studies to enhance the current collectors' design employing porous electrodes and optimized the performance of the MFC. In the same year, Herlambang et al. performed numerical simulations to evaluate air-breathing MFC using direct formic acid. The outcome of the simulation revealed valuable data on the MFCs' operational efficiency [36]. Additionally, Wang et al. carried out a comparative analysis for the co-flow and counter-flow configurations of MFCs by experimental and numerical approaches. The study showed the advantages and limitations of each of the flow configurations (co-flow and counter flow) [37].

10.3.6 Multiphysics and AI integration

In 2018, Li et al. studied the importance of the resistance of the porous electrode in order to improve the overall performance of the MFC. This study was focused on the properties of material and design to achieve higher efficiency and reliability in fuel cell applications [38]. The recent research adopts multidimensional approaches and AI-driven optimization. Wei et al. explored the multi-layered electrode structure and showed that the methanol crossover was effectively controlled, but the flow resistance for discharge products was increased, while the variations in the temperature are linked to the distribution of current density. Sensitivity analysis showed the effect of cell parameters (methanol concentration, porosity, and volume fraction of CO_2) on polarization characteristics [39]. A dual-electrolyte protocol for counter-flow-based MFC was proposed for diverse electrochemical applications and showed a significant advancement in the performance of the device. This approach explored the integration of optimal flow configurations to enhance electrochemical processes. The methodology adopted in this work addressed the challenges associated with the fluid dynamics and electrochemical interactions [40]. The analysis of inlet locations and the multiple compartment configurations was conducted by Tanveer and Kim, whereas Rösing et al. adopted a passive control technique, Dean vortices, to manage boundary layer concentrations [41]. Ouyang et al. analyzed force fields and the influence of design parameters, exergy efficiency, and utilization of fuel in MFCs by employing advanced models. the research explores the interplay of various design factors, providing insights to enhance the performance of MFC [42]. Additionally, a new approach in modeling was developed to understand the coupling of vibrations in MFCs, as well as hydrodynamics, electrochemical kinetics, mass transport, and vibration fields. The study revealed superior resistance to vibrations compared to flow-over designs, providing optimal anti-vibration performance [43]. Fu et al. examined advanced pore-scale modeling for the transport of oxygen transport [44]. Zuria et al. studied numerically the evolution of mixed-reactant in MFCs [45].

Tanveer and Kim proposed the hollow flow channel in a membraneless MFC with porous flow-through electrodes and evaluated its performance. The proposed hollow structure showed a peak power density when compared to the other channel shapes such as H-shaped, simple bridge-shaped, and rectangular channels, by 6%, 18%, and 82%, respectively. The study also showcased its potential for innovative design and improved efficiency [46]. Zhi et al. explored the performance of an H-shaped air-breathing microfluidic fuel cell stack and provided insights into the operational efficiency and design optimization of this configuration. The work also highlighted its potential for enhanced utilization of fuel and conversion of energy [3].

Ouyang evaluated the anti-vibration strategies employing phase field theory [47]. Fu et al. probed into a pore-scale model to understand mass transport in the air-breathing cathode of membraneless MFCs. The model considered transport of oxygen within the porous cathode and provided insights into the performance of fuel cell by focusing the limitations in diffusion of oxygen and reaction kinetics [48].

In recent times, optimization techniques driven by AI have emerged as a powerful tool. Oh et al. optimized the performance of an MFC employing a double-bridge flow channel both computationally and experimentally. The authors presented the enhanced utilization of fuel and reduced fuel and oxidant mixing in the double-bridge configuration, resulting in improved energy conversion. The findings also underline the importance of innovative channel designs in advancing the MFCs for practical applications [49]. Nguyen et al. employed deep learning techniques to explore the optimization of MFC. The study incorporates a data-driven approach with a three-dimensional Multiphysics simulation to optimize the design and operational

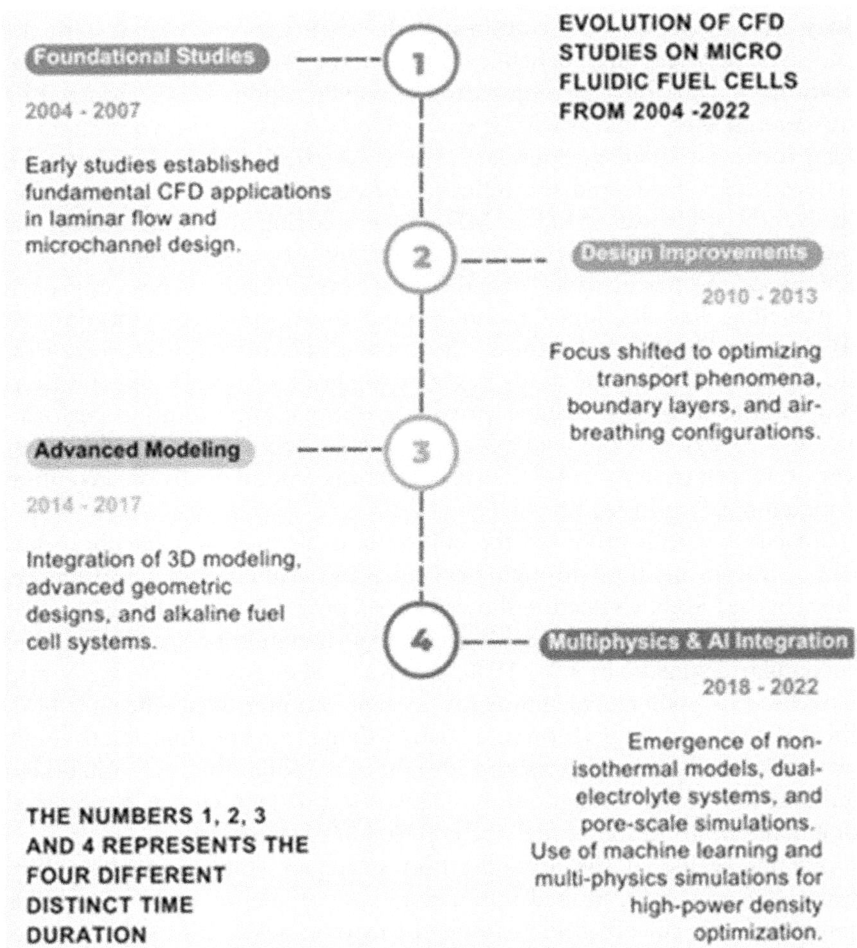

Figure 10.4 The timeline of evolution of CFD studies with of key developments.

Table 10.3 Chronological categorization of computational fluid dynamics (CFD) studies in microfluidic fuel cells (MFCs) from 2004 to 2022 highlighting key advancement, focus areas and the contributions

Time line	*Year*	*Authors*	*Focus area*	*Key contribution*
Foundational Studies (2004–2007)	2004 and 2005	[11,13]	Foundational studies on membraneless liquid-fuel MFCs	Early CFD modeling of laminar flow and fuel utilization in membraneless MFCs.
	2006	[14]	Laminar flow in Y-shaped microchannels	Analysis of the effect of channel geometry on performance.
	2007	[15]	Optimization using transport principles	Development of CFD-based optimization strategies for MFCs.
Design Improvements (2010–2013)	2010	[16]	Air-breathing MFCs	Experimental and numerical investigations of porous planar anodes.
		[17]	Geometric design	Identification of successful and unsuccessful designs for membraneless MFCs.
	2011	[18]	Flow-through porous electrodes	CFD modeling of MFCs with porous electrodes for improved flow management.
		[20,22]	Chaotic flow and air cathode modeling	Counter-flow MFCs and transport behavior predictions.
		[21]	Hydrodynamic focusing	Solutions for breaking the trade-off between fuel utilization and current density.
		[19]	Diffuse charge effects	Modeling of laminar flow and diffuse charge effects in microfluidic fuel cells.
	2012	[23,24]	Gravity effects and critical conditions	Insights into parasitic currents and performance under critical conditions.
	2013	[25–27]	Fuel concentration, air-breathing MFCs, and exergy analysis	Enhanced oxygen performance and energy efficiency studies.
		[28]	Boundary layer control	Improved concentration boundary layers using CFD.
		[29]	Geometric design improvements	Numerical studies to optimize flow geometries in membraneless MFCs.

(continued)

Table 10.3 (Cont.)

Time line	*Year*	*Authors*	*Focus area*	*Key contribution*
Advanced Modeling (2014–2017)	2014	[30]	Air-breathing MFC modeling	3D CFD simulations of flow-over and flow-through anodes.
		[51]	Bridge-shaped microchannels	CFD modeling of cross-section designs for laminar flow MFCs.
	2015	[31,32]	Porous electrodes and sensitivity analysis	Optimization of porous electrodes and parametric sensitivity analysis.
		[52]	Alkaline air-breathing MFCs	CFD modeling of cylinder anode configurations.
		[33]	Counter-flow formic acid MFCs	Achieved high fuel utilization exceeding 90%.
	2016	[34]	Multi-layer porous electrodes	CFD simulation and experimental validation of vanadium-based designs.
	2017	[35]	Channel and electrode geometry	Exploration of geometric variations' impacts on performance.
		[53]	Current collector optimization	Improved porous electrode current collector designs.
		[36]	Air-breathing direct formic acid MFCs	Numerical performance simulations of air-breathing cells.
		[37]	Flow configuration comparison	Comparative CFD study of co-flow and counter-flow configurations.
		[38]	Porous electrode resistance	Analysis of geometry and electrical resistance effects on performance.
Multiphysics & AI Integration (2018–2022)	2018	[39]	Non-isothermal modeling	3D CFD modeling of methanol-based MFCs.
		[40]	Dual-electrolyte protocols	CFD modeling for multiple electrochemical applications.
		[54]	Inlet location optimization	Performance analysis of inlet configurations in MFCs.
	2019	[41]	Passive boundary layer control	Utilization of Dean vortices to manage concentration boundary layers.
	2020	[46]	Hollow flow channels	Enhanced performance with hollow channels and porous electrodes.
		[42]	Force field effects	Evaluation of force fields on exergy efficiency and fuel utilization.
		[43]	Vibration modeling	Coupling vibration effects for enhanced stability in MFCs.

Table 10.3 (Cont.)

Time line	*Year*	*Authors*	*Focus area*	*Key contribution*
		[44]	Pore-scale oxygen transport	Advanced CFD modeling for oxygen transport in air-breathing cathodes.
		[45]	Mixed-reactant MFCs	Numerical modeling of membraneless mixed-reactant fuel cells.
	2021	[47]	Air-breathing MFC stacks	CFD modeling of H-shaped air-breathing fuel cell stacks.
	2022	[48]	Anti-vibration evaluation	Phase field theory for evaluating MFC anti-vibration performance.
		[49]	Mass transport in air-breathing cathodes	Pore-scale CFD modeling of mass transport.
		[49]	Double-bridge flow channels	Optimization of power density using CFD.
		[50]	Deep learning optimization	Data-driven multiphysics CFD modeling for enhanced power density.

parameters of the MFC. Authors explained the potential of AI-driven optimization for MFCs, making them more effective and flexible [50].

10.4 KEY CONTRIBUTIONS OVER TIME

The advancements and key contributions in the design, modeling, and optimization of MFCs over the past two decades employing the CFD studies are presented in this section. The evolution of the CFD simulations between 2004 and 2022 based on available literature is presented in Figure 10.4.

The timeline is grouped into four key phases: (1) Foundational Studies (2004–2007), (2) Design Improvements (2010–2013), (3) Advanced Modeling (2014–2017), and (4) Multiphysics & AI Integration (2018–2022). Each time phase shows the unique contributions and enabled the evolution of MFCs as a practical and efficient energy conversion system. The details of the focus area and the key contributions are presented in Table 10.3.

10.5 CHALLENGES IN CFD STUDIES

Table 10.4 presents an overview of key challenges in CFD studies, including computational complexity, validation and accuracy, integration of

Table 10.4 The challenges and key points in CFD studies

S. No.	*Challenge*	*Description*	*Key points*
1	Computational complexity	High-fidelity CFD models demand extensive computational resources due to the need for detailed spatial and temporal resolutions. Simplified models can mitigate this [50].	Simplified approaches like depth-averaged models balance accuracy and computational feasibility.
2	Validation and accuracy	The accuracy of CFD simulations relies heavily on experimental validation to address discrepancies in reaction kinetics, material properties, and operating conditions (B. Zhang et al. [30]).	Assumptions in models often lead to differences; experimental integration improves reliability.
3	Multiphysics integration	Unifying thermal, mechanical, and electrochemical processes into a single model is challenging, especially for phenomena like vibration coupling and fluid-structure interaction [43,47].	Advanced techniques are required to model interactions like structural dynamics and fluid flow interplay.
4	Hydrodynamic performance	Optimized channel geometries and flow configurations improve reactant mixing, fuel utilization, and overall performance [3,40,41].	Serpentine and H-shaped channels enhance mixing via Dean vortices, while counter-flow setups boost efficiency.
5	Electrochemical performance	Advanced electrode designs and catalyst layer optimizations improve reaction kinetics and minimize concentration polarization [39,49].	Porous electrode designs and uniform catalyst distribution are key innovations for better performance.
6	Mass transport phenomena	Diffusion limitations and concentration boundary layers significantly affect transport processes, particularly at high current densities [28,29,48].	Pore-scale modeling and techniques like hydrodynamic focusing mitigate transport bottlenecks effectively.

Multiphysics interactions, hydrodynamic, electrochemical performances, and the mass transport phenomena.

10.6 FUTURE DIRECTIONS: *MACHINE LEARNING* AND OPTIMIZATION TECHNIQUES

Computational modeling and simulation play an important role in optimizing MFC designs for maximum efficiency. Traditionally, Computational fluid dynamics has been the primary approach for analysis and optimization of the performance of MFCs. However, CFD studies can become prohibitively

Table 10.5 Advantages and limitations of twin models (machine learning) in optimization of MFC systems [49,50]

S. No.	*Factor*	*Advantages*	*Limitations*
1	Computation time	Significantly reduces computation time during parametric optimization compared to CFD methods.	Training the twin model requires substantial time and computational resources for dataset generation.
2	Accuracy	Achieves high accuracy, often with negligible error compared to CFD results (e.g., $R^2 = 0.999$).	Accuracy depends on the quality and quantity of the CFD training datasets.
3	Optimization scope	Enables simultaneous optimization of multiple design parameters, overcoming traditional limitations.	Limited by the need for an initial CFD dataset for training, which can be resource-intensive.
4	Flexibility	Applicable to various parameters, including geometrical and operational aspects of MFC design.	The effectiveness of the model depends on the choice of optimization algorithms and training data.

time-consuming when multiple design parameters are considered, limiting their feasibility for optimization tasks on a large scale. The introduction of machine learning, surrogate models, also known as twin models, emerged as a transformative solution to these challenges [49,50].

The integration of the machine learning approach offers an alternative to traditional CFD studies. The data generated by the CFD simulations are trained using twin models. During the optimization process, these trained models may supersede the CFD models and enable faster computations without losing accuracy. A comparison between the advantages and limitations of using twin models in optimization of MFCs is given in Table 10.5.

10.7 CONCLUSION

Computational fluid dynamics has proven to be an invaluable tool for investigating the design and performance of microfluidic fuel cells. Modeling and simulating transport processes using CFD tools provides insights into the underlying mechanisms governing MFC operations. CFD studies have contributed to advancements in MFCs by evaluating microchannel configurations, electrode designs, and fuel utilization strategies. Despite these advances, challenges persist in fully capturing microscale transport phenomena, especially in complex Multiphysics environments. Capturing real-time scenarios entails high computational costs, and extensive experimental validation is required for CFD modelling outcomes before field applications. However, recent developments, such as hybrid modelling approaches and the integration of artificial intelligence, offer promising solutions to overcome existing limitations. Looking ahead, future research should focus on enhancing computational efficiency, refining algorithms, and developing robust simulation frameworks that effectively complement experimental work. CFD studies are expected to play a key role in bridging the gap between theoretical research and practical implementation of MFCs for applications in energy systems, wearable sensors, biomedical devices, and portable electronics.

REFERENCES

[1] O. A. Ibrahim, M. Navarro-Segarra, P. Sadeghi, N. Sabaté, J. P. Esquivel, and E. Kjeang. Microfluidics for electrochemical energy conversion. *Chemical Reviews* (2022). doi:10.1021/acs.chemrev.1c00499.

[2] *Microfluidic Fuel Cells with Different Types of Fuels: A Prospective Review*. Elsevier Ltd., 2021.

[3] P. Zhi, Z. Liu, K. Jiao and Q. Du, Numerical study on the performance of the H-shaped air-breathing microfluidic fuel cell stack, *Electrochim Acta* 392 (2021), 139024.

[4] E. Kjeang, N. Djilali, and D. Sinton. Microfluidic fuel cells: A review. *Journal of Power Sources* (2009). doi:10.1016/j.jpowsour.2008.10.011..

[5] M. Tanveer and K.Y. Kim, Flow configurations of membraneless microfluidic fuel cells: A review, *Energies (Basel)* 14 (2021), 3381.

[6] B. Ho and E. Kjeang. Microfluidic fuel cell systems. *Central European Journal of Engineering* (2011). doi:10.2478/s13531-011-0012-y.

[7] S. K. S. Cheng,, T. Li, S. S. Meena, Q. Cao, B. Li, B. K. Kosgei, T. Cheng, P. Luo, Q. Liu, and G. Zhu. Advances in Microfluidic Technologies for Energy Storage and Release Systems. In *Advanced Energy and Sustainability Research*. John Wiley & Sons Inc., 2022. doi:10.1002/aesr.202200060.

[8] M.J. Torres, J. Hervas-Ortega, B. Oraá-Poblete, A.B. de Quirós, A.A. Maurice, D. Perez-Antolin et al., *Membraneless Micro Redox Flow Battery: From Vanadium to Alkaline Quinone, Batter Supercaps* 7 (2024), e202400331.

[9] R. Ferrigno, A.D. Stroock, T.D. Clark, M. Mayer and G.M. Whitesides, Membraneless vanadium redox fuel cell using laminar flow, *J Am Chem Soc* 124 (2002), 12930–12931.

[10] R. Choban, L.J. Markoski, J. Stoltzfus, J.S. Moore and J.A. Kenis, Microfludic fuel cells that lack a PEM, *Power Sources Proc* 40 (2002), 317–320.

[11] A. Bazylak, D. Sinton and N. Djilali, Improved fuel utilization in microfluidic fuel cells: A computational study, *J Power Sources* 143 (2005), 57–66.

[12] *Computational Modeling Studies on Microfluidic Fuel Cell: A Prospective Review*. Elsevier Ltd., 2024.

[13] A. Bazylak, D. Sinton and N. Djilali, *Membraneless Liquid-Fuel Microfluidic Fuel Cells: A Computational Study*. ASME 2004 International Mechanical

Engineering Congress and Exposition, Anaheim, CA, November 13–19, 2004.
[14] M.H. Chang, F. Chen and N.S. Fang, Analysis of membraneless fuel cell using laminar flow in a Y-shaped microchannel, *J Power Sources* 159 (2006), 810–816.
[15] J. Lee, G.L. Keng, G.T.R. Palmore and A. Tripathi, Optimization of microfluidic fuel cells using transport principles, *Anal Chem* 79 (2007), 7301–7307.
[16] S.A.M. Shaegh, N.T. Nguyen and S.H. Chan, An air-breathing microfluidic formic acid fuel cell with a porous planar anode: Experimental and numerical investigations, *J Micromech Microeng* 20 (2010), 105008.
[17] D.H. Ahmed, H.B. Park, K.H. Lee and H.J. Sung, The geometrical design of membraneless micro fuel cells: Failure and success, *Int J Energy Res* 34 (2010), 878–896.
[18] D. Krishnamurthy, E.O. Johansson, J.W. Lee and E. Kjeang, Computational modeling of microfluidic fuel cells with flow-through porous electrodes, *J Power Sources* 196 (2011), 10019–10031.
[19] I.B. Sprague and P. Dutta, Modeling of diffuse charge effects in a microfluidic based laminar flow fuel cell, *Numeri Heat Transf A Appl* 59 (2011), 1–27.
[20] J. Xuan, D.Y.C. Leung, M.K.H. Leung, H. Wang and M. Ni, Chaotic flow-based fuel cell built on counter-flow microfluidic network: Predicting the over-limiting current behavior, *J Power Sources* 196 (2011), 9391–9397.
[21] J. Xuan, M.K.H. Leung, D.Y.C. Leung, M. Ni and H. Wang, Hydrodynamic focusing in microfluidic membraneless fuel cells: Breaking the trade-off between fuel utilization and current density, *Int J Hydrogen Energy* 36 (2011), 11075–11084.
[22] H. Wang, D.Y.C. Leung and J. Xuan, Modeling of an air cathode for microfluidic fuel cells: Transport and polarization behaviors, *Int J Hydrogen Energy* 36 (2011), 14704–14718.
[23] J. Xuan, M.K.H. Leung, D.Y.C. Leung and H. Wang, Laminar flow-based fuel cell working under critical conditions: The effect of parasitic current, *Appl Energy* 90 (2012), 87–93.
[24] J. Xuan, M.K.H. Leung, D.Y.C. Leung and H. Wang, Towards orientation-independent performance of membraneless microfluidic fuel cell: Understanding the gravity effects, *Appl Energy* 90 (2012), 80–86.
[25] J. Xuan, D.Y.C. Leung, H. Wang, M.K.H. Leung, B. Wang and M. Ni, Air-breathing membraneless laminar flow-based fuel cells: Do they breathe enough oxygen? *Appl Energy* 104 (2013), 400–407.
[26] H. Zhang, M.K.H. Leung, J. Xuan, H. Xu, L. Zhang, D.Y.C. Leung et al., Energy and exergy analysis of microfluidic fuel cell, *Int J Hydrogen Energy* 38 (2013), 6526–6536.
[27] H. Zhang, J. Xuan, H. Xu, M.K.H. Leung, D.Y.C. Leung, L. Zhang et al., Enabling high-concentrated fuel operation of fuel cells with microfluidic principles: A feasibility study, *Appl Energy* 112 (2013), 1131–1137.
[28] R. Hassanshahi and M. Fathipour, Improved control of concentration boundary layer in microfluidic fuel cells, *Int J Renew Energy Res* 3 (2013), 901–905.

[29] R.A. García-Cuevas, I. Cervantes, L.G. Arriaga and I.A. Diaz-Diaz, Toward geometrical design improvement of membraneless fuel cells: Numerical study, *Int J Hydrogen Energy* 38 (2013), 14791–14800.

[30] B. Zhang, D.D. Ye, P.C. Sui, N. Djilali and X. Zhu, Computational modeling of air-breathing microfluidic fuel cells with flow-over and flow-through anodes, *J Power Sources* 259 (2014), 15–24.

[31] L. Li, K. Zheng, M. Ni, M.K.H. Leung and J. Xuan, Partial modification of flow-through porous electrodes in microfluidic fuel cell, *Energy* 88 (2015), 563–571.

[32] L. Li, W. Fan, J. Xuan and M.K.H. Leung, Dimensionless parametric sensitivity analysis of microfluidic fuel cell with flow-through porous electrodes, *Electrochim Acta* 187 (2016), 636–645.

[33] H. Xu, H. Zhang, H. Wang, D.Y.C. Leung, L. Zhang, J. Cao et al., Counter-flow formic acid microfluidic fuel cell with high fuel utilization exceeding 90%, *Appl Energy* 160 (2015), 930–936.

[34] L. Li, G. Nikiforidis, M.K.H. Leung and W.A. Daoud, Vanadium microfluidic fuel cell with novel multi-layer flow-through porous electrodes: Model, simulations and experiments, *Appl Energy* 177 (2016), 729–739.

[35] M. Tanveer and K.Y. Kim, Effects of geometric configuration of the channel and electrodes on the performance of a membraneless micro-fuel cell, *Energy Convers Manag* 136 (2017), 372–381.

[36] Y.D. Herlambang, J.C. Shyu and S.C. Lee, Numerical simulation of the performance of air-breathing direct formic acid microfluidic fuel cells, *Micro Nano Lett* 12 (2017), 860–865.

[37] Y. Wang, D.Y.C. Leung, H. Zhang, J. Xuan and H. Wang, Numerical and experimental comparative study of microfluidic fuel cells with different flow configurations: Co-flow vs. counter-flow cell, *Appl Energy* 203 (2017), 535–548.

[38] L. Li, S. Bei, Q. Xu, K. Zheng and Y. Zheng, Role of electrical resistance and geometry of porous electrodes in the performance of microfluidic fuel cells, *Int J Energy Res* 42 (2018), 1277–1286.

[39] L. Wei, X. Yuan and F. Jiang, A three-dimensional non-isothermal model for a membraneless direct methanol redox fuel cell, *J Power Sources* 385 (2018), 130–140.

[40] X. Lu, Y. Wang, D.Y.C. Leung, J. Xuan and H. Wang, A counter-flow-based dual-electrolyte protocol for multiple electrochemical applications, *Appl Energy* 217 (2018), 241–248.

[41] W. Rösing, T. Schildhauer, J. König and C. Cierpka, Passive control of the concentration boundary layer in microfluidic fuel cells using Dean vortices, *Microfluid Nanofluidics* 23 (2019). https://doi.org/10.1007/s10404-019-2274-2

[42] T. Ouyang, J. Chen, F. Zhou, J. Lu, P. Xu, N. Chen et al., Effects of force field and design parameters on the exergy efficiency and fuel utilization of microfluidic fuel cells, *Int J Hydrogen Energy* 45 (2020), 12443–12458.

[43] T. Ouyang, F. Zhou, J. Chen, J. Lu and N. Chen, A novel approach for modelling microfluidic fuel cell coupling vibration, *J Power Sources* 450 (2020), 227728.

[44] Y.L. Fu, B. Zhang, X. Zhu, D.D. Ye, P.C. Sui and N. Djilali, Pore-scale modeling of oxygen transport in the catalyst layer of air-breathing cathode in membraneless microfluidic fuel cells, *Appl Energy* 277 (2020), 115536.
[45] A.M. Zuria, J.C. Abrego-Martinez, S. Sun and M. Mohamedi, Prospects of membraneless mixed-reactant microfluidic fuel cells: Evolution through numerical simulation, *Renew Sustain Energy Rev* 134 (2020), 110045.
[46] M. Tanveer and K.Y. Kim, A membraneless microfluidic fuel cell with a hollow flow channel and porous flow-through electrodes, *Int J Energy Res* 45 (2021), 8536–8550.
[47] T. Ouyang, J. Chen, W. Liu, P. Xu, J. Lu and Z. Zhao, A comprehensive evaluation for microfluidic fuel cells from anti-vibration viewpoint using phase field theory, *Renew Energy* 189 (2022), 676–693.
[48] Y.-L. Fu, B. Zhang, X. Zhu, D. ding Ye, P.C. Sui, N. Djilali et al., Pore-scale modeling of mass transport in the air-breathing cathode of membraneless microfluidic fuel cells, *Int J Heat Mass Transf* 188 (2022), 122590.
[49] J.H. Oh, T.D. Vuong and K.Y. Kim, Optimization of a membraneless microfluidic fuel cell with a double-bridge flow channel, *Energies (Basel)* 15 (2022), 973.
[50] D.D. Nguyen, T. Quy Duc Pham, M. Tanveer, H. Khan, J.W. Park, C.W. Park et al., Deep learning–based optimization of a microfluidic membraneless fuel cell for maximum power density via data-driven three-dimensional multiphysics simulation, *Bioresour Technol* 348 (2022), 126794.
[51] P.O. Lopez-Montesinos, A. V. Desai and P.J.A. Kenis, A three-dimensional numerical model of a micro laminar flow fuel cell with a bridge-shaped microchannel cross-section, *J Power Sources* 269 (2014), 542–549.
[52] D.D. Ye, B. Zhang, X. Zhu, P.C. Sui, N. Djilali and Q. Liao, Computational modeling of alkaline air-breathing microfluidic fuel cells with an array of cylinder anodes, *J Power Sources* 288 (2015), 150–159.
[53] L. Li, W. Fan, J. Xuan, M.K.H. Leung, K. Zheng and Y. She, Optimal design of current collectors for microfluidic fuel cell with flow-through porous electrodes: Model and experiment, *Appl Energy* 206 (2017), 413–424.
[54] M. Tanveer and K.Y. Kim, Performance analysis of microfluidic fuel cells with various inlet locations and multiple compartments, *Energy Convers Manag* 166 (2018), 328–336.

Chapter 11

Investigating the process flowsheet for converting cellulose into hexane

A simulation approach

Krishna Srihari Bonasi, Brajesh Kumar, Balasubramanian Sivasamy, Ashish Kapoor, and Dan Bahadur Pal

11.1 INTRODUCTION

In the world of industrialization and urbanization, the consumption of conventional sources has increased in recent decades, due to which the difference between demand and supply of energy is raising concern in today's scenario worldwide, which affects the environment and economy altogether (Kumar et al., 2017). Limited availability, pollution, high cost, etc., are drawbacks of these non-renewable sources, which cannot provide sustainable development solutions and do not fulfil future energy needs. Because of this, many researchers are trying to contribute to research and development at their levels to explore various alternatives for these conventional sources (Demirbas, 2009).

Biomass can be selected as an alternative to renewable sources to provide a sustainable pathway toward the world's development due to its wide availability as fresh material and waste (Guo et al., 2022). The estimated biomass production annually in the global market is expected to exceed 56.8 billion tonnes, with lignocellulose accounting for approximately 80% of this total. Since that time, the applications of cellulose have expanded significantly, ranging from paper production to the development of biosensors (Ummartyotin and Manuspiya., 2015). Using appropriate catalysts, cellulose can be converted into fine chemicals and molecules for wide industrial and other applications. The process of enzymatic hydrolysis of cellulose is frequently luxurious, selective, and eco-friendly. Furthermore, only 35% of the used-up enzyme is recoverable and reprocessed (Higgins and Ho, 1982). In recent times, solid acid catalysts like Ir – ReO_x/SiO_2, Rh-ReO_x, and H-ZSM-5 are identically significant for the hydrogenolysis of the C–O bond (Chen et al., 2013). Baharintasari et al. (2019) exhibited that the movement of Y-zeolite (a solid acid catalyst) for hydrolysis rises with temperature. The biggest demerit of solid acid catalysis is the catalyst-to-substrate mass ratio, around 1:1. Acid hydrolysis of cellulose repeatedly primes to by-product development (xylose, arabinose) (Orozco et al., 2017). The hydrolysis of cellulose surveys a pseudo-first-order reaction, which encourages the

 DOI: 10.1201/9781003581536-11

dehydration of glucose to hydroxylmethyl furfural (HMF) (Op de Beek et al., 2015).

$$\text{Cellulose} + H_2O \xrightarrow{HCl/H_2SO_4} 2\text{Glucose} \xrightarrow{HCl/H_2SO_4} 3H_2O + \text{HMF}$$

The HMF is a complex of high manufacturing position and fuel worth. It is used in the invention of polyesters and plastics, as well as in the drug industry and liquid petrol-like fuels. Plentiful traditions of alteration of HMF to alkanes, alcohols, and aromatic compounds exist. In the contemporary study, hydrogenation and hydrogenolysis of HMF on appropriate catalysts to produce *n*-hexane were preferred (Op de Beeck et al., 2015). For the high concentrations of HCl the hydrolysis is far better and dehydrating capability than H_2SO_4, even though cellulose conversion to glucose is profligate in the case of H_2SO_4 (Hutomo et al., 2015).

$$\text{HMF} + H_2 \xrightarrow[\text{Hydrogenation}]{} 2,5\text{DHMF} \xrightarrow[\text{Hydro-deoxygenation and Hydrogenation}]{} \text{Hexane}$$

When HCl is used as a gas for pretreating the cellulose, HCl is more effective than simple HCl acid treatment, and the reaction proceeds more rapidly. HCl forms a constant boiling azeotrope with water at 108°C, making it easy to separate from the cellulose, HMF, HCl, and water mixture. Others have explored cellular conversion to useful products via modeling and software applications. The population-balance equations of the hydrolysis of cellulose can be used to model the kinetics of the evolution of polydisperse chain lengths (Griggs et al., 2011). Cellulosic feedstock modeling and software application. Researchers use these tools to predict reaction kinetics, assess enzyme performance, optimize process conditions, and generate high-value products, including biofuels and biochemicals (Nzediegwu and Dumont, 2020; Rinaldi and Schüth, 2009). By focusing on mechanistic details and inhibition by various compounds, these models will help develop either more efficient or more economically viable cellulose conversion technologies.

11.1.1 Cellulose

Cellulose is found in abundant quantities in natural sources. Cellulose is recognized as a natural polymer that is decomposed by the action of microorganisms like certain bacteria, algae, and fungi. It is the most abundant organic material found on Earth. Cellulose plays a crucial role in plant growth during photosynthesis and has a variety of uses, such as supporting structural plant support in the textile industry, paper manufacturing, and the generation of diverse biofuels (Saravanan et al., 2023). It is a white, fibrous, odorless solid with a hygroscopic nature. It possesses high tensile strength, low elasticity, high durability, and low thermal expansion. Its eco-friendly nature and renewable availability make it a valuable resource across various sectors (Zhang et al., 2025; Vanderfleet and Cranston, 2021). Cellulose has its own advantages, disadvantages, and applications before being converted into other products. Some are discussed below (Manyatshe and Sibali, 2025; Akshay et al., 2021; Yaowen et al., 2021).

11.2 APPLICATIONS OF CELLULOSE

11.2.1 Reaction mechanism of cellulose to valuable products

Cellulose, a polymer made up of β-1→4 linked D-glucose units, can be transformed into valuable products through a variety of reaction pathways. The pyrolysis process of cellulose occurs in two phases: primary pyrolysis and subsequent secondary reactions. At temperatures extending from 300°C to 350°C, the performance of cellulose during pyrolysis alters significantly, resulting in intermediates that can be converted into gas, liquid, or solid products (Op De Beeck et al., 2015; Kawamoto, 2016). Acid-catalyzed hydrolysis is another key method used to degrade cellulose. Due to their powerful acidity, good solubility, and substantial thermal stability, heteropolyacids (HPAs) have been strong green catalysts for this (Luo et al., 2020). Of particular emphasis has been the improved catalytic efficiency of cellulose conversion through the use of metal catalysts in conjunction with ionic liquids. For example, a catalyst consisting of 1-(4-sulfonic acid) butyl-3-methylimidazolium hydrogen sulfate (IL-1) and $FeCl_2$ induced up to 84% of conversion of microcrystalline cellulose (MCC) into significant levels of 5-hydroxymethyl furfural (HMF) as well as furfural (Tao et al., 2010). A promising alternative is a solid-state reaction with mechanical assistance treatment (Meine et al., 2012), where cellulose is treated with a strong acid, and then milled polymeric chains are converted into water-soluble oligosaccharides. Acid-catalyzed hydrolysis is another important mechanism for degrading cellulose. As they present outstanding acidities, strong solubility in water, and remarkable thermal and chemical stability, heteropolyacids (HPAs) have attracted attention as highly efficient green catalysts for cellulose hydrolysis (Luo et al., 2020). These characteristics enable HPAs to enhance the conversion of cellulose to simple sugars that can be utilized as precursors for an extensive variety of bio-derived chemicals and fuels. Some ionic liquids, when mixed with metal catalyzers, can significantly enhance the catalytic activity of cellulose. For instance, microcrystalline cellulose (MCC) can be produced at a remarkable 84% conversion when utilizing 1-(4-sulfonic acid) butyl-3-methylimidazolium hydrogen sulfate (IL-1) alongside $FeCl_2$ as the catalyst. As a result of this reaction, they obtained 5-hydroxymethyl furfural (HMF) and furfural with high yields; these are valuable intermediates for preparing numerous renewable chemicals and materials (Tao et al., 2010).

In addition, mechanically assisted solid-state reaction provides a new way to perform cellulose processing. This method consists of the introduction of a strong acid into cellulose and mechanical milling. This method has opened up cellulose for chemical reactions and yielded favorable products of cellulose to water-soluble oligosaccharides, which have applications in beverage and drug industries (Meine et al., 2012).

11.2.2 Hexane

Hexane is a hydrocarbon belonging to the alkane family with the chemical formula C_6H_{14}. It is a colorless liquid at room temperature with a characteristic mild odor. Hexane is widely used as a non-polar solvent in various industries, particularly in oil extraction, adhesives, and chemical synthesis. Due to its volatility and low boiling point, it quickly evaporates, making it a valuable component in many industrial and laboratory processes (Cravotto et al., 2022; Writdhama et al., 2022). It can be produced using cellulose as a raw material with the help of some acids such as H_2SO_4 and HCl. The present study focuses on investigating the production of hexane from cellulose with or without acid treatment. There have been many advantages, disadvantages, and applications of using hexane as a substantial product obtained from various sources (Cravotto et al., 2022; Writdhama et al., 2022; Chemat et al., 2019; Fraterrigo Garofalo et al., 2021).

11.2.3 Hexane production feedstock

Hexane, a colourless, flammable organic compound, has a wide range of uses in many different fields. It can be produced from a wide variety of starting materials, with plant biomass and fossil fuels being the two primary feedstocks. Hexoses not only play an important biochemical role but also can be obtained from renewable plant biomass, consisting primarily of lignocellulose, considered to be a sustainable and promising feedstock for biofuel production, e.g. 5-hydroxymethylfurfural (HMF), an important platform chemical that can be catalytically converted into hexane and other valuable chemicals (Chernyshev et al., 2017). This approach is becoming increasingly important as industries look towards more sustainable and greener production methods.

Additionally, algal biomass also serves as a potential feedstock for biofuel production and has also shown considerable potential for the purposes of hexane production (Haas and Wagner, 2011; Venkatachalam et al., 2021). For example, crude oil is the primary feedstock from which light olefins like butene can be separated and further converted into hexane (Emberru et al., 2024). Emphasizing this reality serves to demonstrate further the continuing competitive landscape between renewable and fossil-influenced chemical feedstocks.

11.2.4 Plant biomass as an alternative to fossil fuels

Utilizing plant biomass to produce hexane brings about a myriad of benefits relative to fossil fuels. In contrast to fossil fuels, which are finite in abundance and frequently obtained with considerable environmental consequences, plant biomass acts as a renewable and sustainable resource (Tschan et al., 2012). Drawing on one of the most abundant types of resources (biomass), we can play an important role in preventing the exhaustion of fossil fuel resources and thereby contribute to a more sustainable energy and chemical industry that is less dependent on non-renewable resources (Csefalvay and Horvath, 2018). The shift to renewable plant materials addresses both short-term energy demands and offers substantial potential for development toward pressing global environmental challenges by promoting sustainable practices in plastics production (Kashparova et al., 2021).

Biomass is a clean and renewable power source, and one of its most astonishing features is the potential to produce fuels and materials with a smaller carbon footprint than their fossil counterparts (Kashparova et al., 2021). This is primarily because bioenergy production has a specific carbon cycle by which plants absorb carbon dioxide from the atmosphere during their growth, and when they are processed, they return it to the atmosphere with nearly null net greenhouse gas emissions (Navarro et al., 2009). This is in contrast to fossil fuels, which release carbon that has been locked away in the earth's crust for millions of years and contribute to climate change and environmental degradation.

Furthermore, biomass can be obtained from a wide variety of living organisms, including specially-grown energy crops, agricultural residue, which would be wasted otherwise, and woody biomass collected from forest waste (Navarro et al., 2009). This flexibility not only allows for the optimized use of otherwise underperforming land but also demonstrates the possibility of converting raw organic waste path by-products, which contributes to a circular economy (Brune et al., 2009). Also, biofuels from microalgae and macroalgae as feedstock have shown good potential. These organisms have an exceptionally high productivity and can grow in several types of cultivation systems, making them a promising sustainable chemical feedstock (Mahmood et al., 2022).

11.2.5 Conversion technologies for biomass compared to fossil fuels hexane production

Biomass conversion technologies include a range of advanced technologies, falling under either thermochemical or biochemical processes, and serve as promising substitutes for fossil fuel production techniques (Shahzad et al.,

2024). Leveraging biomass, which is the latest technology, converts organic materials (the byproducts of agriculture, wood, and other plant materials) into biofuels and other value-added products. So, this could find a more sustainable way of creating energy and producing chemicals than relying on fossil fuels.

These conversion technologies can be highly variable in terms of their efficiencies, and they frequently encounter multiple hurdles. As mentioned above, gasification and direct liquefaction processes are two of the primary conversion methods available among thermochemical processes for biomass to biofuel conversion (Akia et al., 2014). Nanocatalysts are specifically designed to improve product quality and maximize operational conditions so these methods can achieve greater efficiency (Akia et al., 2014). Thermochemical processes, however, potentially suffer from high processing costs and harsh temperature requirements, limiting their broader applicability (Shahzad et al., 2024).

In terms of overall efficiency, biomass processing for the production of biofuels is indeed less productive than the fossil fuel-based procedure. Particularly, Ptasinski (2008) highlights the varying efficiency of biofuels and fossil fuels production methods. Moreover, modern biomass-derived fuels generated from lignocellulosic biomass are anticipated to become economically competitive (Faaij, 2006) within one to two decades. At the moment, this means existing biomass conversion technologies are less efficient than established fossil fuel processes.

11.2.6 Influence on the productivity of the hexane manufacturing

Biomass feedstock types, on the other hand, can significantly influence the efficiency at which conversion technologies produce biofuels and biochemicals because of their distinct chemical composition and physical properties (Morales et al., 2021; Op De Beeck et al., 2015). It has been demonstrated that properties, including carbohydrate content, lignin content, ash content, and moisture content, are primary factors in determining overall conversion performance (Tanger et al., 2013). For example, switchgrass has a large amount of carbohydrates for bioethanol conversion and also has better bioethanol yield than other types of feedstock, such as forest residues (Morales et al., 2021). It should be noted that the effect of feedstock composition varies depending on the conversion process. In thermochemical processes, important factors include carbon content, total ash, specific inorganics, density, and particle size (Edmunds et al., 2018). In contrast, cell wall digestibility is more important in biochemical processes (Torres et al., 2016). Indeed, due to the nature of conversion technologies that depend on feedstock properties, selecting suitable biomass feedstock and enhancing its properties through breeding or genetic modification can significantly improve the conversion technology performance (Tanger et al., 2013; Torres et al., 2016). Moreover, combining different biomass types constitutes a practical approach for dealing with issues arising from feedstock variability and supply steady-state (Edmunds et al., 2018; Op De Beeck et al., 2015). Biomass feedstocks are significant determinants for the conversion efficiencies of biomass-based biofuels and biochemicals because of the heterogeneous chemical compositions and physical properties of biomass materials (Morales et al., 2021; Op De Beeck et al., 2015). Factors such as carbohydrate content, lignin concentrations, ash content, and moisture content of the substrates determine the efficiency of these conversion processes. All of these features contribute to the efficiency with which biomass can be converted to biofuels [(Tanger et al., 2013). This feature renders switchgrass a more potential target for the production of biofuels (Morales et al., 2021). However, the feedstock composition/conversion efficiency relationship is not universal and varies greatly based on the selected conversion route. Attributes like carbon content, total ash content, concentration of individual inorganic elements, bulk density, and particle size are Key performance indicators (KPI) for thermochemical processes (Edmunds et al., 2018). In contrast, the digestibility of the cell wall is a key factor influencing overall efficiency during biochemical processes, which typically use enzymatic digestion (Torres et al., 2016).

Biomass comes in various qualities, and the efficiency of the aforementioned conversion technologies (Tanger et al., 2013; Torres et al., 2016) can be improved by optimizing the properties of biomass through breeding techniques or genetic modification.

One creative strategy to address challenges with feedstock variability and supply stability is to blend multiple forms of biomass. One approach to addressing the inconsistency of feedstock quality and availability is through building stockpiles of feed grains, which could provide end users with a more consistent supply of source material for biofuel production (Edmunds et al., 2018; Op De Beeck et al., 2015). Future research should, therefore, focus on the development of dedicated feedstocks for specific conversion pathways. This approach can potentially increase the process's efficiency and economic feasibility, paving the way for more sustainable biofuel production in the future.

11.3 ASPEN SIMULATION

In the present study, the SRK thermodynamic package is chosen to evaluate the thermodynamic data. NIST ThermoData Engine is used to obtain the temperature-dependent and other pure component parameters. In the case of binary interaction parameters, it is necessary to calculate vapor-liquid equilibrium properties for the mixture, and these parameters can be estimated using the UNIFAC property method. The catalyst used for hydrogenation and hydrogenolysis is not considered suitable for the suggested simulation procedure. In the present study, a total of four cases are simulated for further study. The process flow sheet has been shown in Figure 11.1, which considers cases 1 and 2, and Figure 11.2, which considers cases 3 and 4.

Case study 1: CELLU-RE without recycle and HCl acid is used
Case study 2: CELLU-RE without recycle, but H_2SO_4 acid is used
Case study 3: CELLU-RE with recycle, but HCl acid is used
Case study 4: CELLU-RE with recycle, but H_2SO_4 acid is used

11.3.1 Mixer (MIX)

This is applied to mix the freshly available cellulose as feed and the unreacted cellulose, which is separated from the solid-liquid separator (S-L-Sep). The feed, as fresh cellulose, which is named as CELLULOSE in the aspen flowsheet, consists of 80 kmol/h cellulose and 20 kmol/h water.

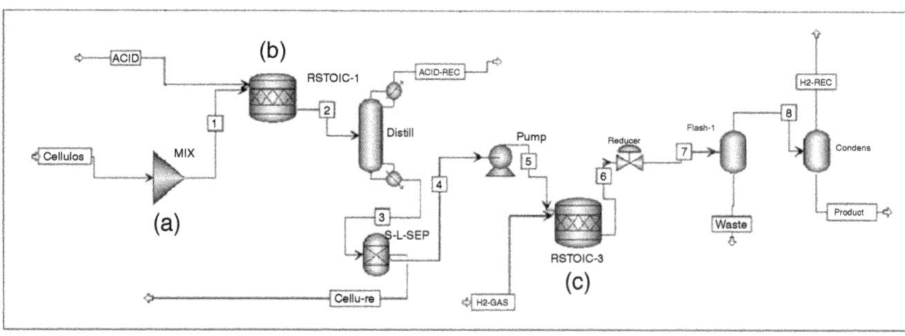

(a) For feeding mixed stream of cellulose and unreacted cellulose; (b) Reactor for conversion of cellulose to hydroxyl methyl furfural (HMF)
(c) Reactor for hydrogenation and hydro-deoxygenation of HMF to produce hexane.

Figure 11.1 Flowsheet of cases 1 and 2.

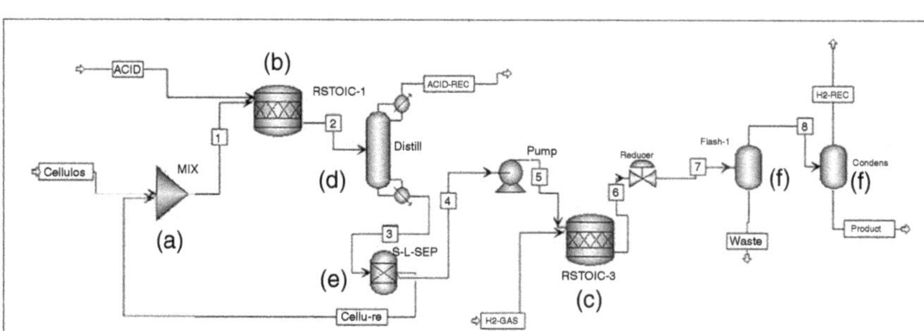

(d) Distillation column to recover the spent acid catalyst; (e) Flash and condenser for separation of hexane and H_2 gas; (f) Flash and condenser for separation of hexane and H_2 gas

Figure 11.2 Flowsheet of cases 3 and 4.

11.3.2 Reactor-1 (RSTOIC-1)

The infeed acid stream composition to the reactor is a 12 M HCl or 18 M H_2SO_4. For cellulose conversion to HMF, the stoichiometric reactor is applied. In this study, it is assumed that the cellulose content is first converted to glucose and subsequently immediately dehydrated to produce HMF. It uses concentrated acid because the acid gets more diluted when the reaction takes place. The following reactions take place in the reactor:

$$Cellulose + H_2O \xrightarrow{Catalyst} 2Glucose \qquad reaction\ 1$$

$$Glucose \xrightarrow{Catalyst} HMF + 3H_2O \qquad reaction\ 2$$

11.3.3 Distillation column (DISTIL)

This distillation column contains ten stages with a reflux of 0.1, and it is used to separate acid from the feed entering the distillation column so that the distillate includes a mixture of acid and water. Because of the high boiling point of cellulose and HMF, they are considered as residue products. The vapor distillate is later cooled down, next concentrated, and thereafter reused.

11.3.4 Separator (S-L-SEP)

This is used for the separation of the unreacted cellulose and acid from the bottom stream 3, from which the pure water and HMF mixture with traces of glucose can be obtained for further processing, or HMF can be collected from this point and utilized on a commercial scale. Despite the current scenario of the present study, this separator can also be used to separate the present solid catalysts (the separator acts like a filter).

11.3.5 Pump and reducer (PUMP and REDUCER)

The pump increases the pressure of stream 4, and the reducer decreases the pressure of stream 6.

11.3.6 Reactor-2 (RSTOIC-2)

Reactor-2 is used for the hydrogenation and hydrogenolysis of HMF to *n*-hexane at a high pressure of 5 bar. The process starts with a hydrogenation reaction where HMF transforms to 2,5-dihydroxymethylfurfural (2,5-DHMF), and then 2,5-DHMF is further hydrogenolysed and forms hexane. During the reaction, the pure hydrogen gas is sent to the reactor with the HMF water mixture. The reaction followed in this reactor is

$$HMF + H_2 \xrightarrow{Catalyst} 2,5\text{-}DHMF \qquad reaction\ 3$$

$$2,5\text{-}DHMF + 6H_2 \xrightarrow{Catalyst} n\text{-}Hexane \qquad reaction\ 4$$

11.3.7 Flash drum (Flash-1)

It separates H_2O and C_6H_{14}, H_2 mixture. The flash drum is operated at ambient conditions like room temperature, i.e., 25°C.

11.3.8 Condenser (CONDENS)

It is used to condense the C_6H_{14} and separate it from hydrogen gas. The condenser was provided with operating conditions for process variables such as a temperature of 25° C and a pressure of 15 bar to achieve a clear separation between the desired product and H_2 as recovered gas. The operating conditions for the different unit operations used in the simulation are tabulated in Table 11.1.

Table 11.1 Operating conditions

Stream/Block	*Parameter*	*Values*
Cellulose	Flowrate	100 (kmol/h)
	Temperature	25°C
	Pressure	1 bar
Acid	Flowrate	100 (kmol/h)
	Temperature	50°C
	Pressure	1 bar
H_2-Gas	Flowrate	1,500 (kmol/h)
	Temperature	25°C
	Pressure	1 bar
Mix	Temperature	25°C
	Pressure	1 bar
RSTOIC-1	Temperature	50°C
	Pressure	1 bar
Distill	Stages	10
	Feed entering stage	5
	Reflux	0.1
	Distillate rate	50 (kmol/h)
S-L-SEP	Split fraction of stream 4	1 for water and HMF
Pump	Discharge pressure	5 bar
RSTOIC-3	Temperature	25°C
	Pressure	5 bar
Reducer	Outlet pressure	1 bar
Flash-1	Temperature	25°C
	Pressure	1 bar
Condens	Temperature	25°C
	Pressure	15 bar

11.4 RESULTS AND DISCUSSION

The simulation was conducted to convert cellulose into hexane for various applications in the process industry, benefiting humanity. Aspen Plus software has been deployed to simulate by developing essential flowsheets with input parameters, suitable assumptions, and precision. The procedure has been divided into four cases: cellulose conversion into hexane using HCl acid without and with recycle stream (Cases 1 and 3) and cellulose conversion into hexane using H_2SO_4 acid without and with recycle stream (Cases 2 and 4). Each case performed with the required parameters has been simulated and explained in detail in the subsequent sections.

Table 11.2 provides case-wise results for cellulose conversion into hexane, including input and output results. Theoretically obtained results dictate that the complete conversion of reactant participated reactions was found in all the categorized cases. No significant difference was found in the design parameters with and without recycle streams using HCl acid for the required process. The changes or variations in other related properties in all four selected cases have also been reported in Table 11.2. For instance, SRK as a property method has been chosen for all specified cases with a constant number of stages as 10. The phrase has been selected as partial vapor for the conditions of the condenser.

Table 11.2 Distillation column summary at 100% conversion for reaction 1

	Case 1	*Case 3*	*Case 2*	*Case 4*
Name	DISTILL	DISTILL	DISTILL	DISTILL
Property method	SRK	SRK	SRK	SRK
Number of stages	10	10	10	10
Condenser	Partial-vapor	Partial-vapor	Partial-vapor	Partial-vapor
Reboiler	KETTLE	KETTLE	KETTLE	KETTLE
Number of phases	2	2	2	2
Top stage pressure [bar]	1	1	1	1
Specified reflux ratio	0.1	0.1	0.1	0.1
Specified distillate rate [kmol/h]	50	50	50	50
Calculated molar reflux ratio	0.1	0.1	0.1	0.1
Calculated bottoms rate [kmol/h]	630	630	630	28,599.00
Calculated boilup rate [kmol/h]	149.84	149.84	173.34	10,744.68
Calculated distillate rate [kmol/h]	50	50	50	50
Condenser/top stage temperature [°C]	90.29	90.29	102.09	102.08
Condenser/top stage pressure [bar]	1	1	1	1
Condenser/top stage heat duty [cal/s]	−14,556.07	−14,556.07	−14,225.89	−14,223.82
Condenser/top stage reflux rate [kmol/h]	5	5	5	5
Reboiler pressure [bar]	1	1	1	1
Reboiler temperature [°C]	101.15	101.15	102.51	288.78
Reboiler heat duty [cal/s]	406,073.47	406,073.47	470,837.22	98,613,121.8
Recovery percentage	100	100	Nil	Nil

11.4.1 Case study 1

In case study 1 (HCl acid and no recycling of unused cellulose), the plot between reaction conversion and the mole fraction of components, such as hydrogen, water, and hexane, has been drawn to check the effect of specified operating conditions. From Figure 11.3, it is clear that with the increase in reaction conversion, the amount of hexane production gradually increased in the final output. Even though the same amount of water was present for all the conversions, the larger amount of hexane in the product would have decreased the fraction of water in the product stream, as shown in Figure 11.3.

11.4.2 Case study 2

The same trend is found in case 2 as in case 1, while the graph has been plotted between mole fractions of hydrogen, water, and hexane concerning reaction conversion. Even with H_2SO_4 as the hydrolysis agent, a gradual increase in the mole fraction of hexane conversion was obtained. In contrast, a gradual decrease in water concentration is observed, as shown in Figure 11.3 (case study 2). It is clear from Figure 11.3 (case study 2) that H_2SO_4 acid in case 2 possesses the same properties as HCl acid in case 1 when cellulose as feed has been converted into the desired product hexane, which has not been recycled.

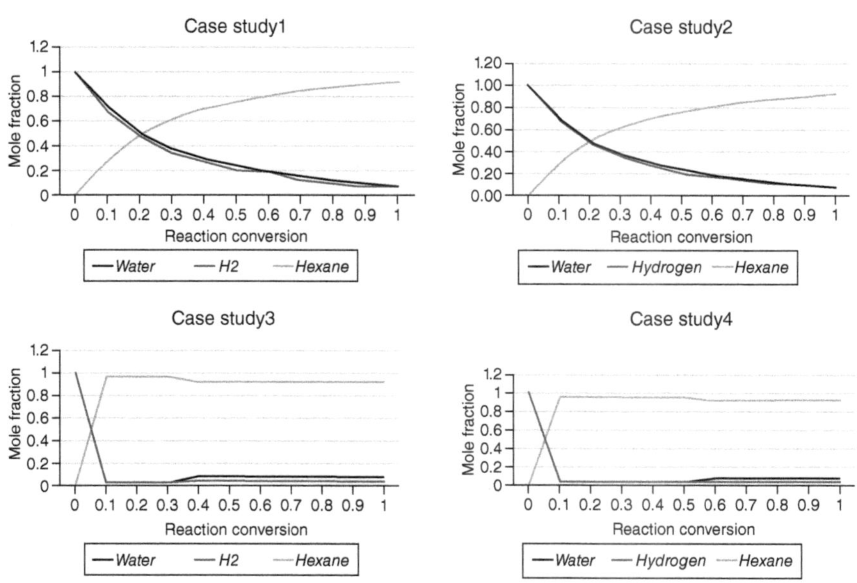

Figure 11.3 Effect of product flow composition for reaction conversion of Cases 1–4.

11.4.3 Case study 3

In case 3, the simulation on the recycling of cellulose was carried out using HCl to produce hexane. From the results obtained, the plot between the mole fraction of components such as hexane, water, and hydrogen was drawn against reaction conversion, and different results were found in cases 1 and 2. With recycling the unused cellulose, the presence of HCl rapidly increased the hexane concentration in the product stream, even at low conversion percentages up to 0.2. Increasing the conversion beyond 0.3 resulted in a slight decrease in the mole fraction of hexane due to the slight increase in water percentage with conversions, as shown in Figure 11.3 (case study 3). Hydrogen decreases at a lower value of reaction conversion, as 0.1, and shows a constant for the preceding values of reaction conversion and shows similar results as water formation.

11.4.4 Case study 4

Case 4 depicts the plot between mole fraction of the same specified components as hexane, water, and hydrogen concerning reaction conversion after a simulation on converting a recycled stream of cellulose into the desired product as hexane with H_2SO_4. With recycling unreacted cellulose and H_2SO_4 as the hydrolysis agent, a rapid increase in the concentration of hexane is observed even at low conversion rates. However, a significant jump in condenser and reboiler duty occurred when HCl was replaced with H_2SO_4. With the recycling of cellulose, hexane production was rapid due to the increase in reactant concentration, as shown in Figure 11.3 (case study 4).

From Table 11.2, it is clear that the recovery of HCl is far easier from the reaction mixture through distillation when HCl is recovered than the recovery of H_2SO_4 from a mixture in the same conditions. The boiler and condenser duty is very low for the separation of HCl, and almost pure separation is possible. This is due to its boiling point and the non-volatile nature of H_2SO_4, but HCl can easily be recovered in low concentrations. After recovery, the acid must be concentrated for further use, so continuous recycling of acid decreases the pH of the reacting mixture, which is not desired.

11.5 CONCLUSION

In the present work, cellulose conversion into hexane has been investigated using simulation. The whole study has been bifurcated into four cases: cellulose conversion into hexane using HCl acid without and with recycle stream (Cases 1 and 3) and cellulose conversion into hexane using H_2SO_4 acid without and with recycle stream (Cases 2 and 4). After conducting a simulation and further analysis based on the obtained theoretical results, some conclusions are drawn, which are as follows:

- The product separation will be high (the product will not have or will have less concentrated Water) when the hydrolysis and dehydration of the cellulose is high.
- HCl is less reactive than H_2SO_4, but it is more useful for the successful recovery and reuse of cellulose and acid. It also reduces the chances of exposure and corrosion.

REFERENCES

Akia, M., Yazdani, F., Motaee, E., Han, D., & Arandiyan, H. (2014). A review on conversion of biomass to biofuel by nanocatalysts. *Biofuel Research Journal*, 01(01), 16–25.

Baharintasari, D. R., Asrori, M. R., Prakasa, Y.F., & Sumari. (2019). Hydrolyze cellulose in banana stem midrib as a glucose source assisted by ultrasonic and activated Y-zeolite catalyst. *Journal of Environmental Sciences*, 10(9), 784–789.

Brune, D. E., Lundquist, T. J., & Benemann, J. R. (2009). Microalgal biomass for greenhouse gas reductions: Potential for replacement of fossil fuels and animal feeds. *Journal of Environmental Engineering*, 135(11), 1136–1144.

Chemat, F., Abert Vian, M., Ravi, H. K., Khadhraoui, B., Hilali, S., Perino, S., and Fabiano Tixier, A. S. (2019). Review of alternative solvents for green extraction of food and natural products: Panorama, principles, applications, and prospects. *Molecules*, 24(16), 3007.

Chen, K., Tamura, M., Yuan, Z., Nakagawa, Y., & Tomishige, K. (2013). One-pot conversion of sugar and sugar polyols to n-alkanes without C_C dissociation over the Ir-ReO_x/SiO_2 catalyst combined with H-ZSM-5. *ChemSusChem*, 6(4), 613–621.

Chernyshev, V. M., Kravchenko, O. A., & Ananikov, V. P. (2017). Conversion of plant biomass to furan derivatives and sustainable access to the new generation of polymers, functional materials and fuels. *Russian Chemical Reviews*, 86(5), 357–387.

Cravotto, C., Fabiano-Tixier, A.-S., Claux, O., Abert-Vian, M., Tabasso, S., Cravotto, G., & Chemat, F. (2022). Towards substituting hexane as extraction solvent of food products and ingredients with no regrets. *Foods*, 11, 3412.

Cséfalvay, E., & Horváth, I. T. (2018). Sustainability assessment of renewable energy in the United States, Canada, the European Union, China, and the Russian Federation. *ACS Sustainable Chemistry & Engineering*, 6(7), 8868–8874.

Demirbas, A. (2009). Political, economic and environmental impacts of biofuels: A review. *Applied Energy*, 86, S108–S117.

Edmunds, C. W., Reyes Molina, E. A., Hamilton, C., André, N., Kelley, S. S., Labbé, N., Tumuluru, J. S., Fasina, O., Adhikari, S., Rials, T. G., & Park, S. (2018). Blended feedstocks for thermochemical conversion: Biomass characterization and bio-oil production from switchgrass-pine residues blends. *Frontiers in Energy Research*, 6, 00079.

Emberru, R. E., Mujtaba, I. M., John, Y. M., & Patel, R. (2024). A review of catalyst modification and process factors in the production of light olefins from direct crude oil catalytic cracking. *Sci*, 6(1), 11.

Faaij, A. (2006). Modern biomass conversion technologies. *Mitigation and Adaptation Strategies for Global Change*, 11(2), 343–375.

Fraterrigo Garofalo, S., Tommasi, T., & Fino, D. (2021). A short review of green extraction technologies for rice bran oil. *Biomass Conversion and Biorefinery*, 11 , 569–587.

Griggs, A. J., Lischeske, J. J., & Stickel, J. J. (2011). A mechanistic model for enzymatic saccharification of cellulose using continuous distribution kinetics I: Depolymerization by EGI and CBHI. *Biotechnology and Bioengineering*, 109(3), 665–675.

Guo, Y., Liu, Y., Guan, M., Tang, H., Wang, Z., Lin, L., & Pang, H. (2022). Production of butanol from lignocellulosic biomass: Recent advances, challenges, and prospects. *RSC Advances*, 12, 18848–18863.

Haas, M. J., & Wagner, K. (2011). Simplifying biodiesel production: The direct or in situ transesterification of algal biomass. *European Journal of Lipid Science and Technology*, 113(10), 1219–1229.

Higgins, F.J., & Ho, G. E. (1982). Hydrolysis of cellulose using HCl: A comparison between liquid phase and gaseous phase processes. *Agricultural Wastes*, 4(2), 97–116.

Hutomo, G. S., Rahim, A., & Kadir, S. (2015). The effect of sulfuric and hydrochloric acid on cellulose degradation from Pod Husk Cacao. *International Journal of Current Microbiology and Applied Sciences*, 4(10), 89–95.

Kashparova, V. P., Chernysheva, D. V., Kravchenko, O. A., Smirnova, N. V., Klushin, V. A., & Andreeva, V. E. (2021). Furan monomers and polymers from renewable plant biomass. *Russian Chemical Reviews*, 90(6), 750–784.

Kawamoto, H. (2016). Review of reactions and molecular mechanisms in cellulose pyrolysis. *Current Organic Chemistry*, 20(23), 2444–2457.

Kumar, B., Kumar, S., & Kumar, S. (2017). Thermodynamic and energy analysis of renewable butanol–ethanol fuel reforming for the production of hydrogen. *Journal of Environmental Chemical Engineering*, 5(6), 5876–5890.

Liu, Y., Ahmed, S., Sameen, D. E., Wang, Y., Lu, R., Dai, J., Li, S., & Qin, W. (2021). A review of cellulose and its derivatives in biopolymer-based for food packaging application. *Trends in Food Science & Technology*, 112, 532–546.

Luo, X., Wu, H., Li, C., Li, Z., Li, H., Zhang, H., Li, Y., Su, Y., & Yang, S. (2020). Heteropoly acid-based catalysts for hydrolytic depolymerization of cellulosic biomass. *Frontiers in Chemistry*, 8, 580146.

Mahmood, T., Iqbal, H. M. N., Shahbaz, A., Mulla, S. I., Hussain, N., & Bilal, M. (2022). Sustainable production of biofuels from the algae-derived biomass. *Bioprocess and Biosystems Engineering*, 46(8), 1077–1097.

Mankar, A. R., Pandey, A., Modak, A., & Pant, K. K. (2021). Pretreatment of lignocellulosic biomass: A review on recent advances. *Bioresource Technology*, 334, 125235.

Manyatshe, A., & Sibali, L. L. (2025). A review on the recovery of cellulose, lignin, and hemicellulose biopolymers from the same source of lignocellulosic biomass – Methodology, characterization and applications. *Journal of Water Process Engineering*, 70, 107037.

Meine, N., Schüth, F., & Rinaldi, R. (2012). Solvent-free catalytic depolymerization of cellulose to water-soluble oligosaccharides. *ChemSusChem*, 5(8), 1449–1454.

Morales, M., Arvesen, A., & Cherubini, F. (2021). Integrated process simulation for bioethanol production: Effects of varying lignocellulosic feedstocks on technical performance. *Bioresource Technology*, 328, 124833.

Navarro, R. M., Fierro, J. L. G., Valle, F. D., Sánchez-Sánchez, M. C., & Alvarez-Galvan, M. C. (2009). Hydrogen production from renewable sources: Biomass and photocatalytic opportunities. *Energy & Environmental Science*, 2(1), 35–54.

Nzediegwu, E., & Dumont, M.-J. (2020). Chemo-catalytic transformation of cellulose and cellulosic-derived waste materials into platform chemicals. *Waste and Biomass Valorization*, 12(6), 2825–2851.

Op de Beeck, B., Dusselier, M., Geboers, J., Holsbeek, J., Morré, E., Oswald, S., Giebeler, L., & Sels, B. F. (2015). Direct catalytic conversion of cellulose to liquid straight-chain alkanes. *Energy & Environmental Science*, 8(1), 230–240.

Orozco, A., Ahmad, M., Rooney, D., & Walker, G. (2017). Dilute acid hydrolysis of cellulose and cellulosic bio-waste using a microwave reactor system. *Process Safety and Environmental Protection*, 85(5), 446–449.

Prasad, W., Wani, A. D., Khamrui, K., Hussain, S. A., & Khetra, Y. (2022). Green solvents, potential alternatives for petroleum-based products in food processing industries. *Cleaner Chemical Engineering*, 3, 100052.

Ptasinski, K. J. (2008). Thermodynamic efficiency of biomass gasification and biofuels conversion. *Biofuels, Bioproducts and Biorefining*, 2(3), 239–253.

Rinaldi, R., & Schüth, F. (2009). Acid hydrolysis of cellulose as the entry point into biorefinery schemes. *ChemSusChem*, 2(12), 1096–1107.

Saravanan, A., Yaashikaa, P. R., Senthil Kumar, P., Thamarai, P., Deivayanai, V. C., & Rangasamy, G. (2023). A comprehensive review on techno-economic analysis of biomass valorization and conversational technologies of lignocellulosic residues. *Industrial Crops, and Products*, 200(Part A), 116822.

Shahzad, H. M. A., Asim, Z., Khan, S. J., Almomani, F., Mahmoud, K. A., Mustafa, M. R. U., & Rasool, K. (2024). Thermochemical and biochemical conversion of agricultural waste for bioenergy production: An updated review. *Discover Environment*, 2(1), 134.

Tanger, P., Defoort, M. W., Jahn, C. E., Leach, J. E., & Field, J. L. (2013). Biomass for thermochemical conversion: targets and challenges. *Frontiers in Plant Science*, 4(50), 00218.

Tao, F., Song, H., & Chou, L. (2010). Hydrolysis of cellulose by using catalytic amounts of $FeCl_2$ in ionic liquids. *ChemSusChem*, 3(11), 1298–1303.

Torres, A. F., Van Boxtel, A. J. B., Trindade, L. M., Slegers, P. M., Vlaswinkel, L., Visser, R. G. F., Dolstra, O., & Noordam-Boot, C. M. M. (2016). Maize feedstocks with improved digestibility reduce biomass pre-treatment and saccharification costs and environmental impacts. *Biotechnology for Biofuels*, 9(1), 63.

Tschan, M. J.-L., Haquette, P., Brulé, E., & Thomas, C. M. (2012). Synthesis of biodegradable polymers from renewable resources. *Polymer Chemistry*, 3(4), 836–851.

Ummartyotin, S., & Manuspiya, H. (2015). A critical review on cellulose: From fundamental to an approach on sensor technology. *Renewable and Sustainable Energy Reviews*, 41, 402–412.

Vanderfleet, O. M., & Cranston, E. D. (2021). Production routes to tailor the performance of cellulose nanocrystals. *Nature Reviews Materials*, 6, 124–144.

Venkatachalam, C. D., Ravichandran, S. R., & Sengottian, M. (2021). Lignocellulosic and algal biomass for bio-crude production using hydrothermal liquefaction: Conversion techniques, mechanism and process conditions: A review. *Environmental Engineering Research*, 27(1), 200555–0.

Zhang, L., Ren, N., Zhang, J., Gao, X., Cui, M., Qi, W., Huang, R., Lam, H. L., & Su, R. (2025). Cellulose nanocrystals: Sustainable production and emerging fruit coating applications. *Chemical Engineering Journal*, 509, 161190.

Chapter 12

Sustainable materials for photovoltaic systems

Uplabdhi Tyagi and Neeru Anand

12.1 INTRODUCTION

As environmental challenges continue to escalate, the demand for sustainable energy solutions has become increasingly critical. Among the various renewable energy technologies, photovoltaic (PV) systems play a vital role in advancing the vision of a low-carbon future. PV technology uses sunlight to generate electricity, which has drastically reduced greenhouse gas emissions, and develops a more sustainable landscape in the energy sector. PV systems directly convert sunlight into electricity, providing clean, renewable energy that is becoming increasingly cost-effective. Their ability to meet diverse energy demands from large-scale solar farms to residential rooftop installations puts PV systems at the heart of efforts to limit greenhouse gas emissions and to wean off finite fossil fuels. As the scale of PV adoption grows, the environmental impact of manufacturing, maintaining, and eventually disposing of PV modules has received much attention, setting off a global movement towards more sustainable materials and practices in PV technology. Sustainability of PV systems goes beyond their operational efficiency. PV modules are mainly composed of semiconducting materials, with silicon being the most widely used. Silicon-based PV cells have the attractive advantages of efficiency, but energy-intensive purification and production processes are usually linked to major environmental impacts. Other alternative technologies, such as cadmium telluride (CdTe) or copper indium gallium selenide (CIGS), have their own problems with resource availability and toxicity (Dada and Popoola, 2023). Therefore, the transition towards truly sustainable PV systems must consider the complete lifecycle, from raw material extraction and manufacturing processes through to waste disposal and recycling. As the industry expands, there is a need to adopt sustainable materials and methods to ensure solar energy remains a green alternative with minimal ecological impact. Among the most critical aspects of sustainability in PV systems is end-of-life management. Solar panels have an average lifespan of 25–30 years; after this, they experience efficiency decline, such that they need replacement or repurposing. Without proper recycling

 DOI: 10.1201/9781003581536-12

and disposal options, of course, high volumes of PV waste will indeed pose an environmental burden. On the other hand, some recent studies on closed-loop recycling and modular design raise optimism. The need to advance more in recycling technology, as well as in the development of PV modules with higher readiness for easy dismantling, can help to significantly reduce waste and resource depletion.

In recent years, researchers and manufacturers have turned their attention toward alternative materials that may alleviate the environmental impact of traditional PV systems. Emerging materials like perovskites, OPVs, and DSSCs have potential for high efficiencies and may reduce the environmental footprint of solar technology. New materials could even be used to cut reliance on scarce metals and facilitate lower-temperature processing that further decreases energy input into the production process. However, the key challenges in stability, efficiency, and scalability still prevail. Understanding the potential of such materials requires an integrated view with performance, environmental considerations, and economic feasibility at the same level. A holistic approach to societal and larger implications of sustainable PV systems is equally appealing: increasing energy security, the opportunity for job creation, and greater social equity since technology becomes more sustainable (Henda et al., 2023). It would be the PVs that make a difference, especially in sunny regions where other sources of energy are scarce. Such systems would contribute to sustainability in multiple ways, meaning that they would support both environmental goals and socio-economic development and resilience.

This chapter provides a structured exploration of sustainable materials for photovoltaic systems, emphasizing both current advancements and future prospects. The discussion begins with a classification of photovoltaic materials, laying the groundwork for understanding the diverse types used in PV technology.

12.2 *CLASSIFICATION OF PHOTOVOLTAIC* MATERIALS

Photovoltaic (PV) materials play a crucial role in converting sunlight into electricity, and their classification is essential for understanding their diverse functionalities, efficiencies, and applications. Additionally, the comparative efficiency of materials (such as organic photovoltaics (OPVs), Perovskite solar cells, dye-sensitized solar cells (DSSCs), silicon-based materials, cadmium telluride (CdTe), and bio-based materials) for photovoltaic applications is a crucial advancing factor in solar technology, where each material has different prospects and limitations based on their energy conversion efficiency, the cost, and environmental impact (Table 12.1). Among these materials, DSSCs have excellent photovoltaic performance because of a specific operational strategy. The basic DSSC has a TCO glass substrate and a TiO_2 layer coated with a dye sensitizer that absorbs sunlight,

Table 12.1 Comparative efficiency of various materials for photovoltaic applications

Material	*Efficiency (%)*	*Properties*	*Advantages*	*Disadvantages*	*References*
Organic photovoltaics (OPVs)	18.2	Bandgap: 1.8–2.0 eV; Absorption: $\lambda <$ 600 nm	Low-cost synthesis, solution processability	Generally lower power conversion efficiency (PCE)	Li et al. (2022)
Perovskite solar cells	25.5	Bandgap: 1.55 eV; Absorption: $\lambda <$ 800 nm	High power conversion efficiency, low material cost potential	Stability concerns and lead toxicity issues	Aldabesh (2023)
Dye-sensitized solar cells (DSSCs)	13.4	Bandgap: ~3.2 eV (TiO_2); Absorption: $\lambda <$ 700 nm	Cost-effective fabrication methods, semi-transparency	Limited long-term stability and efficiency degradation under light exposure	Korir, Kibet and Ngari (2024)
Silicon-based materials	26.7	Bandgap: 1.12 eV; Absorption: $\lambda <$ 1,100 nm	Established technology, high reliability and efficiency	High energy (approx. 200 MJ/kg) and resource intensity in production	Besseau et al. (2023)
Cadmium telluride (CdTe)	22.1	Bandgap: 1.5 eV; Absorption: $\lambda <$ 800 nm	Efficient light absorption, established manufacturing processes	Environmental concerns related to cadmium toxicity, recycling challenges	Katepalli, Wang and Shi (2023)
Bio-based materials	3–10	Bandgap:varies;biodegradable; Absorption: $\lambda <$ 600 nm	Environmentally sustainable, low carbon footprint, renewably sourced	Generally lower efficiency compared to conventional materials; variable performance	Rahimi et al. (2023)

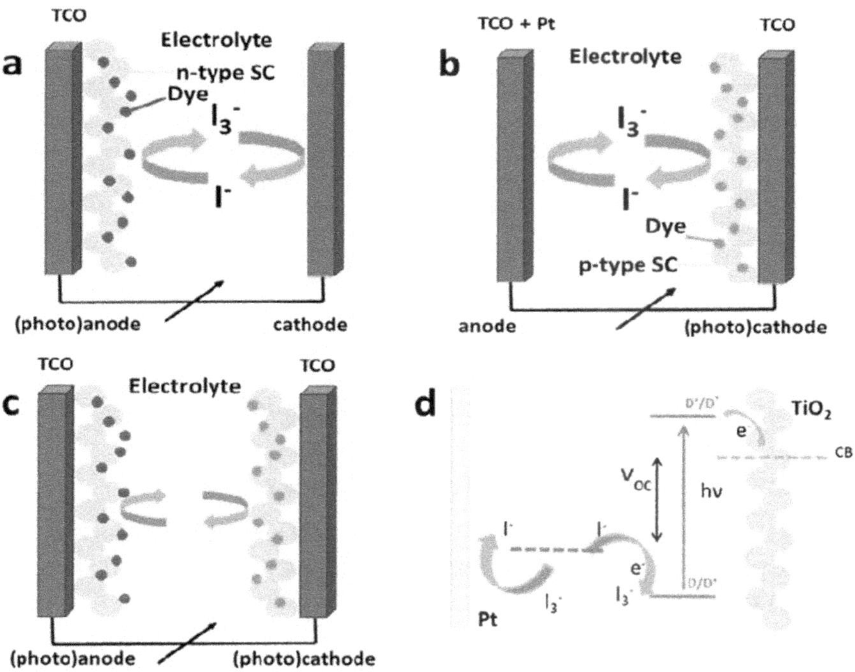

Figure 12.1 Diagram illustrating: (a) the standard design of a dye-sensitized solar cell (DSSC), (b) the inverted DSSC configuration, (c) a tandem device, and (d) primary electronic processes occurring within a traditional DSSC during operation. (Reproduced from an open-access source Mariotti et al., 2020.)

exciting electrons into TiO_2 and allowing them to flow through an external circuit, while an electrolyte regenerates the dye. The excited dye molecules release electrons into TiO_2, which is collected by the TCO electrode, while the electrolyte ensures continuous operation by restoring the dye (as shown in Figure 12.1). Furthermore, the photovoltaic materials can be classified into mainly two categories- conventional and non-conventional, which vary in further technologies.

12.2.1 Conventional photovoltaic materials

Conventional photovoltaic materials predominantly consist of silicon-based solar cells, which have become the backbone of the solar energy industry due to their high efficiency, reliability, and established manufacturing processes. Of these, c-Si solar cells are widely used and have accounted for over 90% of the world's solar cell production. The crystalline silicon cells can further be divided into two categories, namely mono-crystalline and poly-crystalline silicon. These mono-crystalline silicon cells are made up of a single-crystal structure and can reach a staggering efficiency of above 25% while the range starts at 15%. This is due to their uniform crystal lattice, thus minimizing defects and ensuring perfect electron mobility. Although the production costs for mono-crystalline cells are higher than for all other types, the extended life of these cells, exceeding 25 years, makes them a viable, cost-effective solution in the long run (Olorunshogo et al., 2024). Poly-crystalline silicon cells include multiple silicon crystals and usually have reduced efficiencies, from about 13%–20%. CdTe solar cells are especially notable for their low manufacturing costs and exemplary performance in low-light conditions, achieving efficiencies around 9%–11%. The main variation under thin-film technology relates to solar cells prepared by CIGS. Thin layers of CIGS material are deposited onto substrates, which also render an application with flexibility for such purposes (Zhang et al., 2023).

12.2.2 Sustainable photovoltaic materials

In contrast, sustainable photovoltaic materials are gaining traction as alternatives to conventional options, focusing on environmental sustainability and lower production costs. Organic photovoltaics (OPVs) form a significant part of this category. OPVs utilize organic semiconductors, which can be processed at relatively low temperatures, enabling the production of flexible and lightweight solar cells (Figure 12.2). While efficiency is usually low (from 5% to 15%), the potential for high throughput and large-scale production by roll-to-roll techniques, along with their flexibility, makes them interesting candidates for applications ranging from large displays to thin-film photovoltaic devices. The ability of organic materials to vary the band gap is another attractive aspect, as this variability can optimize light absorption for any particular application over the relevant wavelength range. The perovskite solar cells have proved to be one of the most promising technologies within the sustainable PV category. These materials, characterized by a unique crystal structure, have achieved efficiencies greater than 25% in a lab test (Nyamathulla and Chittathuru, 2023). Their band gap can be tuned to optimize solar energy absorption, and they are available using low-cost solution processing routes, making them attractive for use in the commercial sector. The most significant challenges remain particularly in their long-term stability and potential environmental impacts related to lead-based perovskites. Researchers have been working to enhance their durability, find alternatives without lead, and develop encapsulation techniques to improve their resistance to moisture and UV exposure (Table 12.2). The other technology in the sustainable category is DSSCs. DSSCs operate under the mechanism of a dye chemisorbed onto the semiconductor TiO_2 to convert light to electrical energy. These dyes are easy to produce and fabricate with the general efficiencies running

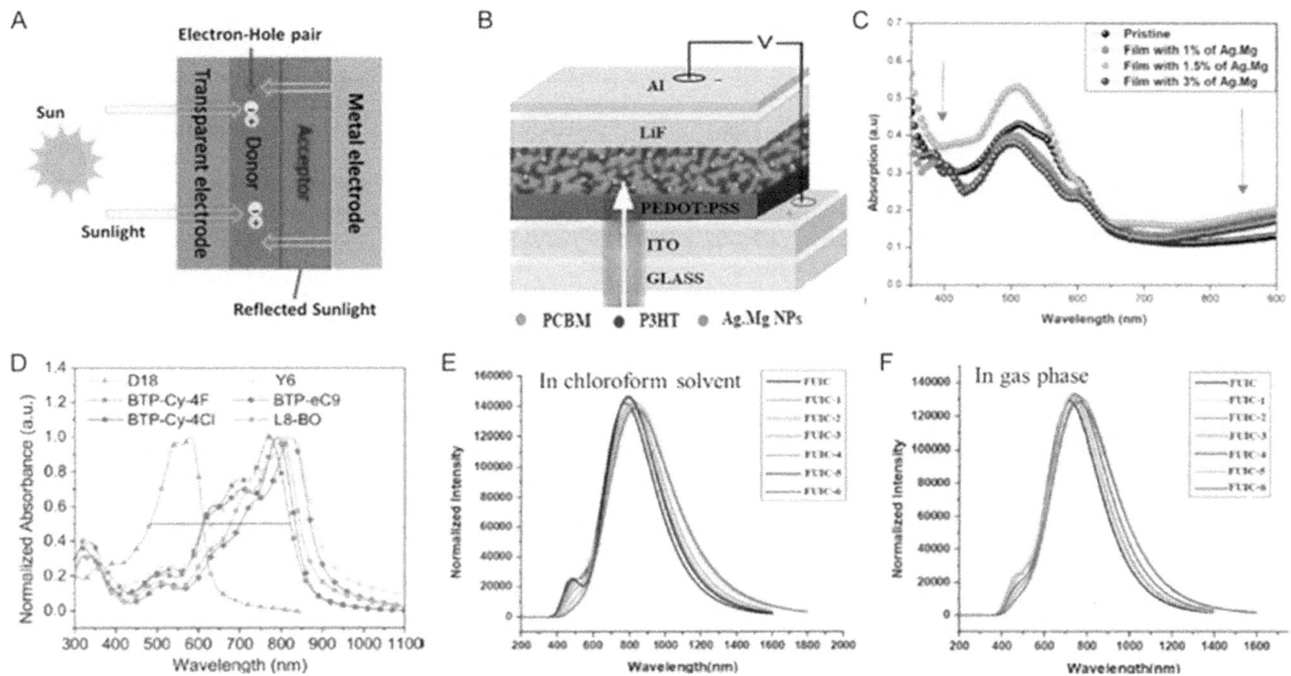

Figure 12.2 Diagrammatic representation of the light absorption mechanism in an organic photovoltaic (OPV) cell along with its performance parameters. (Reproduced from an open access source Ramoroka et al., 2024.)

Table 12.2 Recent innovations in sustainable photovoltaic materials

Innovation	*Description*	*Efficiency (%)*	*Materials*	*Power output (W/m^2)*	*Potential impact*	*References*
Bifacial solar panels	Capture sunlight on both sides, utilizing reflected light from surfaces below.	Up to 27.3	Monocrystalline silicon, glass	400–450	Increased energy yield by 10%–20%, reducing land use for solar farms	Ernst et al. (2024)
Transparent conductors	Use in building-integrated photovoltaics (BIPV), maintaining aesthetic appeal while generating power.	15–20	Indium tin oxide (ITO), graphene	50–100	Allows for multifunctional building materials, integrating power generation seamlessly	Woods-Robinson et al. (2024)
Nanostructured surfaces	Enhance light trapping and absorption through structures at the nanoscale.	20–25	TiO_2, plasmonic nanoparticles	300–350	Higher efficiency and reduced active material requirements, potentially lowering production costs	Josephine, Ikponmwosa and Ikhioya, (2023)
Perovskite tandem cells	Combination of perovskite with silicon to leverage strengths of both materials.	30+	$MAPbI_3$, crystalline silicon	500–600	Significantly higher efficiency potential, enhancing energy conversion in diverse applications	Aydin et al. (2024)
Recyclable solar modules	Incorporation of recyclable materials, improving sustainability throughout the life cycle.	18–22	Bio-based polymers, recycled silicon	150–200	Reduced environmental impact, promoting a circular economy and minimizing waste	Divya et al. (2023)

between 7% and 11%. Owing to increasing global demands for renewable energy, there has been tremendous input in conventional and sustainable materials. Their efforts will be critical in developing their respective efficiencies, reducing costs, and maximizing sustainability (Habib et al., 2023). The future of photovoltaic technology is likely to be characterized by a synergy between high-efficiency conventional materials and innovative sustainable alternatives, paving the way for more accessible and environmentally friendly solar energy solutions. The evolution of PV technology not only addresses the pressing energy needs but also contributes to the global transition towards sustainable energy systems (Table 12.2).

12.3 GREEN ALTERNATIVES TO TOXIC ELEMENTS IN PV SYSTEMS

As the demand for clean energy rises worldwide, there is growing concern about the materials used in solar cells. While toxic elements like cadmium and lead can improve solar cell performance, they also pose serious risks to the environment and human health. This would be critically important for the sustainability of solar energy technologies and for diminishing the environmental impacts of their manufacturing and disposal. Research opportunities in these areas continue to promise to advance solar technology whilst remaining consistent with ecological principles. The transition to cadmium-free and lead-free materials, along with the reduction of rare earth metal usage, represents a significant step forward in creating safer, more sustainable photovoltaic systems.

12.3.1 Cadmium-free materials for thin-film solar cells

Thin-film solar cells are an increasingly important component of the photovoltaic market, prized for their lightweight and flexible designs, which allow for diverse applications beyond traditional solar panels. The problem, however, remains with the use of cadmium in some thin-film technologies, which has been proven to be a toxic heavy metal. In that regard, there is need for more research on the cadmium-free materials with regard to developing solar energy technology in the future. Cadmium-free materials include the copper indium gallium selenide, which is shortened as CIGS. The cells for CIGS have achieved significant efficiencies, and at times, lab-scale versions have gained more than 20%. The most interesting aspect is the ability to optimize the composition of CIGS by adjusting the ratio of copper, indium, gallium, and selenium. This enables designers to come up with cells that can effectively capture a broad spectrum of sunlight. In addition, substitution with zinc for cadmium sulfide (CdS) as a buffer layer also shows that higher efficiency is possible while drastically decreasing environmental concerns (Ahmad et al., 2023). The primary issue at this point is scaling these technologies to industrial-level production for commercial purposes. The costs and availability of indium and gallium can severely impede the spread of such photovoltaic technology. Finally, another area of development includes organic photovoltaics. Their fabrication processes commonly involve the use of abundant, nontoxic materials that make them very suitable for sustainable energy. Currently, efficiencies stand at about 10%–12%. Still, the main research focuses on the stabilization and long-term stability of organic materials. One of the major challenges to environmental degradation is exposure to moisture and heat. Encapsulation techniques and novel material development that are stronger in resistance to degradation improve the overall performance and marketability of OPVs. Along with this, cadmium-free alternatives in a-Si solar cells continue the discussion. Though less efficient than the counterpart of crystalline, specific advantages are low cost for the manufacturing process and a smaller carbon footprint. This potential for deposition on flexible substrates makes possible integrative opportunities in applications as varied as building-integrated photovoltaics and portable electronics (Tseberlidis et al., 2024).

12.3.2 Lead-free perovskites

Perovskite solar cells have gained significant attention due to their exceptional light absorption capabilities and ease of fabrication. However, the reliance on lead poses serious concerns, prompting extensive research into lead-free alternatives. Particularly promising is the tin-based perovskite, like $CH_3NH_3SnI_3$. Tin may potentially replace lead;. However, efficiencies of tin perovskites have already reached over 10%, stability and oxidation problems need to be overcome for long-term viability. Research is focused on stabilizing tin perovskites (Table 12.3) by using innovative additives and coatings that protect the material from environmental degradation. Besides tin, there are also bismuth and antimony compounds under study for lead-free perovskites. For example, such compounds as $Bi_{1-x}Cs_xI_3$ and Sb-based perovskites are not only less toxic but also exhibit unique properties in optoelectronic applications that may be advantageous in enhancing the efficiency of the material (Chauhan et al., 2023). The chief challenge in realizing this material is to achieve efficiencies comparable to those of the parent lead-based materials, whose synthesis and processing methods still require further optimization based on the studies. The latest developments in this area demonstrate that the layer-stacking process with different materials can effectively capture a broader solar spectrum, thereby enhancing performance. Surface passivation is now increasingly used as a method for stabilizing and improving the efficiency of lead-free perovskites (Zhu et al., 2023).

Table 12.3 Lead-free perovskites for photovoltaic application

Material	*Composition*	*Efficiency (%)*	*Synthesis method*	*Advantages*	*Disadvantages*	*References*
Tin-based perovskites	$CH_3NH_3SnI_3$	14.5	Solution processing	Non-toxic, high absorption, good efficiency	Lower stability, high cost of tin	Abate (2023)
Bismuth-based perovskites	$Cs_3Bi_2I_9$	9	Hydrothermal method	Environmentally friendly, good thermal stability	Lower efficiency, limited absorption	Chen et al. (2023)
Double perovskites	$Cs_2AgBiBr_6$	10	Solvothermal method	Improved stability, potential for diverse cation use	Moderate efficiency, complex synthesis	Buvaneswaran et al. (2023)

12.3.3 Reducing the use of rare earth metals

The reliance on rare earth metals (REMs) in photovoltaic technologies is under increasing scrutiny due to their environmental impact and the geopolitical complexities associated with their extraction. Although REMs are used to enhance the performance of certain solar cell materials, their use must be reduced for more environmentally friendly photovoltaic systems. A material substitution approach has proved crucial in reducing dependence on REMs. Hydrometallurgical processes, among others, advanced recycling methods ensure the efficient removal of REMs from disposed of solar panels with minimized resultant wastes and environmental impact by mining and refining processes. A wide scope on recycling places useful resources without losing value in the lifetime of the solar technology uses (Khan et al., 2023). Proper design of such devices will, therefore, help prevent an increased demand for rare earth materials. Innovative designs, such as tandem solar cells, can accommodate multiple layers to achieve high efficiency with minimal quantum of rare earth used on each layer. A sustainable and environmentally responsible future for solar energy would thus require the exploration of green alternatives to toxic elements in photovoltaic systems. Investigations into cadmium-free materials for thin-film solar cells, the development of lead-free perovskites, and reducing the reliance on rare earth metals would all help the solar industry address pressing ecological concerns while enhancing the efficiency and longevity of solar technologies (Sahajwalla and Hossain, 2023).

12.4 NATURAL AND BIO-BASED MATERIALS IN PHOTOVOLTAICS

The urgent need for sustainable energy sources has catalyzed a comprehensive exploration of natural and bio-based materials in photovoltaic technologies. Traditional photovoltaic systems, particularly silicon-based solar cells, are efficient but rely on nonrenewable resources and incur high environmental costs during manufacturing and disposal. This transition involves the use of plant-derived dyes, biodegradable substrates, and polymers sourced from renewable resources, thereby minimizing environmental impact while maintaining performance.

12.4.1 Plant-based and bio-derived dyes for DSSCs

Dye-sensitized solar cells (DSSCs) are a noteworthy area of interest within photovoltaic technology that leverages natural materials. In the case of DSSCs, organic dyes absorb the sun's rays and emit electrons, rather than inorganic semiconductors, which are usually found in traditional solar cells. Bio-derived and plant-based materials (Figure 12.3) are promising substitutes for synthetic dyes, usually derived from petrochemicals, and are

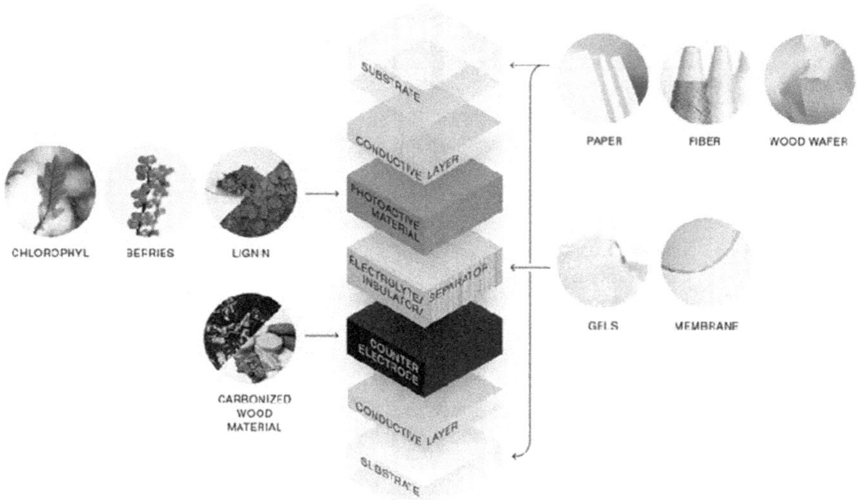

Figure 12.3 Plant-based materials for photovoltaic applications. (Reproduced from an open access source Miettunen et al., 2024.)

considered an ecological hazard. Natural dyes derived from fruits, flowers, and vegetables may efficiently utilize the energy derived from sunlight (Table 12.4). For example, chlorophyll, anthocyanins, and carotenoids are some of the most abundant pigments, which also show effective light absorption in the visible spectrum (Kumar et al., 2023). These dyes can be extracted with a low environmental footprint using eco-friendly methods. Studies have shown that these natural dyes absorb specific wavelengths of light, thereby allowing selective absorption characteristics to be tailored for maximum energy conversion efficiency. In addition to being renewable, these dyes can be chemically modified to achieve higher performance. Such techniques as co-sensitization, where two or more dyes are used in conjunction, help broaden the absorption spectrum, hence increase the harvesting of light. Also, DSSCs can have their efficiency significantly improved using nanomaterials like TiO_2 and ZnO by enhancing electron transport. Particularly, newly developed dyes, unless they are stabilized as well as durable enough to keep a persistent color upon prolonged light exposure as well as in most environmental conditions, thus remain a problem (Arunkumar, Wilson and Lee, 2023).

12.4.2 Biodegradable substrates and encapsulation

The substrate and encapsulation materials in photovoltaic devices are critical for protecting the active components from environmental stressors, such as

Table 12.4 Bio-based materials for photovoltaic applications

Material	*Properties*	*Applications*	*Limitations*	*References*
Plant-based dyes	• High light absorption (up to 95% in certain wavelengths) • Non-toxic (non-carcinogenic) • Tunable properties (can modify dye structure for specific absorption peaks)	• Dye-sensitized solar cells (DSSCs) • Photovoltaic devices	• Stability of < six months under standard conditions • Efficiency typically <10% compared to synthetic dyes • Limited color range for optimal absorption (typically requires optimization)	Miettunen et al. (2024)
Biodegradable substrates	• Composed of renewable materials (e.g., PLA, starch-based polymers) • Lightweight (density ~1.25 g/cm^3) • Compatibility with various PV technologies (can be blended with other materials)	• Module substrates and packaging • Support structures for PV cells	• Mechanical strength may decline under humidity (>70% RH) • Thermal stability may limit use (degrades >60°C)	El Hassani et al. (2023)
Renewable source polymers	• Biocompatible, flexible (elongation at break >300%) • Good barrier properties (O_2 permeability < 2.0 cc/m^2/day) • High thermal stability (glass transition temperature > 50°C)	• Encapsulation and adhesives • Protection against moisture	• Variability in mechanical properties based on source (modulus range: 300–800 MPa) • Higher cost (15%–30% more than traditional polymers) • Potential degradation over time in UV exposure	Alhodaib et al. (2024)

moisture, UV radiation, and mechanical damage. Traditional substrates are mostly rigid and are not biodegradable materials; they can be very polluting to the environment. Biodegradable materials made from natural polymers, such as polylactic acid (PLA), cellulose, or starch, are more ideal. These materials provide lots of support for solar cells and easily decompose into minimal waste at the end of their life cycle. Additionally, the substrate could influence the mechanical properties as well as the overall efficiency of solar cells. Thus, the use of biodegradable flexible substrates can enable the development of ultra-lightweight portable solar panels, which are implemented in building facades as well as clothing. Such freedom in application might open entirely new markets and applications (Dintcheva, Morici and Colletti, 2023).

12.4.3 Role of polymers from renewable sources

Polymers derived from renewable resources are integral to advancing the sustainability of photovoltaic technologies. These polymers can find numerous applications in solar cells, including as binders, conductive materials, and encapsulants. Some of the natural polymers available include cellulose, chitosan, and proteins, all of which have excellent properties suitable for photovoltaic applications. Synthetic biobased polymers include PLA and PBS, which are already made from renewable feedstocks and valued for their versatility and environmental benefits. Recent findings showed that inclusions of conductive additives like carbon nanotubes or graphene can significantly enhance the electrical conductivity of matrices based on biopolymers, reducing or perhaps fully bridging the gap toward conventional material performance (Awad et al., 2023). Despite the benefits, several challenges need to be overcome to realize the full potential of utilizing bio-based polymers in solar cells. The processing conditions required may be drastically different from those needed for conventional polymers, as are the necessary manufacturing techniques. The solar energy industry, for example, would drastically reduce its impacts on the environment by the implementation of plant-based dyes, biodegradable substrates, and renewable-source polymers in relation to the enhanced performance and lifespan of solar cells (Byun et al., 2023).

12.5 NANOMATERIALS AND SUSTAINABLE NANOCOMPOSITES FOR PV

Nanomaterials are increasingly recognized as vital components in the advancement of photovoltaic (PV) technology, significantly enhancing the efficiency and sustainability of solar cells. Some other types of nanomaterials include quantum dots, nanowires, graphene, and carbon nanotubes, which provide distinct properties to improve light absorption, charge transport, and the overall performance of solar cells. For example, quantum dots can be synthesized to absorb specific wavelengths of light, resulting in higher energy conversion efficiencies, while nanowires enhance charge collection and decrease recombination losses. Moreover, the integration of these nanomaterials in sustainable nanocomposites takes a step forward in the advancement of PV technologies (Elnozahy, Abd-Elbary and Abo-Elyousr, 2024). In addition, silicon and graphite nanocomposites exhibit excellent electrochemical performance (Figure 12.4), positioning them as promising materials for photovoltaic applications. In addition, the recyclable nature of nanocomposites follows the circular economy model of waste minimization.

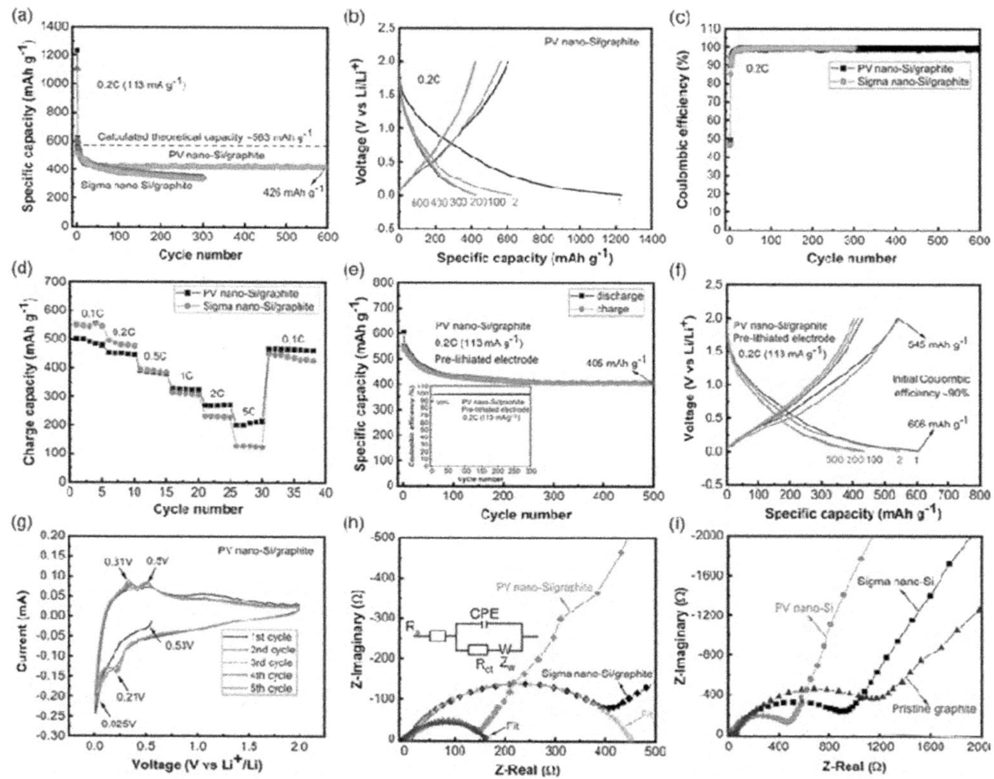

Figure 12.4 Electrochemical performance analysis: (a) the cycling stability of PV nano-Si/graphite, (b) charge-discharge profiles, (c) coulombic efficiency, (d) rate capability, (e) the cycle stability, (f) charge-discharge profiles for the prelithiated PV nano-Si/graphite electrode, (g) cyclic voltammetry (CV) data for the PV nano-Si/graphite electrode during initial cycles, and (h, i) electrochemical impedance spectroscopy (EIS) results and the equivalent circuit model. (Reproduced from an open access source Rahman et al., 2021.)

Furthermore, green synthesis pathways in the synthesis of nanomaterials, like the sol-gel method and plant-based synthesis processes, help reduce the environmental burden arising from the use of toxic chemicals and energy. Nanomaterials, with the support of the newly developing sustainable nanocomposites (Table 12.5), together form that single key that is finally going to make solar solutions to global challenges much more efficient and ecologically sound (Rajamony, Paw, et al., 2024).

12.5.1 Nanoparticle-based absorbers and conductors

Nanoparticle-based absorbers have emerged as a transformative approach to enhancing the efficiency of photovoltaic devices. Unique properties of

Table 12.5 Nanocomposite materials for photovoltaic applications

Nanomaterial type	*Functionality*	*Efficiency (%)*	*Thermal stability (°C)*	*Key applications*	*References*
Nanoparticle absorbers	Enhance light absorption through increased surface area and light scattering effects.	~30	150–200	Photovoltaic cells, particularly in organic and dye-sensitized solar cells (DSSCs).	Öztürk, Yüksel and Çiftçi (2024)
Graphene	Acts as a conductive material with excellent electron mobility and flexibility.	10–15	>300	Transparent conductive layers, electrodes in organic photovoltaics.	Sheikholeslami, Khalili and Mousavi (2023)
Carbon nanotubes	Provide enhanced charge transport properties and mechanical strength.	5–10	>500	Transparent electrodes, conductive composites, and additives in photovoltaic films.	Rajamony, Pandey, et al. (2024)
Nanostructured surfaces	Improve efficiency and durability by enhancing surface area for charge carrier generation.	15–20	10–100	Thin-film solar cells, an enhancement of traditional PV technologies.	Mo et al. (2023)

nanoparticles are due to size, shape, and surface characteristics in such a way that they may be engineered to suit desirable optical responses. The performance parameters of flexible photovoltaic (PV) devices are significantly influenced by the bandgap energy of the absorber layer (Figure 12.5), which plays a crucial role in determining the efficiency and overall energy conversion capabilities of these devices. For instance, semiconductor materials like quantum dots (QDs) exhibit a phenomenon called quantum confinement, where the electronic properties change as the size of the particle decreases. Inclusion of metal nanoparticles within the photovoltaic matrix promises the possibility of improving the absorption of light due to LSPR (Tasnim et al., 2023). Researchers are also working on the functionalization of CNTs for better compatibility with polymer matrices, which may further enhance organic photovoltaic performance. Semiconductor nanoparticles combined with conductive polymers are gaining more attention due to synergistic effects. Such composites could take advantage of the ability of nanoparticles to absorb light and combine it with the flexibility and processability of polymers (Kanani et al., 2024). This approach can result in lightweight, flexible solar cells suitable for a wide range of applications, including wearable electronics and building-integrated photovoltaics.

12.5.2 Graphene and carbon nanotubes in PV systems

Graphene and carbon nanotubes are at the forefront of research in photovoltaic technologies due to their exceptional physical and electronic properties. This material has a two-dimensional structure that allows for outstanding levels of electrical conductivity. It also features optical transparency at very high levels that will enable its use in transparent conductive electrodes in solar cells. For example, it has been demonstrated that integrating graphene with perovskite solar cells results in excellent performance by inhibiting charge recombination, thereby improving stability and efficiency. Another interesting use of graphene in PV is the development of flexible solar panels (Jain et al., 2024). Additionally, it has been shown that the application of CNTs in hybrid solar cells improves the mechanical properties, leading to increased robustness and performance under various environmental conditions. The use of graphene and CNTs within composite materials offers the opportunity to take advantage of the strong points of both materials. The next generations of photovoltaic technologies may thus be prepared for by optimizing light absorption and charge transport in such hybrid nanocomposites (Fikri et al., 2023).

12.5.3 Nanostructuring for efficiency and sustainability

Nanostructuring plays a critical role in maximizing the efficiency of photovoltaic devices by enhancing light management and charge separation.

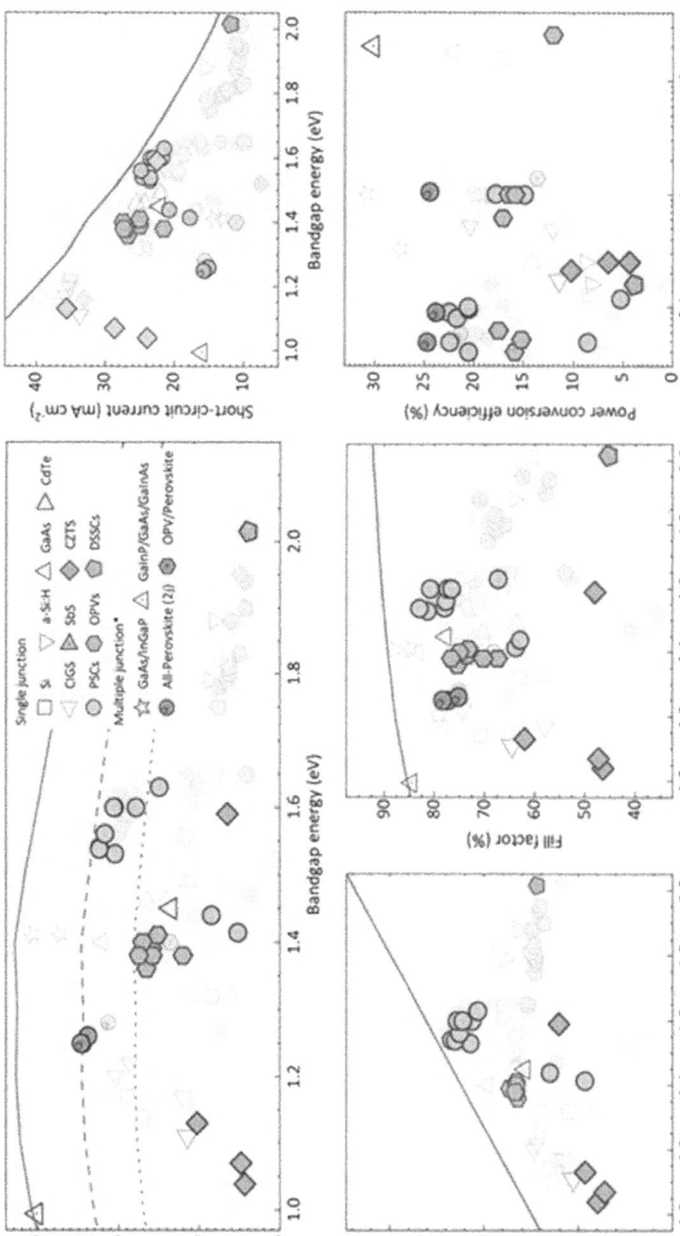

Figure 12.5 Performance parameters of flexible photovoltaic devices vary according to the bandgap energy of the absorber layer. (Reproduced from an open access source Almora et al., 2023.)

Scientists in research can design the solar cell surface morphologies to a nanoscale, which dramatically improves light trapping. Beyond this, nanostructuring might also be applied to a heterojunction in solar cells. Here, in particular, combining two dissimilar materials having different bandgaps would result in creating junctions that provide maximum charge separation with consequent reductions in recombination loss (Shah et al., 2024). For example, nanostructured layers of titanium dioxide (TiO_2) can be used in perovskite solar cells as electron transport layers, thereby significantly enhancing charge extraction and stability. The unique properties of nanoparticles, graphene, and carbon nanotubes have the potential to bring forth a new level of enhancement in light absorption, charge transport, and overall device efficiency (Ali et al., 2023).

12.6 CONCLUSION

The exploration of sustainable materials for photovoltaics (PV) has become increasingly essential in the context of the global push for renewable energy solutions. This chapter has showcased great strides in materials science to transform the face of solar energy generation. New alternatives, such as perovskite, organic photovoltaics, and tandem structures, are emerging alternatives to conventional silicon-based solar cells with improved efficiencies and potentially reduced production costs. Such developments not only make PV technology more viable but also contribute to sustainability goals by reducing resource consumption and further minimizing environmental impact.

REFERENCES

Abate, A. (2023) 'Stable tin-based perovskite solar cells', *ACS Energy Letters*, 8(4), pp. 1896–1899. Available at: https://doi.org/10.1021/acsenergylett.3c00282.

Ahmad, N. et al. (2023) 'Cadmium-free kesterite thin-film solar cells with high efficiency approaching 12%', *Advanced Science*, 10(26). Available at: https://doi.org/10.1002/advs.202302869.

Aldabesh, A.D. (2023) 'A numerical study of $CsSnI_xBr_{3-x}$ perovskite material as an electron transport layer (ETL), in the perovskite solar cell of a photovoltaic system by molecular dynamics method with LAMMPS software: The effects of external convective heat transfer', *Engineering Analysis with Boundary Elements*, 147, pp. 51–58. Available at: https://doi.org/10.1016/j.enganabound.2022.11.032.

Alhodaib, A. et al. (2024) 'Sustainable coatings for green solar photovoltaic cells: Performance and environmental impact of recyclable biomass digestate polymers', *Scientific Reports*, 14(1), p. 11221. Available at: https://doi.org/10.1038/s41598-024-62048-5.

Ali, J. et al. (2023) 'Green synthesized zinc oxide nanostructures and their applications in dye-sensitized solar cells and photocatalysis: A review', *Materials Today*

Communications, 36, p. 106840. Available at: https://doi.org/10.1016/j.mtcomm.2023.106840.

Almora, O. et al. (2023) 'Device performance of emerging photovoltaic materials (Version 3)', *Advanced Energy Materials*, 13(1). Available at: https://doi.org/10.1002/aenm.202203313.

Arunkumar, T., Wilson, H.M. and Lee, S.J. (2023) 'Waste are in the limelight: cost-effective waste materials for sustainable solar desalination', *Clean Technologies and Environmental Policy*, 25(9), pp. 2805–2830. Available at: https://doi.org/10.1007/s10098-023-02574-4.

Awad, M. et al. (2023) 'Design and analysis of photovoltaic/wind operations at MPPT for hydrogen production using a PEM electrolyzer: Towards innovations in green technology', *Plos One*, 18(7), p. e0287772. Available at: https://doi.org/10.1371/journal.pone.0287772.

Aydin, E. et al. (2024) 'Pathways toward commercial perovskite/silicon tandem photovoltaics', *Science*, 383(6679). Available at: https://doi.org/10.1126/science.adh3849.

Besseau, R. et al. (2023) 'An open-source parameterized life cycle model to assess the environmental performance of silicon-based photovoltaic systems', *Progress in Photovoltaics: Research and Applications*, 31(9), pp. 908–920. Available at: https://doi.org/10.1002/pip.3695.

Buvaneswaran, S. et al. (2023) 'Design of ferroelectric double perovskite oxides as photovoltaic materials', *The Journal of Physical Chemistry C*, 127(31), pp. 15486–15499. Available at: https://doi.org/10.1021/acs.jpcc.3c02094.

Byun, J. et al. (2023) 'Reliability, economic, and environmental analysis of fuel-cell-based hybrid renewable energy networks for residential communities', *Energy Conversion and Management*, 284, p. 116964. Available at: https://doi.org/10.1016/j.enconman.2023.116964.

Chauhan, V. et al. (2023) 'Prospects for lead free perovskite for photovoltaic applications and biological impacts: Challenges and opportunities', *Inorganic Chemistry Communications*, 157, p. 111421. Available at: https://doi.org/10.1016/j.inoche.2023.111421.

Chen, X. et al. (2023) 'Recent progress and challenges of bismuth-based halide perovskites for emerging optoelectronic applications', *Advanced Optical Materials*, 11(3). Available at: https://doi.org/10.1002/adom.202202153.

Dada, M. and Popoola, P. (2023) 'Recent advances in solar photovoltaic materials and systems for energy storage applications: A review', *Beni-Suef University Journal of Basic and Applied Sciences*, 12(1), p. 66. Available at: https://doi.org/10.1186/s43088-023-00405-5.

Dintcheva, N.T., Morici, E. and Colletti, C. (2023) 'Encapsulant materials and their adoption in photovoltaic modules: A brief review', *Sustainability*, 15(12), p. 9453. Available at: https://doi.org/10.3390/su15129453.

Divya, A. et al. (2023) 'Review on recycling of solar modules/panels', *Solar Energy Materials and Solar Cells*, 253, p. 112151. Available at: https://doi.org/10.1016/j.solmat.2022.112151.

Elnozahy, A., Abd-Elbary, H. and Abo-Elyousr, F.K. (2024) 'Efficient energy harvesting from PV panel with reinforced hydrophilic nano-materials for eco-buildings', *Energy and Built Environment*, 5(3), pp. 393–403. Available at: https://doi.org/10.1016/j.enbenv.2022.12.001.

Ernst, M. et al. (2024) 'Accurate modelling of the bifacial gain potential of rooftop solar photovoltaic systems', *Energy Conversion and Management*, 300, p. 117947. Available at: https://doi.org/10.1016/j.enconman.2023.117947.

Fikri, M.A. et al. (2023) 'Enhanced thermal properties of phase change materials through surfactant-functionalized graphene nanoplatelets for sustainable energy storage', *Energies*, 16(22), p. 7668. Available at: https://doi.org/10.3390/en16227668.

Habib, S. et al. (2023) 'Assessment and optimization of carport structures for photovoltaic systems: A path to sustainable energy development', *Energy Conversion and Management*, 295, p. 117617. Available at: https://doi.org/10.1016/j.enconman.2023.117617.

El Hassani, S. et al. (2023) 'Towards rural net-zero energy buildings through integration of photovoltaic systems within bio-based earth houses: Case study in Eastern Morocco', *Solar Energy*, 259, pp. 15–29. Available at: https://doi.org/10.1016/j.solener.2023.05.007.

Henda, M.B. et al. (2023) 'High-efficiency perovskite photovoltaic system performance by molecular dynamics method: Optimizing electron transport thicknesses, hole transport, and anti-reflector layers of the sustainable energy materials', *Engineering Analysis with Boundary Elements*, 150, pp. 120–126. Available at: https://doi.org/10.1016/j.enganabound.2023.02.004.

Jain, P. et al. (2024) 'Recent advances in graphene-enabled materials for photovoltaic applications: A comprehensive review', *ACS Omega* [Preprint]. Available at: https://doi.org/10.1021/acsomega.3c07994.

Josephine, E.N., Ikponmwosa, O.S. and Ikhioya, I.L. (2023) 'Synthesis of SnS/SnO nanostructure material for photovoltaic application', *East European Journal of Physics*, (1), pp. 154–161. Available at: https://doi.org/10.26565/2312-4334-2023-1-19.

Kanani, H. et al. (2024) 'Design of graphene-coated silver nanoparticle based on numerical solution to enhance the absorption of the thin-film solar cell', *Plasmonics* [Preprint]. Available at: https://doi.org/10.1007/s11468-024-02231-6.

Katepalli, A., Wang, Y. and Shi, D. (2023) 'Solar harvesting through multiple semi-transparent cadmium telluride solar panels for collective energy generation', *Solar Energy*, 264, p. 112047. Available at: https://doi.org/10.1016/j.solener.2023.112047.

Khan, M. et al. (2023) 'Improving the efficiency of dye-sensitized solar cells based on rare-earth metal modified bismuth ferrites', *Scientific Reports*, 13(1), p. 3123. Available at: https://doi.org/10.1038/s41598-023-30000-8.

Korir, B.K., Kibet, J.K. and Ngari, S.M. (2024) 'A review on the current status of dye-sensitized solar cells: Toward sustainable energy', *Energy Science & Engineering*, 12(8), pp. 3188–3226. Available at: https://doi.org/10.1002/ese3.1815.

Kumar, K. et al. (2023) 'Biomass waste-derived carbon materials for sustainable remediation of polluted environment: A comprehensive review', *Chemosphere*, 345, p. 140419. Available at: https://doi.org/10.1016/j.chemosphere.2023.140419.

Li, Y. et al. (2022) 'Semitransparent organic photovoltaics for building-integrated photovoltaic applications', *Nature Reviews Materials*, 8(3), pp. 186–201. Available at: https://doi.org/10.1038/s41578-022-00514-0.

Mariotti, N. et al. (2020) 'Recent advances in eco-friendly and cost-effective materials towards sustainable dye-sensitized solar cells', *Green Chemistry*, 22(21), pp. 7168–7218. Available at: https://doi.org/10.1039/D0GC01148G.

Miettunen, K. et al. (2024) 'Bio-based materials for solar cells', *WIREs Energy and Environment*, 13(1). Available at: https://doi.org/10.1002/wene.508.

Mo, Y. et al. (2023) 'Improved photovoltaic performance of GaAs/carbon nanotube heterojunction solar cells with a multifunctional nafion/PEDOT:PSS layer', *ACS Applied Nano Materials*, 6(17), pp. 15588–15596. Available at: https://doi.org/10.1021/acsanm.3c02451.

Nyamathulla, S. and Chittathuru, D. (2023) 'A review of multilevel inverter topologies for grid-connected sustainable solar photovoltaic systems', *Sustainability*, 15(18), p. 13376. Available at: https://doi.org/10.3390/su151813376.

Ogundipe, O.B., Okwandu, A.C. and Abdulwaheed, S.A. (2024) 'Recent advances in solar photovoltaic technologies: Efficiency, materials, and applications', *GSC Advanced Research and Reviews*, 20(1), pp. 159–175. Available at: https://doi.org/10.30574/gscarr.2024.20.1.0259.

Öztürk, M., Yüksel, C. and Çiftçi, E. (2024) 'Investigation of a photovoltaic–thermal solar dryer system with double-pass solar air collectors and absorber surfaces enhanced with graphene nanoparticles', *Arabian Journal for Science and Engineering*, 49(8), pp. 11425–11443. Available at: https://doi.org/10.1007/s13369-024-08717-z.

Rahimi, M. et al. (2023) 'Experimental study on photovoltaic panels integrated with metal matrix sheets and bio-based phase change materials', *Energy*, 262, p. 125371. Available at: https://doi.org/10.1016/j.energy.2022.125371.

Rahman, M.M. et al. (2021) 'End-of-life photovoltaic recycled silicon: A sustainable circular materials source for electronic industries', *Advanced Energy and Sustainability Research*, 2(11). Available at: https://doi.org/10.1002/aesr.202100081.

Rajamony, R.K., Pandey, A.K., et al. (2024) 'Heat transfer and energy performance analysis of photovoltaic thermal system using functionalized carbon nanotubes enhanced phase change material', *Applied Thermal Engineering*, 243, p. 122544. Available at: https://doi.org/10.1016/j.applthermaleng.2024.122544.

Rajamony, R.K., Paw, J.K.S., et al. (2024) 'Energizing the thermophysical properties of phase change material using carbon-based nano additives for sustainable thermal energy storage application in photovoltaic thermal systems', *Materials Today Sustainability*, 25, p. 100658. Available at: https://doi.org/10.1016/j.mtsust.2023.100658.

Ramoroka, M.E. et al. (2024) 'Advances in organic photovoltaic cells: Fine-tuning of the photovoltaic processes', *Solar RRL*, 8(7). Available at: https://doi.org/10.1002/solr.202300982.

Sahajwalla, V. and Hossain, R. (2023) 'Rethinking circular economy for electronics, energy storage, and solar photovoltaics with long product life cycles', *MRS Bulletin*, 48(4), pp. 375–385. Available at: https://doi.org/10.1557/s43577-023-00519-2.

Shah, R. et al. (2024) 'Role of quantum dots and nanostructures in photovoltaic energy conversion', *E3S Web of Conferences*, 552, p. 01096. Available at: https://doi.org/10.1051/e3sconf/202455201096.

Sheikholeslami, M., Khalili, Z. and Mousavi, S.J. (2023) 'Influence of self-cleaning coating on performance of photovoltaic solar system utilizing mixture of phase change material and graphene nanoparticle', *Journal of Building Engineering*, 77, p. 107540. Available at: https://doi.org/10.1016/j.jobe.2023.107540.

Tasnim, S. et al. (2023) 'Electrical and optical performance evaluation of plasmonic nanoparticle-based organic photovoltaic cells', *Applied Solar Energy*, 59(4), pp. 420–434. Available at: https://doi.org/10.3103/S0003701X23600236.

Tseberlidis, G. et al. (2024) 'Cd-free kesterite solar cells: State-of-the-art and perspectives', *Sustainable Materials and Technologies*, 41, p. e01003. Available at: https://doi.org/10.1016/j.susmat.2024.e01003.

Woods-Robinson, R. et al. (2024) 'From design to device: Challenges and opportunities in computational discovery of?-Type transparent conductors', *PRX Energy*, 3(3), p. 031001. Available at: https://doi.org/10.1103/PRXEnergy.3.031001.

Zhang, J.-W. et al. (2023) 'Aging phenomena of backsheet materials of photovoltaic systems for future zero-carbon energy and the improvement pathway', *Journal of Materials Science & Technology*, 153, pp. 106–119. Available at: https://doi.org/10.1016/j.jmst.2022.12.063.

Zhu, H. et al. (2023) 'Environmentally friendly perovskite-based photovoltaics and luminescence: Lead less or lead free', *Solar RRL*, 7(24). Available at: https://doi.org/10.1002/solr.202300754.

Chapter 13

Carbon-based nanomaterials utilization in biofuel cells

Avanish Kumar, Adarsh Kumar Arya, Sudeep Yadav, Amit Kumar Tiwari, Sumit Kumar Jana, and Dan Bahadur Pal

13.1 INTRODUCTION

Biofuel cells (BFCs) are an electrochemical technique that produces electricity with the help of catalytic activity of biological mechanisms such as enzymes or microorganisms, which transform chemical energy into electrical energy [1]. Generally, all biofuel systems are operated at mild temperatures and pressures, which makes them more environmentally friendly and makes them a substitute for traditional fuel cells. As we know, with the increasing global demand for sustainable and green energy sources, biofuel cells are considered a favourable technology for various applications, including portable power sources and wastewater treatment. In the present scenario, nanotechnology plays a crucial role in enhancing the efficiency, stability, and electron transfer rates in biofuel cells, which provide a high surface area and excellent conductivity and electron transport between bio-catalysts and electrodes, leading to improved overall performance [2]. Additionally, nanostructured catalysts such as metal nanoparticles and carbon-based nanomaterials enhance reaction kinetics and increase power density, making biofuel cells more effective for energy generation. Nanotechnology also aids in enzyme and microbial immobilization by stabilizing these biocatalysts on electrodes, thereby extending their operational lifespan [3]. In the present time, the advancements in the nano-field, the contribution of biofuel is more useful in the biomedical field. Collectively, contribution increases the functionality and practical applications of biofuel cells in the sustainable energy and healthcare fields. Modern carbon-based nanomaterials, such as carbon nanotubes, graphene, and carbon nanofibres, have exceptional properties that enhance the performance and durability of biofuel [4]. Their high electrical conductivity facilitates quick electron transport, improving overall efficiency. Additionally, their large surface area permits greater biocatalyst loading and interaction, which increases catalytic activity. These nanomaterials also show good chemical stability, which makes them resistant to degradation in bioelectrochemical environments. Their biocompatibility makes them ideal for applications in medical and

DOI: 10.1201/9781003581536-13

environmental systems, including wastewater treatment [5]. The carbon-based nanomaterials provide significant functionalization potential, as they can be modified with biomolecules to further increase catalytic activity and selectivity. These combined advantages make carbon-based nanomaterials a favoured choice for advancing biofuel cell technology.

The recent development in biofuel cell technology has significantly improved the efficiency, functionality, and application scope of biofuel cells. The major developments of hybrid biofuel cells, which combine enzymatic and microbial biofuel cells to achieve better efficiency and improved electron transfer rates [6]. Additionally, the application of 3D nanostructured electrodes made from porous and graded nanomaterials has boosted the performance of catalytic activity. Another promising innovation is the incorporation of self-powered biosensors where biofuel cells are integrated with medical diagnostic devices to provide uninterrupted and defensible energy sources [7]. The development of flexible and wearable biofuel cells has also gained attention in smart textiles and wearable electronics for real-time health checking. Besides, the researchers are discovering waste-to-energy systems that utilize organic waste as a fuel source, which would contribute to sustainable energy generation. These advancements continue to push the boundaries of biofuel cell technology, making them more viable for real-world applications in energy, healthcare, and environmental sustainability [8].

13.1.1 Working principles of biofuel cells

In biofuel cells (BFCs), chemical energy from organic substrates is transformed into electrical energy through biocatalysts. The fundamental principle behind the biofuel involves the oxidation-reduction (redox) reactions facilitated by biological catalysts, such as enzymes or microorganisms, at the electrodes [9]. At the anode, the biofuel, such as glucose, ethanol, or organic waste, undergoes oxidation, releasing electrons and protons. The electrons move through an external circuit to generate electrical power, while the protons move through an electrolyte to the cathode shown in Figure 13.1: At the cathode, these electrons and protons contribute in a reduction reaction and form water as an additional product [10]. The efficiency of electron transfer between the biocatalyst and electrodes significantly influences the overall performance of the BFC. The working of the biofuel cell is shown in Figure 13.1.

Figure 13.1 shows the working principle of a BFC. The fundamental working principle of biofuel cells revolves around bioelectrochemical reactions, where a biocatalyst, such as microorganisms or enzymes, facilitates the oxidation of organic or biochemical substrates to release electrons [11]. These electrons travel through an external circuit and generate an electric current, while protons migrate through an electrolyte to complete the circuit at the cathode, where oxygen or another electron acceptor undergoes reduction. The process takes place in two main compartments:

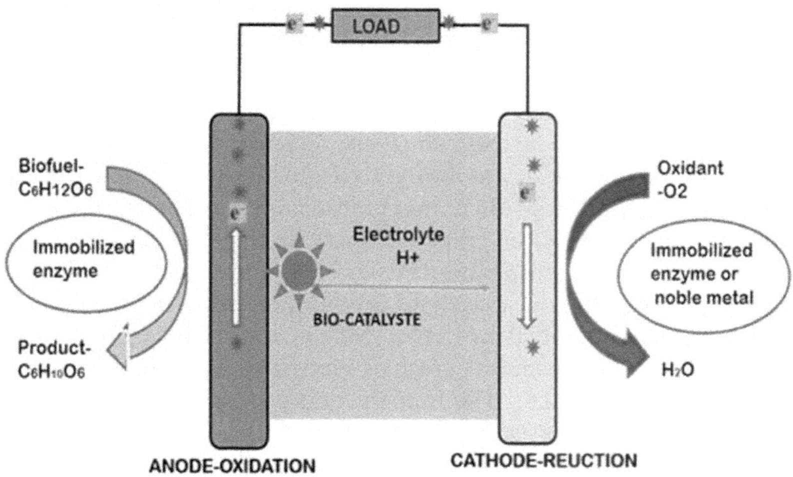

Figure 13.1 A schematic of a biofuel cell.

Anode Chamber: In MFCs, bacteria such as *Geobacter sulfurreducens* and *Shewanella oneidensis* oxidize organic matter, such as glucose or wastewater pollutants, producing electrons and protons. These bacteria use conductive pili (nanowires) or electron mediators to transfer electrons to the anode [12]. In EBCs, enzymes such as glucose oxidase catalyse similar oxidation reactions, leading to electron release. Nano-catalysts like gold nanoparticles and carbon-based nanomaterials (e.g., graphene oxide and carbon nanotubes) are often used to enhance enzyme stability and electron transfer.

Electron Transport and Circuit Flow: The released electrons travel from the anode to the cathode through an external circuit, generating usable electrical power.

Cathode Chamber: Simultaneously, protons move through a proton exchange membrane (PEM) or an electrolyte to reach the cathode, where they combine with oxygen (or another oxidizing agent) and electrons to form water or other harmless byproducts [13]. In EBCs, enzymes such as laccase and bilirubin oxidase, often supported on nano-structured electrodes, catalyse the oxygen reduction reaction (ORR), increasing efficiency.

13.1.2 Applications of biofuel cells in different fields

Biofuel cells (BFCs) are innovative electrochemical devices that generate electricity by harnessing biochemical reactions. They are primarily classified based on the type of biocatalyst involved in electron transfer. The two main types are microbial fuel cells (MFCs) and enzymatic biofuel cells (EBCs) [14].

MFCs employ microorganisms, such as *Shewanella oneidensis*, *Geobacter sulfurreducens*, and *Escherichia coli*, which break down organic matter and transfer electrons to the anode. These fuel cells have gained attention for their dual functionality in wastewater treatment and bioelectricity generation, making them an eco-friendly energy solution. On the other hand, EBCs practice specific enzymes which use glucose oxidase, laccase, and bilirubin oxidase. These act as biocatalysts for oxidation and reduction reactions. These enzymes, often immobilized on nanomaterials such as carbon nanotubes (CNTs) or graphene, increase the electron transfer process, and contribute to improving efficiency [15]. The enzyme-based systems have a limited durability due to enzyme degradation, which affects long-term performance. The microbial fuel cells (MFCs) are utilizing the metabolic activities of microorganisms to produce electricity, including solving the issue of environmental [16]. The fundamental principle of MFCs includes the oxidation of organic matter by bacteria in the anode compartment, releasing electrons that travel through an external circuit to the cathode.

13.1.3 Microbial fuel cells (MFCs) for bio sensing

Microbial fuel cells have also been developed as biosensors to monitor environmental contaminants. Their ability to detect biochemical changes in the surroundings by measuring electrical signals makes them highly active for real-time monitoring. MFC-based biosensors are particularly useful for detecting biochemical oxygen demand (BOD), where the electricity produced correlates with organic pollutant concentrations in water. Heavy metal detection is another essential application, as the presence of toxic metals such as arsenic and cadmium affects bacterial metabolism, causing detectable differences in power output [17]. Furthermore, microbial biosensors can identify pathogens by analysing unique electrochemical signatures produced upon pathogen exposure. These sensors are advantageous due to their low power requirements, cost-effectiveness, and long-term operational stability (Figure 13.2).

13.1.4 Microbial fuel cells (MFCs) for bioelectric generation and wastewater treatment

One of the most significant applications of MFCs is in bioelectricity generation. These systems offer an ecological method of transforming organic waste into electrical energy through microbial respiration. By picking or adopting specific electrode materials such as graphene, carbon nanotubes, and biofilms, researchers have enhanced electron transfer, thereby improving power generation [18]. Additionally, the choice of substrates, including agricultural waste, sewage, and industrial effluents, influences the efficiency of MFCs. The optimization of microbial populations also plays a crucial

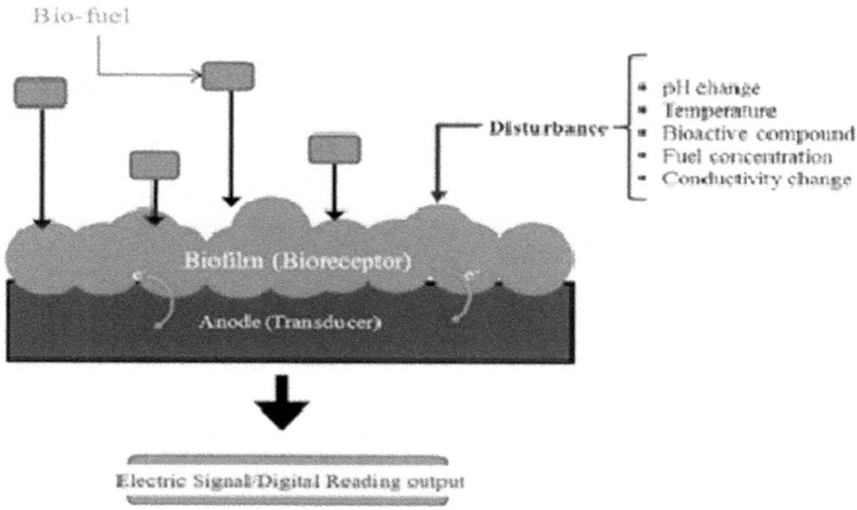

Figure 13.2 Microbial fuel cells (MFCs) for bio-sensing application.

role, with electrogenic bacteria such as *Geobacter* and *Shewanella* demonstrating superior electron transfer capabilities [19]. In spite of their potential, challenges such as low power output keep it up, necessitating further advancements in system design and microbial engineering.

Currently, MFCs have grown considerable attention in wastewater treatment, offering a sustainable alternative to conventional methods. In these systems, bacteria oxidize organic pollutants in wastewater at the anode, generating electrons that travel to the cathode, producing electricity while simultaneously treating the effluent [19]. MFCs are energy-producing along with wastewater treatment technology. They also effectively decrease biochemical oxygen demand (BOD) and chemical oxygen demand (COD), eliminate excess nutrients such as nitrogen and phosphorus, and inactivate pathogens through electrochemical processes [20]. Unlike traditional treatment methods that consume large amounts of energy, MFC-based systems produce power while purifying water [21]. Additionally, they produce minimal sludge, reducing disposal costs.

13.1.4.1 (MFCs) remove heavy metals by microbial reduction

In the field of bioremediation, MFCs break down pollutants, making them practical tools for environmental clean-up. They can be used to remove heavy metals by microbial reduction, where bacteria convert toxic metal ions into less harmful forms. For instance, chromium (Cr^{6+}) can be reduced to Cr^{3+}, which is less toxic and easier to remove from water bodies [22]. Similarly, organic pollutants such as hydrocarbons, pesticides, and pharmaceuticals can be degraded through microbial metabolism. Textile dyes, including methylene blue and Congo red, are also efficiently degraded in MFCs, offering an eco-friendly solution for industrial wastewater treatment [23]. However, improving biodegradation rates remains a challenge, necessitating microbial community optimization and improved reactor configurations.

13.1.4.2 Microbial fuel cells (MFCs) for carbon sequestration and sediment

Another promising application of MFCs is carbon sequestration, where microbial metabolism plays a role in sequestering and storing carbon dioxide. Some bacteria, such as *Clostridium* and *Rhodopseudomonas*, can convert CO_2 into organic compounds, decreasing greenhouse gas emissions [24].

Electroactive biofilms further enhance carbon capture by facilitating microbial electro synthesis, where CO_2 is transformed into biofuels like acetate or methane (Figure 13.3). By integrating MFCs into carbon sequestration systems, researchers aim to develop sustainable methods for

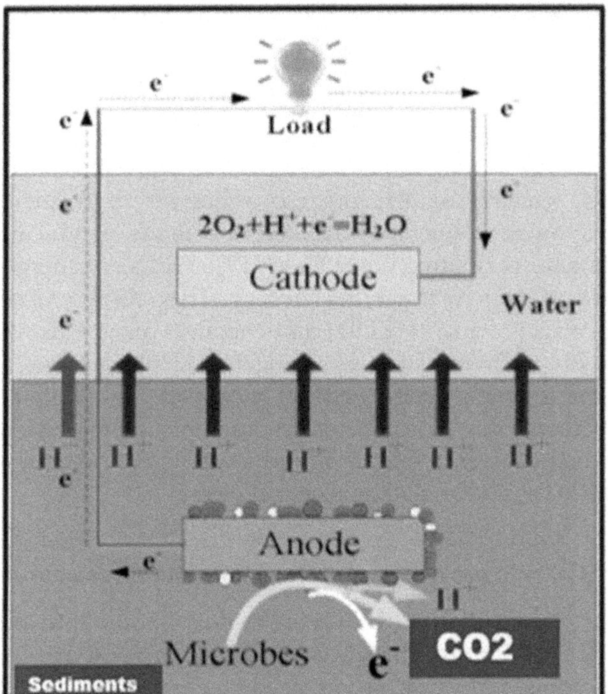

Figure 13.3 Microbial fuel cells (MFCs) for carbon sequestration and sediment.

moderating climate change while generating appreciated bio-based products [25]. The ability of MFCs to simultaneously reduce CO_2 emissions and produce energy makes them highly promising for future sustainable energy solutions. Besides all, sediment microbial fuel cells (SMFCs) represent a particular application of MFCs, mainly in marine and freshwater environments. These systems generate electricity by utilizing natural organic matter in sediments as a fuel source. In SMFCs, the anode is buried in the anaerobic sediment, where bacteria oxidize organic compounds, while the cathode is placed in the aerobic water column, where electrons reduce oxygen [26]. This electron flow creates a continuous power source, making SMFCs ideal for remote environmental monitoring and powering underwater sensors. Figure 13.3 shows the Microbial Fuel Cells (MFCs) for carbon sequestration and sediment.

13.1.4.3 Advantages and disadvantages of biofuel cells

Biofuel cells (BFCs) offer several advantages and disadvantages across different aspects in terms of sustainability. BFCs utilize renewable biofuels, reducing dependence on fossil fuels; however, their large-scale implementation is limited due to low power density [26]. They are eco-friendly, generating minimal harmful emissions and byproducts, but optimal microbial or enzymatic activity requires specific conditions. BFCs excel in waste utilization by converting organic waste into energy, aiding in waste management, though slow electron transfer rates in microbial fuel cells limit efficiency [27]. Operating under mild conditions, such as room temperature and atmospheric pressure, sets BFCs apart from traditional fuel cells; however, the sensitivity of enzymes and microbes to environmental changes affects performance. BFCs have significant application potential in areas like medical implants, biosensors, and remote power generation, but the short lifespan of enzymes in enzymatic BFCs demands frequent replacement. The detailed advantages and disadvantages of biofuel cells are displayed in Table 13.1.

13.2 APPLICATION OF NANOMATERIAL IN BIOFUEL CELL

Nanomaterials, with structural features at the nano scale (1–100 nm), possess unique physical, chemical, and electrical properties due to their high surface area, quantum effects, and enhanced reactivity [31].

These properties make nanomaterials highly beneficial for biofuel cells, which convert chemical energy from organic substances like glucose and ethanol into electricity using biocatalysts. Nanomaterials improve biofuel cell performance by enhancing electron transfer, catalytic activity, and enzyme stability, while reducing the energy required to initiate the reaction [32]. Various methods are used to prepare nanomaterials, including physical methods like ball milling and laser ablation, chemical methods such

Table 13.1 Advantages and disadvantages of biofuel cells

Aspect	*Advantages*	*Disadvantages*
Sustainability	Uses renewable biofuels, reducing reliance on fossil fuels [28].	Limited large-scale implementation due to low power density.
Eco-friendly	Generates minimal harmful emissions and by products.	May require specific conditions for optimal microbial or enzymatic activity.
Waste utilization	Can utilize organic waste as a fuel source, contributing to waste management.	Slow electron transfer rates in microbial fuel cells limit efficiency.
Mild operating conditions	Operates at room temperature and atmospheric pressure, unlike traditional fuel cells [29].	Enzymes and microbes are sensitive to environmental changes, affecting performance.
Application potential	Suitable for medical implants, biosensors, and remote power applications.	Short lifespan of enzymes in enzymatic BFCs requires frequent replacement.
Scalability	Microbial fuel cells have potential for large-scale wastewater treatment and energy production.	High initial costs and challenges in maintaining microbial growth conditions [30].

as sol-gel synthesis and hydrothermal synthesis, and biological methods like microbial and plant-based synthesis [33]. The primary sources of nanomaterials include carbon-based materials (e.g., graphene, carbon nanotubes), metal oxides (e.g., TIO_2, ZnO), metallic nanoparticles (e.g., gold, silver, platinum), and composite materials that combine metal and carbon-based nanostructures. These nanomaterials play important role in improving the efficiency and stability of biofuel cells [34]. Nanomaterials significantly increase the efficiency of biofuel cells due to their exceptional properties. Their high surface area allows for better enzyme immobilization, leading to improved electron transfer efficiency. The carbon-based nanomaterials, such as graphene and carbon nanotubes, enhance conductivity between the electrode and biocatalyst, while metal nanoparticles such as platinum (Pt), gold (Au), and silver (Ag) help as efficient catalysts for oxidation and reduction reactions [35]. Additionally, nanomaterials enhance the stability of enzymes, preventing their denaturation during the response, and reduce the overpotential, which lowers the energy required to initiate the reaction, thereby boosting the overall performance of biofuel cells. Different types of nanomaterials used for biofuel cell are shown in Figure 13.2.

Each of these nanomaterials plays a crucial role in improving the efficiency of biofuel cells by addressing challenges related to electron transfer, enzyme stability, and catalytic efficiency [33]. Table 13.2 shows the different types of nanomaterials used in biofuel cells.

Table 13.2 Different type of nanomaterials used in biofuel cells

Nanomaterial type	*Examples*	*Application in biofuel cells*	*Advantages*	*Disadvantages*
Carbon-based	Carbon nanotubes (CNTs), graphene, carbon quantum dots, carbon aerogels	Electron transfer enhancement • Increased electrode surface area • Bio-electro catalysis support	• High conductivity • Good biocompatibility • Large surface area for enzyme immobilization	• Complex synthesis • Expensive fabrication techniques
Inorganic-based	Metal oxides (TiO_2, ZnO, CeO_2), Silica nanoparticles [36]	• Supports bio-electro catalysts • Enhances enzyme stability • Improves mechanical strength	• High stability • Cost-effective • Biocompatible	• Poor conductivity (some oxides) • Requires modification for better electron transfer
Metal-based	Gold nanoparticles (AuNPs), silver nanoparticles (AgNPs), platinum (Pt), palladium (Pd)	• Direct electron transfer improvement • Enzyme immobilization	• High catalytic efficiency • Strong interaction with biomolecules • Good conductivity	• Expensive • Prone to aggregation • Possible toxicity concerns

13.2.1 Carbon-based nanomaterials for biofuel

Carbon-based nanomaterials (CBNMs) have been considered as essential components in biofuel production due to their physicochemical properties. These materials have improved the efficiency of biofuel cells (BFCs) by improving electron transfer, catalytic activity, and stability [4]. Their structural diversity, high surface area, superior conductivity, and biocompatibility make them ideal for applications such as biofuel cell electrodes, catalyst supports, and bioelectrochemical reactors. Carbon-based nanomaterials basically contribute to biofuel production. Carbon nanotubes (CNTs), which might be single-walled (SWCNTs) or multi-walled (MWCNTs), display exceptional electrical conductivity and mechanical strength, improving electron transfer efficiency in biofuel cells [37]. Graphene, a single layer of carbon atoms in a hexagonal lattice, provides a high surface area and excellent electrical conductivity, aiding enzyme and microbial attachment for efficient biofuel production. Carbon quantum dots (CQDs) are nanoscale carbon particles with quantum confinement effects, holding exclusive optical properties and biocompatibility, making them useful for bio-imaging and enzymatic catalysis improvement [38]. Carbon nanofibres

(CNFs), having excellent mechanical stability and high porosity, efficiently give conductive support materials in microbial and enzymatic fuel cells. Mesoporous carbon, known for its highly porous structure, offers a large surface area for enzyme and microbe immobilization, improving mass transport and diffusion of reactants in biofuel cells [39]. These nanomaterials hold unique properties like a high surface area, which allows for greater catalyst loading, enzyme immobilization, and microbial adhesion, thereby improving bio-electrocatalytic reactions and energy conversion efficiency [40]. The superior electrical conductivity of these materials facilitates rapid electron transfer, reducing resistance in biofuel cells and increasing power density. Their biocompatibility properties ensure compatibility with biological systems, making them suitable for microbial interactions and safe for biomedical and environmental applications [41].

13.2.2 Carbon-based nanomaterials

Carbon nanomaterials shown an important role in improved efficiency of biofuel cells (BFCs) by improving electron transfer, catalytic efficiency, and durability [33].

Carbon nanotubes (CNTs) work as excellent electrode materials due to their high electrical conductivity and mechanical strength. Functionalization strategies such as surface modification with biomolecules or metal nanoparticles enable effective enzyme immobilization, which improves biocatalytic activity [42]. Several case studies demonstrate CNTs' success in microbial and enzymatic fuel cells, improving power density and durability. Graphene and its derivatives offer better conductivity and high surface area, making them ideal for anode and cathode materials in biofuel cells. Hybrid composites of graphene with metal nanoparticles further improved catalytic performance, increasing reaction rates and overall energy conversion efficiency [2]. Carbon quantum dots (CQDs) contribute to bioelectrocatalysis by improving charge transfer kinetics while also exhibiting unique photoluminescence properties that enhance cell performance. These properties make CQDs highly suitable for self-powered biosensors, where they help integrate energy generation with real-time sensing applications [43]. Carbon nanofibres (CNFs) and mesoporous carbon provide structural advantages for enzyme immobilization due to their porous nature and high stability, allowing for prolonged bio-catalytic activity.

13.2.2.1 Carbon nanotubes (CNTs)

Carbon nanotubes (CNTs) are cylindrical nanostructures composed of graphene sheets rolled into tubes [44]. They can be synthesized or prepared through several methods as arc discharge, laser ablation, and chemical vapour deposition (CVD).

Among these, CVD is the most applicable method due to its scalability and control over the growth process. In this method, a carbon-containing gas, such as methane or acetylene, is decomposed on a metal catalyst, typically iron, cobalt, or nickel, at high temperatures. The catalyst particles act as nucleation sites for the growth of CNTs. CNTs has very good mechanical strength, high electrical and thermal conductivity and specific optical properties [45]. They are classified into single-walled (SWCNTs) and multi-walled (MWCNTs) nanotubes. SWCNTs have a diameter of about 0.4–2 nm, while MWCNTs are prepared with multiple concentric graphene cylinders with diameters ranging from 2 to 100 nm. In biofuel production, CNTs are used, which enhance the activity and selectivity of metal catalysts used in the conversion of biomass to biofuels. The CNT-based electrodes increase the efficiency of microbial fuel cells and increase hydrogen production through water splitting [17].

13.2.2.2 Graphene

Graphene, a single layer of carbon atoms organised in a honeycomb lattice, can be prepared through mechanical exfoliation, chemical vapour deposition (CVD), and chemical reduction of graphene oxide [44]. The CVD method involves the decomposition of hydrocarbons on a metal substrate, such as copper or nickel, at high temperatures [46]. This method allows the production of large-area, high-quality graphene sheets. Graphene shows remarkable electrical conductivity and high thermal conductivity, along with mechanical strength. It is highly flexible and transparent with a high surface area and excellent chemical stability [47]. Graphene's two-dimensional structure and high electron mobility make it a promising material for various applications. Graphene's exceptional conductivity and flexibility make it ideal for electronic devices, sensors, and energy storage systems. Graphene-based materials serve as efficient catalysts and catalyst supports in biofuel production [48]. They enhance the activity of metal nanoparticles used in biomass conversion processes. Graphene also improves the performance of electrodes in microbial fuel cells and supercapacitors, contributing to sustainable energy production.

13.2.2.3 Carbon quantum dots (CQDs)

Carbon quantum dots (CQDs) are nano-sized carbon particles with unique photoluminescence properties that can be synthesized through top-down methods, such as laser ablation and chemical oxidation or bottom-up methods, such as hydrothermal and microwave-assisted synthesis [49]. The hydrothermal method involves heating carbon precursors, such as citric acid or glucose, in a high-pressure autoclave. CQDs exhibit strong photoluminescence, high chemical stability, and good biocompatibility. They have a size range of 2–10 nm and possess abundant surface functional groups, which enhance their solubility and reactivity [50]. It has also exhibited excellent electron transfer properties. They are also employed in photocatalysis and optoelectronic devices. Additionally, CQDs are non-toxic and environmentally friendly, making them suitable for sustainable applications. CQDs act as photocatalysts in the conversion of biomass into biofuels. They enhance light absorption and facilitate electron transfer in photochemical reactions [51].

13.2.2.4 Carbon nano fibres (CNFs)

Carbon nanofibres (CNFs) are cylindrical nanostructures with diameters ranging from tens to hundreds of nanometers. They can be synthesized through chemical vapour deposition (CVD), electrospinning, and template-assisted methods [52]. In CVD, carbon-containing gases are decomposed on metal catalysts at high temperatures, leading to the growth of CNFs. CNFs hold high mechanical strength, electrical conductivity, and thermal stability. They have a large surface area and high aspect ratio, which enhances their performance in various applications. CNFs are also chemically stable and resistant to corrosion [53]. CNFs are used in composite materials for structural reinforcement and in energy storage devices for enhanced conductivity. Their high surface area makes them suitable for adsorption and catalytic applications. CNFs also exhibit excellent thermal conductivity, making them ideal for heat management systems [54]. CNFs serve as catalyst supports in biomass conversion processes. They enhance the activity and stability of metal catalysts used in biofuel production. CNFs also improve the performance of electrodes in microbial fuel cells and facilitate hydrogen generation through water electrolysis.

13.2.2.5 Mesoporous carbon

Mesoporous carbon materials are characterized by a highly ordered porous structure with pore sizes ranging from 2 to 50 nm. They can be synthesized through soft-templating and hard-templating methods. The soft-templating method involves the self-assembly of surfactants or block copolymers, while the hard-templating method uses silica or other solid templates [55]. The porous structure allows efficient mass transfer and adsorption. Mesoporous carbon materials are used in energy storage devices, such as supercapacitors and batteries. Their high surface area makes them ideal for adsorption and catalytic applications [56]. Mesoporous carbon is also employed in environmental remediation and gas separation processes. The porous structure facilitates efficient mass transfer and improves the performance of electrodes in microbial fuel cells and other bioelectrochemical systems [57]. A comparison of different types of carbon-based nano materials can be demonstrated in Table 13.3.

Table 13.3 Summary of carbon-based nano material based on their development, uses advantage and disadvantages

Carbon-based nanomaterial	*Development technology*	*Uses in biofuel production*	*Advantages*	*Disadvantages*
Carbon nanotubes (CNTs)	Chemical vapour deposition (CVD), arc discharge, laser ablation	Enhancing electron transfer in biofuel cells, catalyst support, enzyme immobilization	High electrical conductivity, excellent mechanical strength, large surface area	High production cost, difficulty in large-scale synthesis, potential toxicity
Graphene	Mechanical exfoliation, chemical vapour deposition (CVD), reduction of graphene oxide [58]	Supports microbial and enzymatic biofuel cells, enhances bio-catalytic reactions	High surface area, superior electrical conductivity, excellent flexibility	Restacking issue reduces effective surface area, complex and expensive production
Carbon quantum dots (CQDs)	Hydrothermal synthesis, electrochemical methods, chemical oxidation	Enhancing enzymatic catalysis, bio-imaging, biosensors [59]	Biocompatibility, tunable optical properties, good electron donor/ acceptor behaviour	Lower electrical conductivity compared to CNTs and graphene, limited large-scale production
Carbon nanofibers (CNFs)	Electro spinning, CVD, pyrolysis of organic precursors	Conductive support for biofuel cells, microbial and enzymatic fuel cells	High mechanical stability, efficient charge transport, high porosity	Requires complex fabrication techniques, potential structural defects
Mesoporous carbon	Sol-gel process, template-assisted synthesis, pyrolysis	Enzyme and microbe immobilization, improved mass transport in biofuel cells	Large surface area, tunable pore size, excellent adsorption capacity [59]	Poor mechanical strength, requires precise synthesis control

13.3 DIFFERENT CHALLENGES AND LIMITATIONS FOR USING C-BASED NANO MATERIAL

Carbon-based nanomaterials, such as graphene, carbon nanotubes (CNTs), and fullerenes, hold great potential in various applications, including energy storage, sensors, and biomedical devices [60]. However, several challenges and limitations hinder their widespread use. Their tendency to agglomerate affects their dispersion in solutions, impacting performance in composite materials.

Carbon-based electrodes, such as carbon nanotubes (CNTs), graphene, and activated carbon, are primarily in biofuel cells due to their excellent conductivity, large surface area, and biocompatibility [61]. However, their long-term stability and durability remain critical challenges. Over time, biofouling, structural degradation, and surface passivation can reduce electrode performance. In microbial fuel cells (MFCs), the interaction between bacteria and the electrode surface can lead to the formation of biofilms, which may either increase or decrease electron transfer depending on their composition and thickness [62]. Similarly, in enzymatic biofuel cells (EBCs), enzyme degradation and leaching reduce the effectiveness of electron transfer reactions. To improve durability, researchers are exploring surface modifications using metal oxide coatings, polymeric encapsulation, and nanocomposite materials that enhance electrode stability without compromising biocatalyst activity [1]. Additionally, functionalizing carbon electrodes with nanocatalysts such as gold nanoparticles (Au NPs) or transition metal oxides has been indicated to enhance electron transfer rates while reducing electrode degradation.

13.4 CONCLUSION

Biofuel cells (BFCs) signify an encouraging approach to sustainable energy generation, which uses biological catalysts to produce electricity without compromising the environment. The use of carbon-based nanomaterials, such as carbon nanotubes (CNTs), graphene, carbon quantum dots (CQDs), carbon nanofibers (CNFs), and mesoporous carbon, has meaningfully improved the performance of biofuel cells by growing electron transfer, catalytic activity, and stability. Advances in nanomaterial synthesis techniques, including chemical vapour deposition and hydrothermal methods, have given insight for bioelectrochemical applications.

REFERENCES

[1] Nasar, A. and Perveen, R., 2019. Applications of enzymatic biofuel cells in bioelectronic devices: A review. *International Journal of Hydrogen Energy*, *44*(29), pp.15287–15312.

[2] Adeel, M., Bilal, M., Rasheed, T., Sharma, A. and Iqbal, H.M., 2018. Graphene and graphene oxide: Functionalization and nano-bio-catalytic system for enzyme immobilization and biotechnological perspective. *International Journal of Biological Macromolecules*, *120*, pp.1430–1440.

[3] Rubenwolf, S., Kerzenmacher, S., Zengerle, R. and von Stetten, F., 2011. Strategies to extend the lifetime of bioelectrochemical enzyme electrodes for biosensing and biofuel cell applications. *Applied Microbiology and Biotechnology*, *89*, pp.1315–1322.

[4] Zhao, C.E., Gai, P., Song, R., Chen, Y., Zhang, J. and Zhu, J.J., 2017. Nanostructured material-based biofuel cells: Recent advances and future prospects. *Chemical Society Reviews*, *46*(5), pp.1545–1564.

[5] Gude, V.G., 2016. Wastewater treatment in microbial fuel cells: An overview. *Journal of Cleaner Production*, *122*, pp.287–307.

[6] Xiao, X., Xia, H.Q., Wu, R., Bai, L., Yan, L., Magner, E., Cosnier, S., Lojou, E., Zhu, Z. and Liu, A., 2019. Tackling the challenges of enzymatic (bio) fuel cells. *Chemical Reviews*, *119*(16), pp.9509–9558.

[7] Ibrahim, R., Shaari, N. and Mohd Aman, A.H., 2021. Bio-fuel cell for medical device energy system: A review. *International Journal of Energy Research*, *45*(10), pp.14245–14273.

[8] Mathur, V. and Mukherji, R., 2025. Advances in biofuel cell research and future prospects. In Kishor Kumar Sadasivuni and Mithra Geetha (Eds.), *Biofuel Cells and Energy Generation* (pp.27–65). Cambridge, USA: Woodhead Publishing.

[9] Guengerich, F.P. and Yoshimoto, F.K., 2018. Formation and cleavage of C–C bonds by enzymatic oxidation–reduction reactions. *Chemical Reviews*, *118*(14), pp.6573–6655.

[10] Liu, H., Zhang, L., Zhang, J., Ghosh, D., Jung, J., Downing, B.W. and Whittemore, E., 2006. Electrocatalytic reduction of O_2 and H_2O_2 by adsorbed cobalt tetramethoxyphenyl porphyrin and its application for fuel cell cathodes. *Journal of Power Sources*, *161*(2), pp.743–752.

[11] Piera, J. and Bäckvall, J.E., 2008. Catalytic oxidation of organic substrates by molecular oxygen and hydrogen peroxide by multistep electron transfer: A biomimetic approach. *Angewandte Chemie International Edition*, *47*(19), pp.3506–3523.

[12] Ptushenko, V.V., 2020. Electric cables of living cells. II. Bacterial electron conductors. *Biochemistry (Moscow)*, *85*, pp.955–965.

[13] Sharma, V.K., Zboril, R. and Varma, R.S., 2015. Ferrates: Greener oxidants with multimodal action in water treatment technologies. *Accounts of Chemical Research*, *48*(2), pp.182–191.

[14] Wang, M.H. and Wang, L., 2024. Design and performance optimization of enzyme-catalyzed biofuel cells. *Journal of Energy Bioscience*, *15*, 314–325.

[15] Ansari, S.A. and Husain, Q., 2012. Potential applications of enzymes immobilized on/in nano materials: A review. *Biotechnology Advances*, *30*(3), pp.512–523.

[16] Chiranjeevi, P. and Patil, S.A., 2020. Strategies for improving the electroactivity and specific metabolic functionality of microorganisms for various microbial electrochemical technologies. *Biotechnology Advances*, *39*, p.107468.

[17] Li, X., Brejnrod, A.D., Ernst, M., Rykær, M., Herschend, J., Olsen, N.M.C., Dorrestein, P.C., Rensing, C. and Sørensen, S.J., 2019. Heavy metal exposure causes changes in the metabolic health-associated gut microbiome and metabolites. *Environment International*, *126*, pp.454–467.

[18] Ngaw, C.K., Zhao, C.E., Wang, V.B., Kjelleberg, S., Tan, T.T.Y., Zhang, Q. and Loo, S.C.J., 2017. A graphene/carbon nanotube biofilm based solar-microbial fuel device for enhanced hydrogen generation. *Sustainable Energy & Fuels*, *1*(1), pp.191–198.

[19] Logan, B.E., 2005. Simultaneous wastewater treatment and biological electricity generation. *Water Science and Technology*, *52*(1–2), pp.31–37.

[20] Du, J., Waite, T.D., Feng, J., Lei, Y. and Tang, W., 2023. Coupled electrochemical methods for nitrogen and phosphorus recovery from wastewater: A review. *Environmental Chemistry Letters*, *21*(2), pp.885–909.

[21] Huang, Y., Zhao, Y., Tang, C., Yadav, A.K., Abbassi, R., Kang, P., Cai, Y., Liu, A., Yang, A. and Li, M., 2023. A glance of coupled water and wastewater treatment systems based on microbial fuel cells. *Science of the Total Environment*, *892*, p.164599.

[22] Sharma, M., Sharma, S., Paavan, Gupta, M., Goyal, S., Talukder, D., Akhtar, M.S., Kumar, R., Umar, A., Alkhanjaf, A.A.M. and Baskoutas, S., 2024. Mechanisms of microbial resistance against cadmium: A review. *Journal of Environmental Health Science and Engineering*, *22*(1), pp.13–30.

[23] Sivasankar, P., Poongodi, S., Seedevi, P., Sivakumar, M., Murugan, T. and Loganathan, S., 2019. Bioremediation of wastewater through a quorum sensing triggered MFC: A sustainable measure for waste to energy concept. *Journal of Environmental Management*, *237*, pp.84–93.

[24] Sharma, K., Park, Y.K., Nadda, A.K., Banerjee, P., Singh, P., Raizada, P., Banat, F., Bharath, G., Jeong, S.M. and Lam, S.S., 2022. Emerging chemo-biocatalytic routes for valorization of major greenhouse gases (GHG) into industrial products: A comprehensive review. *Journal of Industrial and Engineering Chemistry*, *109*, pp.1–20.

[25] Das, S., Das, S., Das, I. and Ghangrekar, M.M., 2019. Application of bioelectrochemical systems for carbon dioxide sequestration and concomitant valuable recovery: A review. *Materials Science for Energy Technologies*, *2*(3), pp.687–696.

[26] Freguia, S., Rabaey, K., Yuan, Z. and Keller, J., 2008. Sequential anode–cathode configuration improves cathodic oxygen reduction and effluent quality of microbial fuel cells. *Water Research*, *42*(6–7), pp.1387–1396.

[27] Chakma, R. and Hossain, M.K., 2025. Optimizing biofuel cell technology through electrocatalysis. In Kishor Kumar Sadasivuni and Mithra Geetha (Eds.), *Biofuel Cells and Energy Generation* (pp. 211–241). Cambridge, USA: Woodhead Publishing.

[28] Malla, F.A., Bandh, S.A., Wani, S.A., Hoang, A.T. and Sofi, N.A., 2023. Biofuels: Potential alternatives to fossil fuels. In Suhaib A. Bandh, Fayaz A. Malla (Eds.), *Biofuels in Circular Economy* (pp. 1–15). Singapore: Springer Nature Singapore.

[29] Bhatta, R., Shuklaa, P., Mishrab, A. and Bajpaic, A.K. 2023. Emerging applications of nano-modified bio-fuel cells. In Ashok Kumar Nadda, Tuan Anh Nguyen, Swati Sharma, Muhammad Bilal, Ram K. Gupta (Eds.), *Nanotechnology for Advanced Biofuels: Fundamentals and Applications* (pp. 213–242).

[30] Dutta, N., Kundu, P., Lee, J.T.E. and Bhattacharya, S., 2023. Implementation and optimization of algal biomass in value-added products recovery: A step towards algae-based green economy. *Hydrobiology*, *2*(2), pp.326–346.

[31] Onyia, A.I., Ikeri, H.I. and Chima, A.I., 2020. Surface and quantum effects in nanosized semiconductor. *American Journal of Nano Research and Applications*, *8*(3), pp.35–41.

[32] ul Haque, S., Nasar, A., Duteanu, N. and Pandey, S., 2023. Carbon based-nanomaterials used in biofuel cells: A review. *Fuel*, *331*, p.125634.

[33] Vivekanandhan, S., 2019. Combustion process using plant-based fuels for the synthesis of metal-oxide nanostructures. *Chemistry Select*, *4*(27), pp.8026–8042.

[34] Mishra, A., Bhatt, R., Bajpai, J. and Bajpai, A.K., 2021. Nanomaterials based biofuel cells: A review. *International Journal of Hydrogen Energy*, *46*(36), pp.19085–19105.

[35] Shen, X., Liu, W., Gao, X., Lu, Z., Wu, X. and Gao, X., 2015. Mechanisms of oxidase and superoxide dismutation-like activities of gold, silver, platinum, and palladium, and their alloys: A general way to the activation of molecular oxygen. *Journal of the American Chemical Society*, *137*(50), pp.15882–15891.

[36] Anucha, C.B., Altin, I., Bacaksiz, E. and Stathopoulos, V.N., 2022. Titanium dioxide (TiO_2)-based photocatalyst materials activity enhancement for contaminants of emerging concern (CECs) degradation: In the light of modification strategies. *Chemical Engineering Journal Advances*, *10*, p.100262.

[37] Zhou, W., Bai, X., Wang, E. and Xie, S., 2009. Synthesis, structure, and properties of single-walled carbon nanotubes. *Advanced Materials*, *21*(45), pp.4565–4583.

[38] Pérez, J.A.C., Sosa-Hernández, J.E., Hussain, S.M., Bilal, M., Parra-Saldivar, R. and Iqbal, H.M., 2019. Bioinspired biomaterials and enzyme-based biosensors for point-of-care applications with reference to cancer and bio-imaging. *Biocatalysis and Agricultural Biotechnology*, *17*, pp.168–17.

[39] Benzigar, M.R., Talapaneni, S.N., Joseph, S., Ramadass, K., Singh, G., Scaranto, J., Ravon, U., Al-Bahily, K. and Vinu, A., 2018. Recent advances in functionalized micro and mesoporous carbon materials: Synthesis and applications. *Chemical Society Reviews*, *47*(8), pp.2680–2721.

[40] Liu, L. and Pant, D., 2024. Integrative electrochemical and biological catalysis for the mild and efficient utilization of renewable electricity and carbon resources. *Sustainable Energy & Fuels*, *8*(3), pp.460–480.

[41] Mu, Q., Jiang, G., Chen, L., Zhou, H., Fourches, D., Tropsha, A. and Yan, B., 2014. Chemical basis of interactions between engineered nanoparticles and biological systems. *Chemical Reviews*, *114*(15), pp.7740–7781.

[42] Bilal, M., Zhao, Y., Rasheed, T. and Iqbal, H.M., 2018. Magnetic nanoparticles as versatile carriers for enzymes immobilization: A review. *International Journal of Biological Macromolecules*, *120*, pp.2530–2544.

[43] Sun, F., Dong, G., Jiang, F., Wang, X., Diao, B., Li, X., Joo, S.W., Zhang, L., Kim, S.H., Cong, C. and Li, X., 2024. Carbon quantum dot/multiwalled carbon nanotube-based self-powered strain sensors for remote human motion detection. *ACS Applied Nano Materials*, 7(23), pp.27706–27716.

[44] López, V., Sundaram, R.S., Gómez-Navarro, C., Olea, D., Burghard, M., Gómez-Herrero, J., Zamora, F. and Kern, K., 2009. Chemical vapor deposition repair of graphene oxide: A route to highly-conductive graphene monolayers. *Advanced Materials*, *21*(46), pp.4683–4686.

[45] Li, K., Eres, G., Howe, J., Chuang, Y.J., Li, X., Gu, Z., Zhang, L., Xie, S. and Pan, Z., 2013. Self-assembly of graphene on carbon nanotube surfaces. *Scientific Reports*, *3*(1), p.2353.

[46] Narkiewicz, U., Podsiadły, M., Jędrzejewski, R. and Pełech, I., 2010. Catalytic decomposition of hydrocarbons on cobalt, nickel and iron catalysts to obtain carbon nanomaterials. *Applied Catalysis A: General*, *384*(1–2), pp.27–35.

[47] Peng, H., Dang, W., Cao, J., Chen, Y., Wu, D., Zheng, W., Li, H., Shen, Z.X. and Liu, Z., 2012. Topological insulator nanostructures for near-infrared transparent flexible electrodes. *Nature Chemistry*, *4*(4), pp.281–286.

[48] Navalon, S., Dhakshinamoorthy, A., Alvaro, M. and Garcia, H., 2014. Carbocatalysis by graphene-based materials. *Chemical Reviews*, *114*(12), pp.6179–6212.

[49] Triphati, A. and Pirzadah, T.B., 2023. Synthesis methods of nanoparticles and their key applications. In Munir Ozturk, Arpita Roy, Rouf Ahmad Bhat, Fazilet Vardar-sukan, and Fernanda Maria Policarpo Tonelli (Eds.), *Synthesis of Bionanomaterials for Biomedical Applications* (pp.57–76). Amsterdam, Netherlands: Elsevier.

[50] Dubey, R., Dutta, D., Sarkar, A. and Chattopadhyay, P., 2021. Functionalized carbon nanotubes: Synthesis, properties and applications in water purification, drug delivery, and material and biomedical sciences. *Nanoscale Advances*, *3*(20), pp.5722–5744.

[51] Xu, J.Y., Tong, X., Yu, P., Wenya, G.E., McGrath, T., Fong, M.J., Wu, J. and Wang, Z.M., 2018. Ultrafast dynamics of charge transfer and photochemical reactions in solar energy conversion. *Advanced Science*, *5*(12), p.1800221.

[52] Karaman, M., Sarıipek, F., Köysüren, Ö. and Yıldız, H.B., 2013. Template assisted synthesis of photocatalytic titanium dioxide nanotubes by hot filament chemical vapor deposition method. *Applied Surface Science*, *283*, pp.993–998.

[53] Oh, H.S., Kim, K., Ko, Y.J. and Kim, H., 2010. Effect of chemical oxidation of CNFs on the electrochemical carbon corrosion in polymer electrolyte membrane fuel cells. *International Journal of Hydrogen Energy*, *35*(2), pp.701–708.

[54] Zhang, K., Tao, P., Zhang, Y., Liao, X. and Nie, S., 2019. Highly thermal conductivity of CNF/AlN hybrid films for thermal management of flexible energy storage devices. *Carbohydrate Polymers*, *213*, pp.228–235.

[55] Zhang, L., Jin, L., Liu, B. and He, J., 2019. Templated growth of crystalline mesoporous materials: From soft/hard templates to colloidal templates. *Frontiers in Chemistry*, *7*, p.22.

[56] Zhu, Q.L. and Xu, Q., 2016. Immobilization of ultrafine metal nanoparticles to high-surface-area materials and their catalytic applications. *Chem*, *1*(2), pp.220–245.

[57] Logan, B.E., 2012. Essential data and techniques for conducting microbial fuel cell and other types of bioelectrochemical system experiments. *ChemSusChem*, *5*(6), pp.988–994.

[58] Hao, G.P., Jin, Z.Y., Sun, Q., Zhang, X.Q., Zhang, J.T. and Lu, A.H., 2013. Porous carbon nanosheets with precisely tunable thickness and selective CO_2 adsorption properties. *Energy & Environmental Science*, *6*(12), pp.3740–3747.

[59] Zhang, H., Chen, W., Shen, X., Zhang, M., Yang, Y., White, J.C., Tao, S. and Wang, X., 2017. Influence of multi-walled carbon nanotubes and fullerenes on the bioaccumulation and elimination kinetics of phenanthrene in geophagous earthworms (Metaphire guillelmi). *Environmental Science: Nano*, *4*(9), pp.1887–1899.

[60] Veerubhotla, R., Varanasi, J.L. and Das, D., 2018. Biofilm formation within microbial fuel cells. In Patit Paban Kundu, Kingshuk Dutta Location (Eds.), *Progress and Recent Trends in Microbial Fuel Cells*, (pp.231–242). Amsterdam, Netherlands: Elsevier.

[61] Sheng, T., Guan, X., Liu, C. and Su, Y., 2021. De novo approach to encapsulating biocatalysts into synthetic matrixes: From enzymes to microbial electrocatalysts. *ACS Applied Materials & Interfaces*, *13*(44), pp.52234–52249.

[62] Zadeh, P.G., Rezania, S., Fattahi, M., Dang, P., Vasseghian, Y. and Aminabhavi, T.M., 2024. Recent advances in microbial fuel cell technology for energy generation from wastewater sources. *Process Safety and Environmental Protection*.

Chapter 14

Biohydrogen production through food waste materials

Utilization and future prospectives

Ramdendra Pal, Dan Bahadur Pal, and Amit Kumar Rathoure

14.1 INTRODUCTION

The requirement for sustainable and clean energy solutions has put significant pressure on finding alternatives to fossil fuels, which are associated with environmental degradation and greenhouse gas emissions [1]. Hydrogen, particularly "BH", has been a promising solution in the exploration for a clean, renewable, and efficient energy carrier. Unlike conventional H_2 produced from natural gas or coal, BH is generated through biological processes using organic matter, thus aligning with the goals of sustainability and environmental preservation [2]. Among the various feedstocks investigated for BH production, food waste has received significant interest due to its abundance, low cost, and organic-rich composition. The use of wasted food for BH generation represents an innovative approach to addressing two global issues: energy sustainability and waste management.

H_2 is generally detected as a versatile and efficient energy carrier, with applications spanning from transportation to industrial processes. It's pure combustion, yielding only water as a by-product, which positions H_2 as a key player in reducing carbon emissions and combating climate change [3,4]. H_2 can be preserved and applied in fuel cells to generate electricity, making it appropriate for a vast range of energy systems, including portable electronics, vehicles, and large-scale power plants [5]. However, the current global H_2 production is predominantly reliant on fossil fuels, which limits its sustainability and contributes to greenhouse gas emissions [6]. This has led to growing interest in developing alternative, environmentally friendly production methods, such as BH production.

BH production refers to the generation of H_2 through biological processes, primarily involving microorganisms [7]. Unlike thermochemical methods such as steam reforming or gasification, which require high temperatures and the utilization of fossil fuels, BH can be developed under mild conditions using biological systems. There are several pathways for BH production, including DF, PF, and microbial electrolysis, each utilizing different microbial communities and operational conditions [8]. These operations have

 DOI: 10.1201/9781003581536-14

the potential to convert a wide range of organic substrates, including agricultural waste, wastewater, and food waste, into H_2, making them highly adaptable to different regions and waste streams.

The generation of food waste is a developing global issue, with considerable environmental, economic, and social consequences. According to the Food and Agriculture Organization (AFAO), approximately one-third of all food made for human consumption is lost or wasted, resulting in 1.3 billion tons (annually). This waste occurs across the food distribution network, from agricultural production and processing to retail and consumers. The disposal of waste food in landfills results in the generation of CH_4, a powerful greenhouse gas that plays a role in global warming. Additionally, the decomposition of food waste in anaerobic conditions creates leachate, which has the potential to pollute groundwater and soil. Given the vast quantities of organic material present in waste food, it represents an excellent feedstock for BH production. Waste food is high in carbohydrates, proteins, and lipids, which microorganisms can break down during fermentation processes to produce H_2. Moreover, utilizing food waste for BH generation not only supplies an RE source but also offers an environmentally friendly solution to the problem of waste disposal. By diverting food waste from landfills and using it as a resource for bioenergy production, we can reduce methane emissions, lower waste management costs, and contribute to the circular economy. The pressing issues of energy security and environmental sustainability have spurred the search for RE technologies that are both effective and scalable. The consumption of fossil fuel reserves, rising energy demand, and the effect of climate change have heightened the need to shift towards low-carbon and sustainable energy systems. H_2, particularly when produced from renewable sources, has the potential to play a vital role in this transition by providing a clean alternative to fossil fuels. However, the widespread adoption of H_2 as an energy carrier is contingent on the development of cost-effective and environmentally benign production methods [5].

Simultaneously, the management of waste food is an increasingly pressing issue, with profound implications for the environment and public health. In numerous countries, food waste constitutes an important portion of municipal solid waste, contributing to landfill overflow, greenhouse gas emissions, and resource inefficiencies. The conventional methods of food waste disposal, such as landfilling and incineration, are not only environmentally detrimental but also fail to harness the valuable organic material contained in the waste. As such, there is a growing recognition of the need to develop alternative approaches that can convert food waste into useful products, such as biofuels, bio-plastics, and BH. Figure 14.1 illustrates the process of converting biomass and organic solid waste to H_2, storing it, and using it in the energy industry.

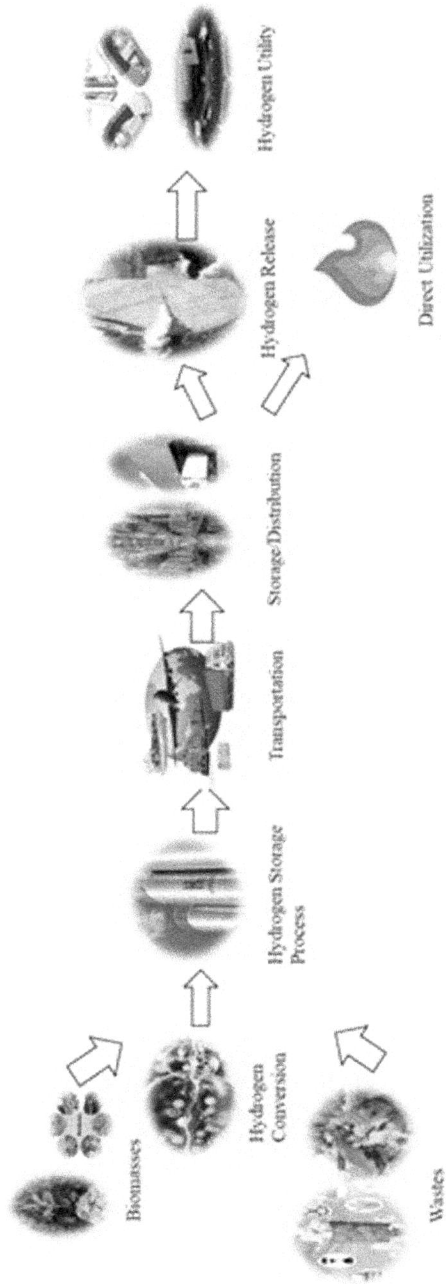

Figure 14.1 Conceptual illustration of the energy system's use of H_2, storage of H_2, and conversion of biomass as well as waste to H_2 [9].

Harnessing food waste for BH production provides significant benefits from both an environmental and economic perspective. First and foremost, food waste is a readily available and low-cost feedstock, making it an attractive option for bioenergy production. In regions with high levels of food waste generation, such as urban centers and food processing industries, food waste can be collected and used as a sustainable feedstock for BH production. This can help to minimize the environmental impact related to waste disposal and mitigate the costs of landfill management.

Additionally, the conversion of waste food into BH can contribute to the circular economy by converting waste into a worthy resource. BH production not only generates clean energy but also produces byproducts, like organic acids and fertilizers, which can be used in agriculture and other industries. This closed-loop approach reduces waste and increases resource efficiency, aligning with the principles of sustainability and resource conservation.

This review seeks to present an overview of the power for BH production using food waste materials. It will explore the biological processes involved in BH generation, including DF and PF, and discuss the suitability of different types of food waste as feedstocks for these processes. The review will also examine the key challenges and opportunities connected with using food waste for BH production, including reactor design, process optimization, and economic feasibility. Through a detailed analysis of current research and case studies, this chapter seeks to highlight the significant potential of waste food as a sustainable resource for BH production. By leveraging the inherent organic content of food waste, we can not only generate RE but also contribute to more enduring waste control practices, ultimately advancing the goals of a low-carbon and circular economy. Figure 14.1 shows the conceptual illustration of the energy system's use of H_2, storage of H_2, and conversion of biomass as well as waste to H_2.

14.2 BASICS OF BIO-HYDROGEN PRODUCTION

BH production is an emerging field in clean energy, offering a sustainable method to generate H_2 fuel using biological processes [2]. As H_2 is deemed a highly promising energy carrier in the transition from fossil fuels to RE systems, BH production holds the potential to perform a critical role in meeting future energy demands while reducing environmental impacts. This section explores the basics of H_2 as an energy carrier, various methods of H_2 production, and the biological pathways involved in BH generation.

14.2.1 Fundamentals of H_2 as an energy carrier

H_2 is mainly mentioned as the "fuel of the future" due to its remarkable energy density, clean combustion, and versatility across various applications

[5]. It is usable in fuel cells to generate electricity, heat, and mechanical power, offering a sustainable alternative to fossil fuels.

14.2.2 Properties of hydrogen and its role in clean energy transitions

H_2 possesses the utmost energy content per unit mass of any fuel—about 142 MJ/kg—approximately three times more than traditional fossil fuels like gasoline [10]. This makes it an excellent choice for applications requiring high energy density, such as transport, power generation, and industrial processes. Furthermore, when used in fuel cells or combusted, H_2 generates only water vapor as a by-product, making it a zero-emission fuel.

H_2 plays a major role in the clean energy transition for several reasons:

(a) *Decarbonization of Industry*: H_2 can replace fossil fuels in industries that are hard to electrify, such as steel production and massive transportation.
(b) *Energy Storage:* H_2 can store surplus electricity built from intermittent RE (wind and solar), stabilizing grids.
(c) *Transport:* H_2 fuel cells are being a viable energy source for vehicles, especially long-haul trucks, buses, and trains, where batteries alone may not be feasible.

However, despite these advantages, H_2 faces challenges, particularly in production and infrastructure development. Most H_2 today is produced using fossil fuels, leading to significant carbon dioxide emissions. The shift to renewable H_2 production, including BH, is crucial for achieving its role in clean energy transitions.

14.2.3 Comparison with other biofuels

In comparison with other biofuels like biogas and bioethanol, H_2 offers distinct advantages [11]:

(a) *Biogas*: Produced by anaerobic digestion of organic matter, biogas mainly consists of CH_4 and CO_2. Although it's renewable, its combustion still results in CO_2 emissions, unlike H_2 clean combustion.
(b) *Bioethanol*: Derived from biomass fermentation, bioethanol is broadly used as a liquid fuel additive. However, its energy content is considerably lower than that of H_2, and its combustion generates carbon emissions.

BH production thus offers a cleaner alternative with higher energy yields and no harmful emissions, positioning it as a superior choice among biofuels for a sustainable energy future.

14.2.4 Methods of hydrogen production

H_2 can be generated using a selection of methods, each with its own advantages, limitations, and environmental impacts. These methods include thermochemical, electrochemical, and biological processes [12]. Figure 14.2 illustrates the several processes used to produce H_2.

14.2.4.1 Thermochemical methods

1. *Steam Methane Reforming (SMR):* This is the most common method for H_2 production today, accounting for about 95% of global H_2 output. SMR involves reacting natural gas, CH_4, with steam at high temperatures to produce H_2 and CO. However, the operation emits substantial amounts of CO_2, making it unsuitable for long-term sustainability unless carbon capture technologies are applied.
2. *Gasification:* Gasification converts carbon-based materials (coal, biomass, or waste) into H_2 and CO by exposing the feedstock to high temperatures in an oxygen-limited environment. While gasification can use renewable biomass as feedstock, the process is still energy-intensive and emits greenhouse gases unless paired with carbon capture and storage.
3. *Pyrolysis:* This method entails the thermal decay of organic materials in the absence of O_2 to produce H_2-rich syngas (a mixture of H_2, CO) and CH_4. Pyrolysis can process biomass, but its efficiency and scalability are challenges that need to be addressed.

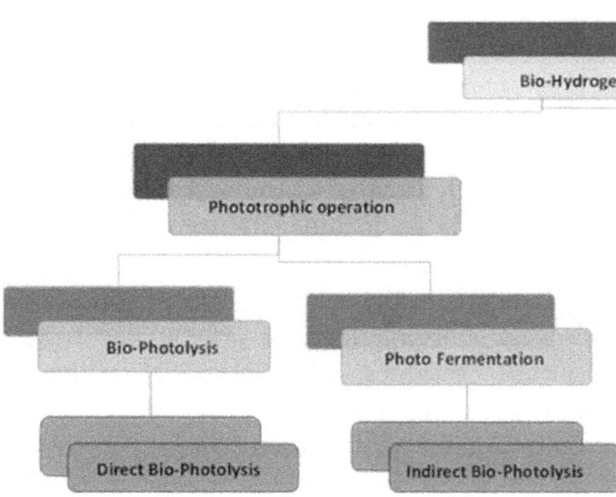

Figure 14.2 The method for producing bio-hydrogen [13].

14.2.4.2 Electrochemical methods

1. *Water Electrolysis:* Water electrolysis splits water into H_2 and O_2 using an electric current. This method is clean, especially when powered by RE sources like wind or solar. The two main types of water electrolysis are:
2. *Alkaline Electrolysis*: AE operates at lower temperatures with a liquid electrolyte.
3. *Proton Exchange Membrane (PEM) Electrolysis:* PEM uses a solid electrolyte and operates at higher efficiency but is costlier.

Water electrolysis is regarded as a key method for green H_2 production, but its widespread adoption is limited by the cost and availability of renewable electricity.

14.2.4.3 Biological methods

Biological methods, or BH production, utilize micro-organisms to produce H_2 from organic materials under mild conditions, offering a sustainable alternative to thermochemical processes. These methods are particularly attractive for converting waste materials like waste food, agricultural residues, and wastewater into H_2.

Biological pathways for H_2 production primarily involve microbial processes that convert organic substrates into H_2. The three major biological pathways are DF, PF, and combined processes such as two-stage fermentation and microbial-electrolysis.

14.2.4.3.1 Dark fermentation

DF is a procedure where anaerobic bacteria decompose carbohydrates and other organic compounds in the absence of light, producing H_2, CO_2, and organic acids as byproducts [14]. This process operates under controlled conditions, with key parameters like pH (typically 5.5–6.5), temperature (30°C–55°C), and retention time influencing efficiency. The process is facilitated by H_2-producing bacteria such as *Clostridium* and *Enterobacter*. An illustration of the various methods to include DF into other post-treatment procedures is provided in Figure 14.3.

DF is advantageous because it operates under relatively simple conditions (no light required) and can process a broad range of organic substrates, including waste food, agricultural residues, and industrial effluents. However, one major limitation is the relatively low H_2 yield, as much of the substrate's energy is converted into organic acids rather than H_2.

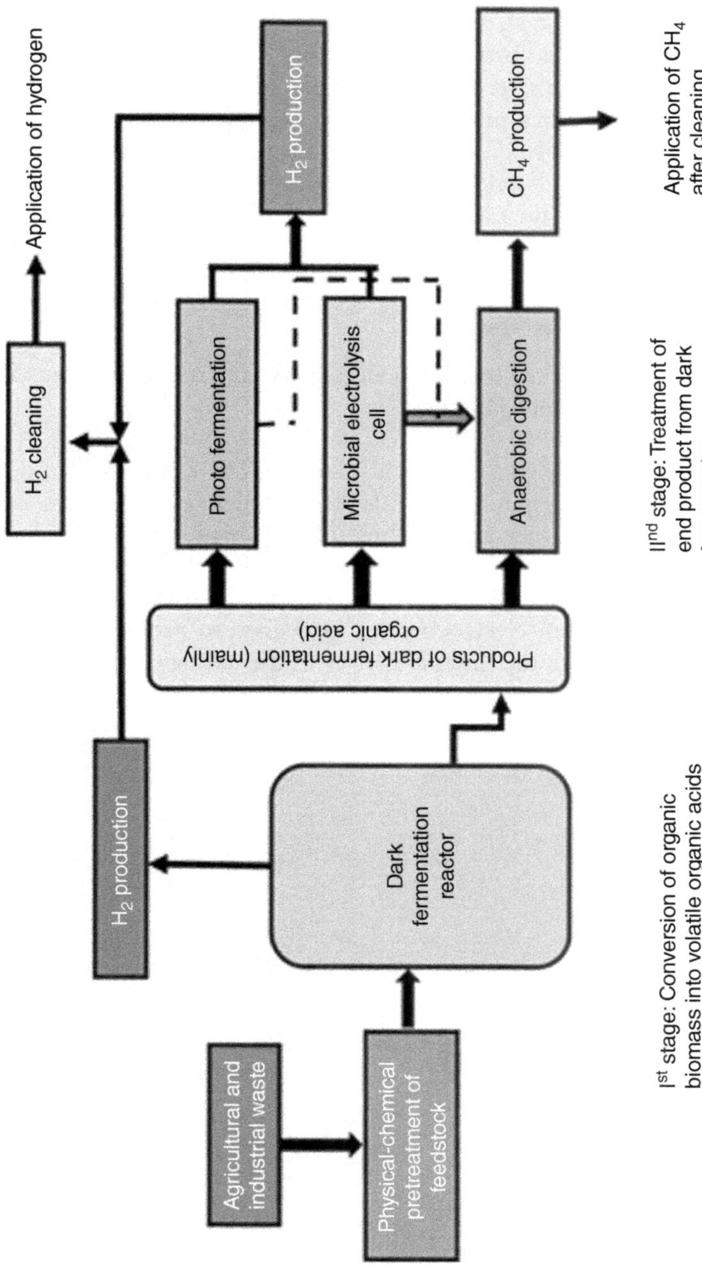

Figure 14.3 Various approaches of incorporating DF into posttreatment procedures for enhanced biofuel production [7].

14.2.4.3.2 Photo-fermentation

In PF, photo-synthetic bacteria, particularly purple non-sulfur bacteria (e.g., *Rhodobacter sphaeroides*), use light energy to metabolize organic acids (produced from DF) and produce H_2 [15,16]. The process is dependent on light and can achieve greater H_2 yields compared to DF [17]. The microbial community dynamics in a mixed PF culture enriched from a digestate from the AD of pig dung showed the prevalence of mostly *R. palustris*. The main disadvantage of PF is its reliance on sunlight or artificial light, which can make large-scale implementation challenging. The diagrammatic model of PF is displayed in Figure 14.4.

14.2.4.3.3 Combined processes

Combining DF with PF or microbial electrolysis can increase H_2 yields and improve process efficiency [19–23].

1. *Two-Stage Fermentation:* In this method, DF is first used to decompose complex organic materials into simpler organic acids [24,25]. These acids are then transformed into H_2 by photosynthetic bacteria in a second stage, improving overall H_2 yield.
2. *Microbial Electrolysis Cells (MECs):* MECs combine microbial metabolism with electrochemical processes to produce H_2 [26,27]. Bacteria oxidize the organic matter in the anode chamber, releasing electrons. A little external voltage is utilized to drive the reduction of protons to H_2 at the cathode. MECs can form H_2 from a wide range

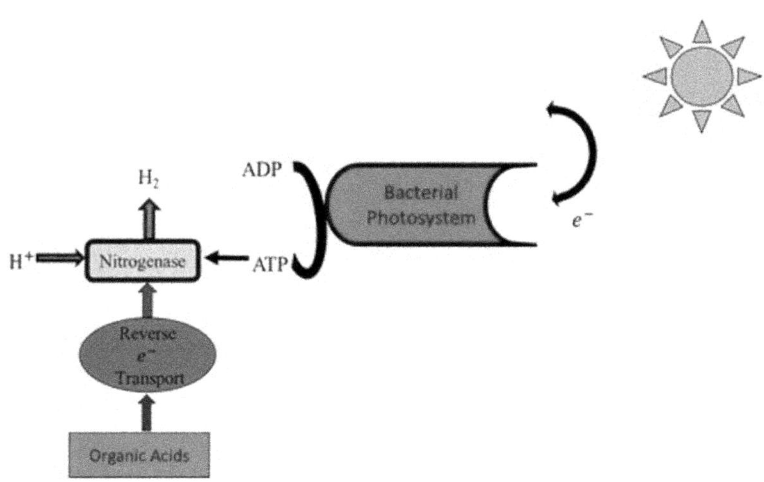

Figure 14.4 Diagrammatic representation of PF [18].

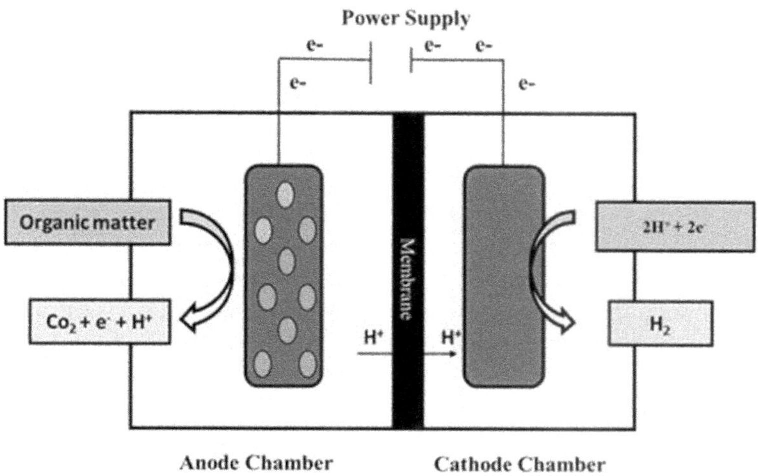

Figure 14.5 The basic idea behind the microbial electrolysis process that produces H_2 [9].

of organic feedstock, including wastewater and food waste, with higher efficiency than DF. Figure 14.5 illustrates the fundamental concept of the H_2-production process from MECs [9].

Researchers [28] presented separate combination strategies for mixing two-step DF and PF processes. Others [29] studied joining DF and PF for BH manufacturing using glucose as the substrate. [30] and [31] attained a total yield of 10.25 mol H_2/mol sucrose and 6.63 mol H_2/mol sucrose, respectively. In a study by [32], a yield of 4.16 mol H_2 mol/mol glucose was obtained from PF of DF effluents using glucose as the substrate, which increased the total yield to 5.48 mol H_2/mol glucose.

14.2.5 Key microorganisms involved in bio-hydrogen production

The progress of BH production relies on the activity of specific groups of microorganisms, each adapted to different metabolic processes.

1. *Anaerobic Bacteria and Archaea in Dark Fermentation*
   Anaerobic bacteria, including *Clostridium*, *Enterobacter*, and *Bacillus*, are the primary microorganisms responsible for H_2 production in DF [29,33]. These bacteria utilize the hydrogenase enzyme to split water or organic compounds, releasing H_2 gas. Other microorganisms, including certain archaea, may also contribute to

the breakdown of complex organic materials, although they do not directly produce H_2.

H_2 production in DF is limited by factors such as substrate composition, pH, and temperature. Thermophilic bacteria (operating at 50°C–60°C) tend to produce higher H_2 yields than mesophilic bacteria (operating at 30°C–40°C) [34].

2. *Cyanobacteria and Purple Non-Sulfur Bacteria (NSB) for PF*
   In PF, H_2 production is carried out by photo-synthetic bacteria, most notably purple non-sulfur bacteria (NSB) (such as Rhodobacter and Rhodopseudomonas species) [29,35]. These bacteria are capable of utilizing light energy to transform organic acids into H_2 through the activity of nitrogenase or hydrogenase enzymes. Cyanobacteria can also be used for biological H_2 production through a process called bio-photolysis, where water is broken down into H_2 and O_2 using sunlight [36]. However, this method is less efficient and more complex to analyze PF with organic substrates.

   BH production offers a promising, sustainable approach to producing H_2 from organic waste materials using biological pathways. By utilizing DF, PF, or a combination of these processes, BH can be produced efficiently under mild conditions with minimal environmental impact. The role of microorganisms, including anaerobic bacteria and photosynthetic bacteria, is central to this process. With ongoing advancements in microbial engineering and process optimization, BH has the power to become a leading player in the future of clean energy transitions.

14.3 WASTE FOOD AS A FEEDSTOCK FOR BIO-HYDROGEN GENERATION

Waste Food is one of the world's most generous and underutilized resources, with millions of tons discarded annually across households, industries, and agriculture [37]. This discarded food material, often viewed as a waste product, contains valuable organic compounds that can be transformed into BH through biological processes. Utilizing waste food for BH generation not only addresses energy demands but also contributes to waste management and environmental sustainability.

This section explores the power of waste food as a feedstock for BH generation, examining its composition, pre-processing methods, and the suitability of various kinds of food waste based on their nutritional content.

14.3.1 Composition of waste food

The formation of waste food varies depending on its source, but it generally contains a mix of carbohydrates, proteins, fats, fibers, vitamins, and minerals. Figure 14.6 illustrates the many types of feedstock utilized for bio-H_2 production. These nutrients serve as feedstocks for microorganisms that generate BH. Understanding the formation of waste food is crucial to determining its compatibility for H_2 production, as different types of waste will favor various biological pathways and yields [38].

14.3.2 Types of food waste

Waste food can be classified into various categories depending on its origin and composition:

1. *Kitchen Waste*: This includes food scraps from households, restaurants, and catering services, such as leftover meals, spoiled fruits and vegetables, and peels. Kitchen waste is typically rich in carbohydrates and starches (e.g., from bread, rice, and potatoes), as well as fats and proteins (from meat, dairy, and oils).
2. *Agricultural Residue*: Post-harvest agricultural activities produce a large amount of food waste, including crop residue, fruit and vegetable peels, leaves, and stalks. Agricultural waste tends to be rich in fibers and lignocellulosic materials, which require more complex degradation processes for BH generation.
3. *Fruit and Vegetable Peels:* These are another significant source of food waste, often discarded during food processing. Fruit peels (such as banana, citrus, and apple) and vegetable peels (like potato and carrot) are rich in sugars, making them highly suitable for H_2 production through fermentation processes.

14.3.3 Nutritional content and its relevance for bio-hydrogen production

The effectiveness of waste food as a raw material for BH production is largely determined by its nutritional content, particularly the carbon, nitrogen, and trace element composition, which affects microbial activity. Table 14.1 shows a comparison of the BH yield for different types of food waste.

1. *Carbohydrates:* Carbohydrates, including sugars, starches, and cellulose, are the most readily degradable components for BH production. Bacteria involved in DF, such as Clostridium species, efficiently convert carbohydrates into H_2. Food waste rich in carbohydrates, like bread, rice, and fruit, typically yield higher amounts of BH [39].
2. *Proteins*: Protein-rich food waste (e.g., from meat and dairy) is more difficult to convert into H_2. Proteins are first decomposed into amino acids and then metabolized by bacteria, but this process often results in the production of ammonia, which inhibits H_2 production [40].

Table 14.1 Comparison of the bio-hydrogen yield on different types of food wastes

Substrates/microorganisms/inoculums	*Pretreatments*	*Operational period/ reactor type*	*PH, temperature & HRT condition*	*H_2 production yield [HPY]*	*References*
Rice waste from food waste	thermal	Anaerobic bio reactor	pH; 5.5, temperature; 55°C	40 mL H_2/g VS	[43]
Waste food/sewage sludge	Heat up to 90°C, 20 minutes	Batch	pH 8.0, temperature; 35°C	2.11	[44]
Food waste (grains and vegetables) or sewage sludge	Heat up to 15 minutes	Continuous reactor	pH 6.5; temperature 35°C	0.33 H_2/mol	[45]
Waste food/heat shock processed anaerobic sludge	NA	Leaching bed reactor type	Temperature 37°C, pH 5.5–7	Max. H_2 310	[46]
Anaerobic digested sludge	Heat stock treatment	NA	Temperature 36°C, pH-7	1.9	[47]
Waste of kitchen like rice, noodle, cake or mixed sludge	NA	I-CST Reactor	pH-4.4; temperature-55°C	2.1	[48]
Vegetable from kitchen waste or kitchen waste break down	NA	Batch type	pH-7.0; temperature-55°C	0.57	[49]
From kitchen rice slurry or anaerobic digester	Heat-100°C, 30 minutes	Batch type	pH-3.7; temperature-45°C	346	[50]
Food industry residue or anaerobic digester sludge	NA	CSTR type	pH-6.3–6.8; temperature-55°C	16.5	[51]
Potato's processing waste or soil extract from potato plant	nd	Batch type	pH-4.0–5.0, temperature-35°C–38°C	2.3	[52]

However, small amounts of proteins can contribute to microbial growth and BH production under optimized conditions.

3. *Lipids:* Lipid-rich food waste, including oils and fats, poses challenges for BH production because lipids are not easily degraded by most H_2-producing bacteria [41]. While some microorganisms can metabolize fatty acids, the BH yield from lipid degradation is relatively low. Lipid-rich waste can also cause issues with reactor operation, such as clogging and foaming.
4. *Other nutrients:* The availability of trace components, vitamins, and ores in waste food can also influence BH production. Certain trace metals, such as iron and nickel, play critical roles in the enzymatic activity of hydrogenases, which are involved in H_2 production by bacteria [42].

14.3.4 Pre-treatment methods for food waste

Before waste food can be utilized as a feedstock for BH generation, it often requires pre-treatment to break down complex organic materials, enhance microbial accessibility, and improve H_2 yields. Various pre-treatment methods are used to prepare food waste for biological processes, including mechanical, thermal, chemical, and biological treatments [53,54].

14.3.4.1 Mechanical pre-treatment

Mechanical pre-processing includes physically altering the structure of food waste to increase the surface area available for microbial digestion. Standard mechanical pre-treatment methods include [54]:

1. *Grinding:* Reducing the particle size of food waste makes it easier for microbes to access the organic matter. Fine grinding also accelerates the breakdown of complex materials, such as lignocellulosic fibers, which are otherwise difficult to degrade.
2. *Shredding:* Shredding food waste into smaller pieces helps to homogenize the material, improving the productivity of BH production by enhancing the availability of fermentable substrates.

Mechanical pre-treatment is typically energy-efficient and straightforward, making it an essential step in processing large volumes of food waste.

14.3.4.2 Thermal pre-treatment

Thermal pre-processing involves heating food waste to break down complex molecules and improve the digestibility of organic matter by microorganisms. Some common thermal pre-treatment methods include [55]:

1. *Heating:* Heating food waste to moderate temperatures (50°C–90°C) can denature proteins and dissolve starches, making them more accessible for fermentation. However, excessive heating can degrade sugars and reduce H_2 yields.
2. *Steam Explosions*: Steam explosions are highly intense thermal processes in which food waste is rapidly heated using high-pressure steam and then abruptly depressurized. This process breaks down lignocellulosic materials, making the cellulose and hemicellulose fractions more accessible to microbes.

Thermal pre-treatment can be particularly useful for agricultural residues and other lignocellulosic materials that are difficult to degrade under normal fermentation conditions.

14.3.4.3 Chemical pre-treatment

Chemical pre-treatment involves using acids, alkalis, or other chemicals to break down the structural components of food waste, particularly lignocellulosic materials, and release fermentable sugars. The most common chemical pre-treatment methods include [56]:

1. *Acid Treatment:* Acid hydrolysis uses strong acids (e.g., H_2SO_4) to decompose cellulose and hemicellulose into simple sugars. Acid pre-treatment can improve H_2 yields, but it also generates inhibitory byproducts, like furfural and phenolic compounds, which must be removed or neutralized before fermentation.
2. *Alkaline Treatment*: Alkaline pre-treatment with substances like sodium hydroxide helps to solubilize lignin and improve the biodegradability of the remaining cellulose and hemicellulose. Alkaline treatment is often less harsh than acid treatment and generates fewer toxic by-products.

14.3.4.4 Biological pre-treatment

Biological pre-processing types include the use of enzymes or microorganisms to diminish complex organic materials in food waste. This operation is often more environmentally friendly and energy-efficient compared to chemical or thermal techniques. Common biological pre-treatment methods include [56,57]:

1. *Enzymatic Treatment:* Enzymes, such as cellulase and amylase, are used to break down cellulose, starches, and other polysaccharides into fermentable sugars. Enzymatic treatment is highly specific and can be tailored to target particular components of food waste.

2. *Microbial Digestion:* Microbial consortia can be used to partially degrade food waste before H_2 production. This process, known as bioaugmentation, involves inoculating food waste with specialized microorganisms that break down complex organic matter, making its accessibility for H_2-producing bacteria [58].

14.3.5 Suitability of various types of waste food

The suitability of different types of waste food for BH generation depends on their composition, particularly the relative quantities of carbohydrates, proteins, and lipids [37].

1. *High Carbohydrate Content Waste:* Food wastes that are rich in carbohydrates, such as bread, rice, pasta, and potatoes, are the most appropriate feedstocks for BH production. Carbohydrates, especially sugars and starches, are easily fermentable by H_2-producing bacteria, leading to high H_2 yields. DF is particularly effective in converting carbohydrate-rich waste into H_2.
2. *Protein-Rich Food Waste:* Protein-rich food waste, including meat, dairy products, and eggs, is less ideal for BH production due to the production of ammonia during protein degradation. Ammonia inhibits microbial activity and can lower H_2 yields. However, small amounts of protein-rich waste can still be processed in combination with carbohydrate-rich waste to improve the overall nutrient balance for microbial growth.
3. *Lipid-Rich Food Waste:* Lipid-rich food waste (LRFW), such as oils, fats, and grease, presents challenges for BH production. Lipids are not easily fermentable, and their degradation, which is the production of long-chain fatty acids, can inhibit H_2 production [19]. Moreover, lipids can cause operational issues in reactors, such as clogging and foaming. However, some microbial strains have been developed to degrade fatty acids, and lipid-rich waste can be co-digested with carbohydrate-rich waste to enhance overall BH yields.

Food waste serves as an abundant and renewable feedstock for BH production, offering a promising solution to both energy and waste management difficulties. By understanding the composition of waste food and applying suitable pre-treatment methods, it is possible to enhance the efficiency of H_2 generation. Wastes rich in carbohydrates are the most suitable for BH production, while protein and lipid-rich wastes may require specialized treatment or co-digestion strategies [59]. Ongoing advancements in pre-treatment technologies and microbial engineering will play a vital role in optimizing the use of food waste for sustainable H_2 production.

14.4 REACTOR DESIGNS FOR BIO-HYDROGEN PRODUCTION FROM FOOD WASTE

Reactor design plays a central role in the efficiency and scalability of BH production from waste food [60]. A well-designed reactor ensures optimal conditions for microbial activity, maximizes H_2 yields, and minimizes operational challenges such as by-product inhibition. Different types of reactors have been explored for BH production, including anaerobic digesters, photo-bioreactors, and microbial electrolysis cells (MECs). Every reactor has specific benefits and restrictions depending on the biological pathway utilized—DF, PF, or microbial electrolysis. The operational parameters, such as temperature, pressure, retention time, and microbial loading, significantly influence reactor performance. Additionally, the operational parameters, such as temperature, pressure, retention time, and microbial loading, plays a crucial role in influence reactor performance.

14.4.1 Anaerobic digesters

Anaerobic digesters are widely used for BH production via DF. DF is a biological process that occurs in the lack of light, where micro-organisms decompose complex organic materials in food waste into H_2, CO_2, and other by-products like volatile fatty acids (VFAs) [61].

14.4.2 Types of digesters suitable for food waste-based bio-hydrogen generation

1. *Continuous Stirred Tank Reactor (CSTR):* The CSTR is one of the most common types of digesters used for DF [62–64]. It is a well-mixed reactor where food waste and microorganisms are continuously fed, while H_2 gas and effluent are continually removed. The key advantage of CSTRs is their ability to maintain homogenous conditions, allowing for efficient substrate degradation. However, due to constant mixing, the dilution effect can limit H_2 yields as excess organic acids can accumulate and inhibit microbial activity.
2. *Upflow Anaerobic Sludge Blanket (UASB):* UASB reactors are mostly used in wastewater treatment and have been adapted for BH production [65,66]. In this design, food waste is identified at the bottom of the reactor, and as it moves upward, it contacts a dense microbial sludge bed where H_2 production occurs. The gas produced is collected at the top. The key advantage of UASB reactors is their ability to retain high microbial biomass concentrations, leading to higher yields. However, substrate composition variability in food waste can disrupt the microbial community and reduce performance.

3. *Packed Bed Reactor (PBR):* PBRs contain solid packing material that provides a surface for microbial biofilm formation [67]. The food waste flows through the reactor, where H_2-producing microbes in the biofilm convert the organic matter into H_2. PBRs are efficient for BH production as they support high cell densities and protect the microbial community from washout. However, biofilm formation may lead to clogging, and the reactor's performance can decrease if the biofilm is not maintained properly.
4. *Fluidized Bed Reactor (FBR):* FBRs use fluidized particles to maintain high mixing rates and mass transfer [68]. Food waste particles are fluidized by the up-flow of liquid and gas, providing excellent contact between the waste and microorganisms. FBRs have been shown to increase H_2 production rates due to their ability to maintain a high surface area for microbial activity. These reactors are more complex than CSTRs and UASB reactors but offer greater control over microbial interactions.
5. *Photo-bioreactors:* Photo-bioreactors are specialized reactors designed for BH production through PF. In this process, photosynthetic bacteria, such as purple non-sulfur bacteria (PNSB), used light energy to convert organic compounds in food waste into H_2 [5,69]. Photo-bioreactors are more complex than anaerobic digesters because they require light exposure, but they offer the potential for higher H_2 yields under optimal conditions.

14.4.3 Designs for maximizing light exposure and efficiency in photo-fermentation

1. *Flat Plate Photobioreactors:* Flat plate photobioreactors comprise shallow, transparent chambers where light penetrates the reactor uniformly. This design maximizes the surface area exposed to light, ensuring efficient absorption by the photosynthetic microorganisms [70]. Flat plate reactors are well-suited for PF as they minimize light limitations and reduce the risk of shading effects, which can reduce H_2 production efficiency. However, the shallow design may limit the volume of food waste that can be processed at a time, requiring larger surface areas for industrial-scale operations.
2. *Tubular Photo Bio-Reactors:* Tubular photo bio-reactors are made up of transparent tubes arranged horizontally or vertically [69]. These tubes provide a large surface area for light exposure while allowing continuous circulation of the food waste and microbial culture. Tubular reactors are more scalable than flat plate reactors and can be designed to maximize both light exposure and reactor volume. The primary challenge with tubular reactors is maintaining

uniform light distribution along the length of the tubes, which can lead to reduced efficiency in areas further from the light source.

3. *Immobilized Photo Bio-Reactors:* In immobilized photo-bioreactors, photo-synthetic bacteria are immobilized on a solid support within the reactor, while the food waste solution flows past them [69]. This design allows for higher microbial concentrations, improving H_2 production rates. Additionally, immobilized systems can help protect the photosynthetic microorganisms from washout. However, like packed bed reactors in DF, immobilized photo-bio-reactors can suffer from clogging or biofilm overgrowth, which reduces efficiency.
4. *Microbial Electrolysis Cells(MECs):* Microbial electrolysis cells (MECs) represent a hybrid approach to BH production, combining biological processes with electrochemical reactions [71]. In an MEC, microorganisms degrade organic matter in food waste, releasing electrons and protons. An exterior voltage is employed to facilitate the electrochemical decrease of protons to H_2 gas at the cathode. MECs have several advantages, including higher H_2 yields and the capability to treat a wide variety of waste food compositions.

14.4.4 Configuration and advantages for bio-hydrogen production

1. *Anode Chamber:* The anode chamber in an MEC houses electroactive microorganisms that oxidize organic compounds from waste food, producing protons and electrons [71]. These microorganisms, such as Geobacter and Shewanella, transfer electrons to the anode either through direct electron transfer or via electron-shuttling compounds. The execution of the MEC is dependent on the efficiency of electron transfer and the availability of organic substrates.
2. *Cathode Chamber:* In the cathode chamber, the protons produced at the anode combine with the electrons supplied by the exterior voltage to generate H_2 gas [71]. The effectiveness of H_2 output at the cathode is affected by the material used for the cathode, the applied voltage, and the proton conductivity between the anode and cathode chambers. MECs typically require lower external voltages than traditional electrolysis, making them more energy-efficient for H_2 production.

14.4.4.1 Advantages

MECs offer several advantages over traditional DF and PF reactors. They can achieve higher H_2 yields, process a wider variety of food waste substrates, and reduce the accumulation of inhibitory by-products. MECs are particularly suitable for mixed food waste streams that include carbohydrates, proteins, and lipids, as the electroactive microorganisms can degrade a broad range of organic materials. Additionally, MECs can operate under ambient conditions, reducing energy requirements for heating or pressurization.

14.4.4.2 *Operational parameters*

The efficiency of reactors used for BH production is influenced by several critical operational parameters, including temperature, pressure, retention time, and microbial loading.

1. *Temperature:* Most H_2-producing microorganisms thrive in mesophilic (30°C–40°C) or thermophilic (50°C–60°C) conditions [72]. Higher temperatures generally enhance microbial activity and H_2 production rates, but they also increase energy requirements for reactor operation. Thermophilic conditions can lead to higher H_2 yields but may require more robust microbial strains capable of withstanding elevated temperatures.
2. *Pressure:* BH production reactors typically operate at atmospheric pressure, but pressurized systems can improve gas collection efficiency. However, maintaining pressure requires additional infrastructure and increases operational costs.
3. *Retention Time:* Retention time indicates the duration that the food waste stays in the reactor. Longer retention times allow for more complete degradation of organic material, leading to higher H_2 yields. However, prolonged retention times may result in the accumulation of inhibitory by-products. Optimizing retention time is essential for balancing H_2 production and operational efficiency.
4. *Microbial Loading:* The concentration of microorganisms in the reactor significantly impacts H_2 production rates. Higher microbial loading increases the rate of substrate degradation and H_2 generation. However, overloading the reactor with microorganisms can lead to competition for resources and reduce overall efficiency.

Ongoing studies focus on enhancing microbial efficiency, optimizing reactor configurations, and improving operational parameters to make BH production from waste food a viable alternative to conventional energy sources.

14.5 ECONOMIC AND ENVIRONMENTAL ANALYSIS

BH production from food waste presents an exciting opportunity to address both energy and waste management challenges. However, like any emerging technology, its economic feasibility and environmental sustainability must be evaluated to ensure it can be widely adopted. This analysis covers the financial costs associated with BH production from waste food, environmental implications using LCA, and social and policy aspects that influence the growth of this technology.

The economic feasibility of BH production from waste food depends on different factors, including the cost of collecting, transporting, and treating food waste, as well as the capital and operational expenditures required for BH production [73]. One of the primary advantages of using food waste as a feedstock is that it is often available at low or no cost, especially if the waste is sourced from municipal waste streams or agricultural industries. However, the logistics of food waste collection-such as separation, transportation, and storage—can incur significant costs, particularly in urban areas with complex waste management systems.

Once collected, food waste requires pre-treatment to decompose complex organic compounds, making them more approachable to H_2-producing microorganisms. These pre-treatments, including mechanical, thermal, or chemical processes, can contribute to the overall cost of the production process. The type of pre-processing used often relies on the composition of the food waste and the specific BH production pathway. For instance, high-carbohydrate food wastes may require less intensive pre-treatment than lipid or protein-rich wastes, which are less susceptible to microbial degradation. The capital costs for BH production technologies, such as the construction of anaerobic digesters, photo-bioreactors, or microbial electrolysis cells (MECs), are another significant economic factor. These costs include reactor materials, energy inputs (e.g., electricity or heat), and infrastructure for gas collection and storage. Operational costs include maintenance, microbial inoculation, and control of process conditions (temperature, pressure, etc.). As the technology matures and scales up, it is predicted that these costs will be minimized, improving the overall economic feasibility.

BH has significant market potential as a renewable fuel. It can be used in fuel cells for power production in different sectors, including transportation, residential energy systems, and industrial applications. H_2-powered fuel cells produce electricity with high effectiveness and release only water as a byproduct, making them ideal for applications where clean energy is prioritized.

The transportation sector represents one of the greatest power markets for BH. As countries seek alternatives to fossil fuels to decrease greenhouse gas emissions, H_2 fuel cells for vehicles (cars, trucks, and buses) offer a promising solution. Furthermore, H_2 can be mixed with natural gas for use in combined heat and power (CHP) systems, further diversifying its market applications. Government incentives and subsidies for renewable fuels, such as those seen in the European Union (EU), the United States (US), and Japan, are likely to drive the growth of in these markets.

14.5.1 Life cycle assessment (LCA)

LCA is a tool used to examine the environmental effects of a product or procedure from "cradle to grave." For BH production from waste food, LCA considers the entire process, from waste collection and treatment to H_2 generation and final use [74]. This assessment provides insights into the sustainability of the technology by accounting for resource inputs (e.g., energy, water, and materials) and environmental outputs (e.g., greenhouse gas emissions, water pollution, and waste generation).

BH production has the potential to considerably minimize environmental impacts compared to traditional H_2 production methods, such as steam methane reforming (SMR) of natural gas. By using food waste as a feedstock, the process redirects organic waste from landfills, reducing methane emissions and helping to mitigate climate change. Moreover, the biological processes involved in BH production typically occur under milder conditions, requiring less energy input than thermochemical methods like gasification or pyrolysis.

14.5.2 Comparative analysis with other waste-to-energy processes (biogas, composting)

When comparing BH production with other waste-to-energy processes, like biogas generation and composting, each technology has its unique advantages and environmental impacts. Biogas production via anaerobic digestion converts waste food into CH_4, which can serve as an RE source. However, CH_4 is a potent greenhouse gas, and its leakage during production, storage, or transportation can contribute to climate change. In contrast, BH, when properly collected, emits no direct greenhouse gases during use, making it a cleaner alternative.

Composting, on the other hand, is an environmentally friendly method of managing organic waste, as it returns nutrients to the soil and improves soil health. However, composting does not provide a direct energy output like biogas or BH. From an LCA perspective, BH production offers a more direct route to RE generation, while composting primarily contributes to soil and ecosystem health.

14.5.3 Social and policy implications

Government policies and regulations are crucial in shaping the development of BH technologies. Several countries have introduced RE policies that incentivize the formation and use of H_2 as part of their strategy to decrease dependence on fossil fuels and reduce carbon emissions. These policies often include subsidies, tax credits, or feed-in tariffs for H_2 producers, as well as stricter regulations on landfill use and organic waste management, which encourage the diversion of food waste to RE processes. In addition to national policies, international agreements, such as the Paris Agreement, place increasing pressure on countries to meet their greenhouse gas reduction targets. This has led to a rise in interest in BH production as part of a broader effort to decarbonize energy systems. The shift to a circular economy, where waste is viewed as a resource rather than a burden, further supports the use of waste food in RE production. Additionally, BH plays a role in achieving net-zero emissions targets set by many governments. As H_2 is a versatile and clean fuel, its development from sustainable sources like food waste can help reduce the carbon footprint of sectors such as transportation, industry, and power generation. With the right policy framework and continued technological advancements, bio-H_2 could become an essential component of renewable and low-carbon energy in the future.

Studies analysing the monetary and ecological feasibility of BH production consistently highlight the potential for this technology to play a substantial role in future energy systems. Research on LCA for BH has shown its environmental superiority over fossil-based H_2 and other waste-to-energy technologies, while financial analyses underscore the need for continued improvements in cost reduction and reactor design optimization to make BH production from waste food commercially viable.

14.6 CHALLENGES AND FUTURE DIRECTIONS

BH production from waste food holds tremendous potential as an RE source, addressing both the progressing rise for sustainable energy and the need for effective waste management. However, significant challenges remain in realizing this potential on a commercial scale. Technical limitations, inefficiencies in microbial processes, and operational complexities hinder the large-scale application of BH technologies. Despite these challenges, advances in microbial biotechnology, synthetic biology, and reactor design present exciting opportunities for improvement. This section outlines the technical challenges, opportunities for innovation, and future research directions in BH production from waste food.

14.6.1 Technical challenges

Most of the primary challenges in BH production are the comparatively low H_2 yields and formation effectiveness associated with biological processes. During DF, for instance, only a fraction of the available organic material in waste food is altered into H_2 [30]. The leftover energy is trapped in byproducts like volatile acetic(fatty) acids (VFAs), alcohols, and gases (e.g., CO_2 and CH_4), which reduce the total energy efficiency of the process. This inefficiency can be attributed to several factors:

1. *Microbial Limitations:* The micro-organisms available in BH production, like Clostridium and Enterobacter, have limited capacity to fully convert substrates into H_2. Their metabolic pathways often prioritize the production of organic acids over hydrogen, leading to suboptimal H_2 yields.
2. *Inhibitors:* Accumulation of by-products like VFAs and ammonia (from protein degradation) can inhibit microbial activity and reduce H_2 production.
3. *Substrate Composition:* Food waste is heterogeneous, containing varying amounts of carbohydrates, proteins, and lipids. This variation affects the consistency and predictability of H_2 production, as different components require different metabolic pathways for conversion into H_2.

Addressing these limitations requires optimizing microbial metabolism and developing integrated processes that can fully utilize the energy in food waste.

14.6.2 Scale-up issues and operational complexities

Scaling up BH production from lab-scale experiments to industrial-scale systems poses significant challenges. Several factors contribute to the complexity of scaling up BH production:

1. *Reactor Design:* BH production requires carefully controlled environments, including anaerobic conditions, optimal pH, and temperature, to maximize microbial activity. Maintaining these conditions consistently at larger scales is challenging and often requires sophisticated control systems.
2. *Substrate Variability:* Food waste is inherently variable in terms of composition, water content, and nutrient balance. As a result, it is difficult to standardize BH production processes for large-scale operations. Variations in substrate composition can affect microbial performance, reactor stability, and H_2 yields.
3. *Gas Separation and Purification:* At the industrial level, separating H_2 from other gases (like CO_2 and CH_4) produced during fermentation is a critical challenge. The clarity of the H gas stream is crucial for its use in fuel cells and other energy applications, but current gas separation technologies are costly and energy-intensive.

14.6.3 Opportunities for improvement

Recent advances in microbial biotechnology offer promising solutions for improving BH production. The detection and engineering of novel microbial strains with improved H_2-producing abilities could lead to higher yields and more efficient conversion processes. Key strategies include:

1. *Selection of Superior Strains*: Screening for naturally occurring bacteria with superior H_2 production abilities can lead to the identification of strains that are more efficient at converting complex organic substrates into H_2.
2. *Metabolic Engineering:* By modifying the metabolic pathways of microorganisms, researchers can direct more of the substrate's energy toward H_2 production. For example, genetic modifications can be made to inhibit the production of competing byproducts (like VFAs) and enhance hydrogenase enzyme activity, increasing H_2 yields [75].

14.6.4 Use of synthetic biology to engineer superior bio-hydrogen-producing organisms

Synthetic biology offers unprecedented opportunities to design and engineer microorganisms with enhanced BH production capabilities. By combining knowledge from systems biology, genetic engineering, and metabolic modeling, researchers can create tailored microbial strains with optimized traits. For example:

1. *Tailored Hydrogenase Enzymes*: Hydrogenase enzymes are crucial for the production of H_2 in microbial systems [75]. Using synthetic biology, researchers can engineer hydrogenases with improved stability, activity, and tolerance to inhibitors (such as oxygen or ammonia), which could significantly boost H_2 production.
2. *Pathway Optimization:* Synthetic biology can be used to rewire microbial metabolic pathways to ensure that more energy is channeled into H_2 production rather than competing processes. For instance, metabolic engineering could block pathways leading to the formation of organic acids, directing more carbon and energy to H_2 generation.

14.6.5 Future research areas

While significant progress has been made in using carbohydrate-rich food waste for BH production, there remains a need to explore underutilized food waste streams. Certain food waste types, such as protein- and lipid-rich waste, have not been fully exploited due to the challenges they present (e.g., ammonia inhibition or the difficulty of lipid degradation).

The design of cost-effective and scalable reactors is critical for the commercial viability of BH production. Current reactors used for lab-scale production may not be suitable for industrial applications due to factors like material costs, energy efficiency, and operational complexity. Key areas for future research in reactor development include:

1. *Integration of BH Production with Other Processes*: Hybrid systems that combine BH production with other waste management or energy production methods (like biogas generation or wastewater treatment) could improve total system productivity and reduce costs.
2. *Advanced Reactor Designs:* Research into novel reactor designs, such as biofilm, fluidized bed reactors (FBR), and Membrane bioreactors, could improve the performance of H_2 production by providing better control over microbial communities and substrate availability.

Research gaps and ongoing innovations in BH production focus on enhancing microbial effectiveness, enhancing reactor designs, and expanding the range of usable feedstocks. Studies have identified the need for integrating multiple biological pathways, improving the scalability of production processes, and reducing the operational costs associated with pre-treatment and gas purification. For example, the use of microbial consortia and advanced biotechnological tools is being explored to address inefficiencies and optimize the transformation of diverse waste food streams in H_2.

14.7 CONCLUSION

BH formation from waste food presents a promising and sustainable strategy to address both energy demand and waste management challenges. This process not only converts organic waste into valuable H_2 gas but also contributes to the diminution of greenhouse gas emissions and landfill usage. By leveraging biological processes, particularly DF and PF, food waste, which is often considered a burden, can be transformed into an RE source. In conclusion, the power of food waste as output for BH production is substantial, offering a sustainable solution to both energy generation and waste reduction. Although challenges remain in terms of efficiency, scalability, and operational costs, advancements in microbial biotechnology, reactor design, and waste pre-treatment methods provide significant opportunities for future improvements. With continued research and technological development, BH production from waste food could become a key component of the global conversion to RE and a circular economy.

ACKNOWLEDGEMENT

The authors are grateful to the Harcourt Butler Technical University, Kanpur, India, for technical facilities.

REFERENCES

[1] Zott C, Amit R, Massa L. The business model: Recent developments and future research. *Journal of Management* 2011;37:1019.

[2] Sharma AK, Ghodke PK, Manna S, Chen W-H. Emerging technologies for sustainable production of biohydrogen production from microalgae: A state-of-the-art review of upstream and downstream processes. *Bioresource Technology* 2021;342:126057.

[3] Pramanik H, Rathoure AK. Electrooxidation study of $NaBH_4$ in a membraneless microfluidic fuel cell with air breathing cathode for portable power application. *International Journal of Hydrogen Energy* 2017;42:5340.

[4] Rathoure AK, Pramanik H. Electrooxidation study of methanol using H_2O_2 and air as mixed oxidant at cathode in air breathing microfluidic fuel cell. *International Journal of Hydrogen Energy* 2016;41:15287.

[5] Zhang L, Jia C, Bai F, Wang W, An S, Zhao K, et al. A comprehensive review of the promising clean energy carrier: Hydrogen production, transportation, storage, and utilization (HPTSU) technologies. *Fuel* 2024;355:129455.

[6] Ishaq H, Dincer I, Crawford C. A review on hydrogen production and utilization: Challenges and opportunities. *International Journal of Hydrogen Energy* 2022;47:26238.

[7] Ghimire A, Frunzo L, Pirozzi F, Trably E, Escudie R, Lens PN, et al. A review on dark fermentative biohydrogen production from organic biomass: Process parameters and use of by-products. *Applied Energy* 2015;144:73.

[8] Yadav VS, Vinoth R, Yadav D. Bio-hydrogen production from waste materials: A review. *MATEC Web of Conferences: EDP Sciences* 2018;192:02020.

[9] Aziz M, Darmawan A, Juangsa FB. Hydrogen production from biomasses and wastes: A technological review. *International Journal of Hydrogen Energy* 2021;46:33756.

[10] Abdin Z, Zafaranloo A, Rafiee A, Mérida W, Lipiński W, Khalilpour KR. Hydrogen as an energy vector. *Renewable and Sustainable Energy Reviews* 2020;120:109620.

[11] Datta A, Hossain A, Roy S. An overview on biofuels and their advantages and disadvantages. *Asian Journal of Chemistry* 2019;31:1851–1458.

[12] Balat M. Possible methods for hydrogen production. *Energy Sources, Part A: Recovery, Utilization, and Environmental Effects* 2008;31:39.

[13] Jain R, Panwar NL, Jain SK, Gupta T, Agarwal C, Meena SS. Bio-hydrogen production through dark fermentation: An overview. *Biomass Conversion and Biorefinery* 2024;14:12699.

[14] Giordano A, Cantù C, Spagni A. Monitoring the biochemical hydrogen and methane potential of the two-stage dark-fermentative process. *Bioresource Technology* 2011;102:4474.

[15] Eroglu E, Melis A. Photobiological hydrogen production: Recent advances and state of the art. *Bioresource Technology* 2011;102:8403.

[16] Adessi A, De Philippis R. Photobioreactor design and illumination systems for H_2 production with anoxygenic photosynthetic bacteria: A review. *International Journal of Hydrogen Energy* 2014;39:3127.
[17] Yanling Y, Zhenmei L, Hang M, Cheng J. Dynamic changes of microbial community diversity in a photohydrogen producing reactor monitored by PCR-DGGE. *Journal of Environmental Sciences* 2008;20:1118.
[18] Hallenbeck PC, Ghosh D. Advances in fermentative biohydrogen production: The way forward? *Trends in Biotechnology* 2009;27:287.
[19] Moreno R, Escapa A, Cara J, Carracedo B, Gómez X. A two-stage process for hydrogen production from cheese whey: Integration of dark fermentation and biocatalyzed electrolysis. *International Journal of Hydrogen Energy* 2015;40:168.
[20] Chookaew T, Prasertsan P, Ren ZJ. Two-stage conversion of crude glycerol to energy using dark fermentation linked with microbial fuel cell or microbial electrolysis cell. *New Biotechnology* 2014;31:179.
[21] Gómez X, Fernández C, Fierro J, Sánchez M, Escapa A, Morán A. Hydrogen production: Two stage processes for waste degradation. *Bioresource Technology* 2011;102:8621.
[22] Liu W, Huang S, Zhou A, Zhou G, Ren N, Wang A, et al. Hydrogen generation in microbial electrolysis cell feeding with fermentation liquid of waste activated sludge. *International Journal of Hydrogen Energy* 2012;37:13859.
[23] Li X-H, Liang D-W, Bai Y-X, Fan Y-T, Hou H-W. Enhanced H_2 production from corn stalk by integrating dark fermentation and single chamber microbial electrolysis cells with double anode arrangement. *International Journal of Hydrogen Energy* 2014;39:8977.
[24] Chen C-Y, Yang M-H, Yeh K-L, Liu C-H, Chang J-S. Biohydrogen production using sequential two-stage dark and photo fermentation processes. *International Journal of Hydrogen Energy* 2008;33:4755.
[25] Redwood MD, Macaskie LE. A two-stage, two-organism process for biohydrogen from glucose. *International Journal of Hydrogen Energy* 2006;31:1514.
[26] Jeremiasse AW, Hamelers HV, Buisman CJ. Microbial electrolysis cell with a microbial biocathode. *Bioelectrochemistry* 2010;78:39.
[27] Logan BE, Call D, Cheng S, Hamelers HV, Sleutels TH, Jeremiasse AW, et al. Microbial electrolysis cells for high yield hydrogen gas production from organic matter. *Environmental Science & Technology* 2008;42:8630.
[28] Redwood MD, Paterson-Beedle M, Macaskie LE. Integrating dark and light bio-hydrogen production strategies: Towards the hydrogen economy. *Reviews in Environmental Science and Bio/Technology* 2009;8:149.
[29] Nath K, Kumar A, Das D. Hydrogen production by Rhodobacter sphaeroides strain OU 001 using spent media of Enterobacter cloacae strain DM11. *Applied Microbiology and Biotechnology* 2005;68:533.
[30] Chen C-Y, Yeh K-L, Lo Y-C, Wang H-M, Chang J-S. Engineering strategies for the enhanced photo-H_2 production using effluents of dark fermentation processes as substrate. *International Journal of Hydrogen Energy* 2010;35:13356.

[31] Tao Y, Chen Y, Wu Y, He Y, Zhou Z. High hydrogen yield from a two-step process of dark-and photo-fermentation of sucrose. *International Journal of Hydrogen Energy* 2007;32:200.

[32] Su H, Cheng J, Zhou J, Song W, Cen K. Combination of dark-and photo-fermentation to enhance hydrogen production and energy conversion efficiency. *International Journal of Hydrogen Energy* 2009;34:8846.

[33] Monlau F, Aemig Q, Trably E, Hamelin J, Steyer J-P, Carrere H. Specific inhibition of biohydrogen-producing Clostridium sp. after dilute-acid pretreatment of sunflower stalks. *International Journal of Hydrogen Energy* 2013;38:12273.

[34] Qiu C, Zheng Y, Zheng J, Liu Y, Xie C, Sun L. Mesophilic and thermophilic biohydrogen production from xylose at various initial pH and substrate concentrations with microflora community analysis. *Energy & Fuels* 2016;30:1013.

[35] Argun H, Kargi F, Kapdan IK. Light fermentation of dark fermentation effluent for bio-hydrogen production by different Rhodobacter species at different initial volatile fatty acid (VFA) concentrations. *International Journal of Hydrogen Energy* 2008;33:7405.

[36] Madamwar D, Garg N, Shah V. Cyanobacterial hydrogen production. *World Journal of Microbiology and Biotechnology* 2000;16:757.

[37] Habashy MM, Ong ES, Abdeldayem OM, Al-Sakkari EG, Rene ER. Food waste: A promising source of sustainable biohydrogen fuel. *Trends in Biotechnology* 2021;39:1274.

[38] Dinesh GK, Chauhan R, Chakma S. Influence and strategies for enhanced biohydrogen production from food waste. *Renewable and Sustainable Energy Reviews* 2018;92:807.

[39] Hawkes FR, Hussy I, Kyazze G, Dinsdale R, Hawkes DL. Continuous dark fermentative hydrogen production by mesophilic microflora: Principles and progress. *International Journal of Hydrogen Energy* 2007;32:172.

[40] Yin D, Liu W, Zhai N, Yang G, Wang X, Feng Y, et al. Anaerobic digestion of pig and dairy manure under photo-dark fermentation condition. *Bioresource Technology* 2014;166:373–380.

[41] Kim H, Moon S, Abug A, Choi S-C, Zhang R, Oh Y-S. Effect of fermentation conditions on biohydrogen production from lipid-rich food material. *International Journal of Hydrogen Energy* 2012;37:15062.

[42] Das D, Dutta T, Nath K, Kotay SM, Das AK, Veziroglu TN. Role of Fe-hydrogenase in biological hydrogen production. *Current Science* 2006;90:1627–1637.

[43] Sattar A, Arslan C, Ji C, Sattar S, Umair M, Sattar S, et al. Quantification of temperature effect on batch production of bio-hydrogen from rice crop wastes in an anaerobic bio reactor. *International Journal of Hydrogen Energy* 2016;41:11050.

[44] Kim D-H, Kim S-H, Kim H-W, Kim M-S, Shin H-S. Sewage sludge addition to food waste synergistically enhances hydrogen fermentation performance. *Bioresource Technology* 2011;102:8501.

[45] Han S-K, Shin H-S. Biohydrogen production by anaerobic fermentation of food waste. *International Journal of Hydrogen Energy* 2004;29:569.

[46] Han S-K, Shin H-S. Performance of an innovative two-stage process converting food waste to hydrogen and methane. *Journal of the Air & Waste Management Association* 2004;54:242.
[47] Wang J, Wan W. Comparison of different pretreatment methods for enriching hydrogen-producing bacteria from digested sludge. *International Journal of Hydrogen Energy* 2008;33:2934.
[48] Wang Y-H, Li S-L, Chen I-C, Tseng I-C, Cheng S-S. A study of the process control and hydrolytic characteristics in a thermophilic hydrogen fermentor fed with starch-rich kitchen waste by using molecular-biological methods and amylase assay. *International Journal of Hydrogen Energy* 2010;35:13004.
[49] Lee Z-K, Li S-L, Lin J-S, Wang Y-H, Kuo P-C, Cheng S-S. Effect of pH in fermentation of vegetable kitchen wastes on hydrogen production under a thermophilic condition. *International Journal of Hydrogen Energy* 2008;33:5234.
[50] Fang HH, Li C, Zhang T. Acidophilic biohydrogen production from rice slurry. *International Journal of Hydrogen Energy* 2006;31:683.
[51] Karlsson A, Vallin L, Ejlertsson J. Effects of temperature, hydraulic retention time and hydrogen extraction rate on hydrogen production from the fermentation of food industry residues and manure. *International Journal of Hydrogen Energy* 2008;33:953.
[52] Van Ginkel SW, Oh S-E, Logan BE. Biohydrogen gas production from food processing and domestic wastewaters. *International Journal of Hydrogen Energy* 2005;30:1535.
[53] Wang J, Wan W. Factors influencing fermentative hydrogen production: A review. *International Journal of Hydrogen Energy* 2009;34:799.
[54] Guo XM, Trably E, Latrille E, Carrère H, Steyer J-P. Hydrogen production from agricultural waste by dark fermentation: A review. *International Journal of Hydrogen Energy* 2010;35:10660.
[55] Guo J, Wang W, Liu X, Lian S, Zheng L. Effects of thermal pre-treatment on anaerobic co-digestion of municipal biowastes at high organic loading rate. *Chemosphere* 2014;101:66.
[56] Rafieenia R, Lavagnolo MC, Pivato A. Pre-treatment technologies for dark fermentative hydrogen production: Current advances and future directions. *Waste Management* 2018;71:734.
[57] Singh L, Wahid ZA. Methods for enhancing bio-hydrogen production from biological process: A review. *Journal of Industrial and Engineering Chemistry* 2015;21:70.
[58] Lü F, Ji J, Shao L, He P. Bacterial bioaugmentation for improving methane and hydrogen production from microalgae. *Biotechnology for Biofuels* 2013;6:1.
[59] Gu S, Xing H, Zhang L, Wang R, Kuang R, Li Y. Effects of food wastes based on different components on digestibility and energy recovery in hydrogen and methane co-production. *Heliyon* 2024;10:e25421.
[60] Jung K-W, Kim D-H, Kim S-H, Shin H-S. Bioreactor design for continuous dark fermentative hydrogen production. *Bioresource Technology* 2011;102:8612.

[61] Fontanille P, Kumar V, Christophe G, Nouaille R, Larroche C. Bioconversion of volatile fatty acids into lipids by the oleaginous yeast Yarrowia lipolytica. *Bioresource Technology* 2012;114:443.
[62] Chen C-C, Chuang Y-S, Lin C-Y, Lay C-H, Sen B. Thermophilic dark fermentation of untreated rice straw using mixed cultures for hydrogen production. *International Journal of Hydrogen Energy* 2012;37:15540.
[63] Zhang M-L, Fan Y-T, Xing Y, Pan C-M, Zhang G-S, Lay J-J. Enhanced biohydrogen production from cornstalk wastes with acidification pretreatment by mixed anaerobic cultures. *Biomass and Bioenergy* 2007;31:250.
[64] Kotsopoulos TA, Fotidis IA, Tsolakis N, Martzopoulos GG. Biohydrogen production from pig slurry in a CSTR reactor system with mixed cultures under hyper-thermophilic temperature (70°C). *Biomass and Bioenergy* 2009;33:1168.
[65] Castelló E, y Santos CG, Iglesias T, Paolino G, Wenzel J, Borzacconi L, et al. Feasibility of biohydrogen production from cheese whey using a UASB reactor: Links between microbial community and reactor performance. *International Journal of Hydrogen Energy* 2009;34:5674.
[66] Carrillo-Reyes J, Celis LB, Alatriste-Mondragón F, Razo-Flores E. Different start-up strategies to enhance biohydrogen production from cheese whey in UASB reactors. *International Journal of Hydrogen Energy* 2012;37:5591.
[67] Zhao R-J, Zhang Z, Yang S-S, Min G, Liu S-J, Qiu X-T, et al. Study on the performance of a new type of combined packing biofilm reactor treating wastewater. *Environmental Technology* 2024;45:4191.
[68] Lu Y, Jin H, Guo L, Zhang X, Cao C, Guo X. Hydrogen production by biomass gasification in supercritical water with a fluidized bed reactor. *International Journal of Hydrogen Energy* 2008;33:6066.
[69] Dasgupta CN, Gilbert JJ, Lindblad P, Heidorn T, Borgvang SA, Skjanes K, et al. Recent trends on the development of photobiological processes and photobioreactors for the improvement of hydrogen production. *International Journal of Hydrogen Energy* 2010;35:10218.
[70] Melis A. Photosynthetic H_2 metabolism in Chlamydomonas reinhardtii (unicellular green algae). *Planta* 2007;226:1075.
[71] Aboelela D, Soliman MA. Hydrogen production from microbial electrolysis cells powered with microbial fuel cells. *Journal of King Saud University-Engineering Sciences* 2024;36:369–374.
[72] Karakashev D, Kotay SM, Trably E, Angelidaki I. A strict anaerobic extreme thermophilic hydrogen-producing culture enriched from digested household waste. *Journal of Applied Microbiology* 2009;106:1041.
[73] Jarunglumlert T, Prommuak C, Putmai N, Pavasant P. Scaling-up biohydrogen production from food waste: Feasibilities and challenges. *International Journal of Hydrogen Energy* 2018;43:634.
[74] Bhandari R, Trudewind CA, Zapp P. Life cycle assessment of hydrogen production via electrolysis: A review. *Journal of Cleaner Production* 2014;85:151.
[75] Constant P, Hallenbeck PC. Hydrogenase. In Ashok Pandey, S. Venkata Mohan, Jo-shu Chang, Patrick C. Hallenbeck, Christian Larroche (Eds.), *Biohydrogen*. Amsterdam, Netherlands Elsevier; 2019, p. 49.

Index

For Product Safety Concerns and Information please contact our EU representative GPSR@taylorandfrancis.com
Taylor & Francis Verlag GmbH, Kaufingerstraße 24, 80331 München, Germany

www.ingramcontent.com/pod-product-compliance
Ingram Content Group UK Ltd.
Pitfield, Milton Keynes, MK11 3LW, UK
UKHW020350280226
468510UK00007B/61
9781032947402